MOTH BORERS OF SUGARCANE

The Author

Dr. S. K. Duttamajumder is a revered sugarcane crop protection specialist of National and International fame. Graduated in Agriculture from the Bidhan Chandra Krishi Viswavidyalaya, Nadia, West Bengal during 1975-79; did both M.Sc. and Ph.D. in Plant Pathology from the Indian Agricultural Research Institute, New Delhi during 1980-85. He joined the Indian Institute of Sugarcane Research, Lucknow as Scientist in 1985 and worked on important bacterial and fungal diseases of sugarcane (Leaf scald, Ratoon stunting disease, Red rot, Smut and Wilt), on the development of ESTs of sugarcane and in the high sugar parent development programme of sugarcane for the subtropical India. He was the key architect in developing the road map for sugarcane research and development to be followed at the Indian Institute of Sugarcane Research, Lucknow. He also served as a member of RCGM, New Delhi for the identification of suitable genetically modified crop varieties for India. Dr. Duttamajumder served as the first Associate Editor (Bacteriology) of the Indian Phytopathological Society, New Delhi. He became Principal scientist in 2006 and finally headed the Division of Crop Protection of IISR, Lucknow from 2009 to 2014. His interest on Sugarcane Entomology started in 2009. He has authored two authoritative books on crop protection of sugarcane. The book *Red rot of Sugarcane* is a landmark publication on this very important disease of sugarcane and incidentally it is the first book on this disease. The other book, *Top borer of Sugarcane* is a well acclaimed publication in the field of Sugarcane Entomology. Dr. Duttamajumder is an ace scientific photographer and these two books bear the testimony.

MOTH BORERS OF SUGARCANE

S. K. Duttamajumder
Former Head
Division of Crop Protection
Indian Institute of Sugarcane Research
Lucknow - 226002

2020
Daya Publishing House
A Division of
Astral International Pvt. Ltd.
New Delhi - 110 002

ISBN: 978-93-89719-44-4

Published by : **Daya Publishing House®**
A Division of
Astral International Pvt. Ltd.
– ISO 9001:2015 Certified Company –
4736/23, Ansari Road, Darya Ganj
New Delhi-110 002
Ph. 011-43549197, 23278134
E-mail: info@astralint.com
Website: www.astralint.com

Digitally Printed at : **Replika Press Pvt. Ltd.**

Acknowledgements

It is my pleasant duty to acknowledge the help and support of many persons, who have contributed knowingly or unknowingly on my journey of learning of Sugarcane Entomology. In this journey, many scholars have enlightened me with their erudite discussions on complex and intricate matters of entomology and their field experiences (not documented in text books) on sugarcane insect pests. I gratefully acknowledge all the help I received.

I take this opportunity to express my gratitude to all my teachers who have instilled in me the requisite confidence to take up any scientific challenge without prejudice and fear. Nobody is born with the knowledge, rather everybody imbibes it through nurturing, self reading, patient hearing, assimilating from teachers, discussions with peers, and "*Learning by doing*".

I am highly indebted to Dr. Ashok Varma, who headed the Division of Entomology, IISR, Lucknow for a long time in 1980's and 1990's. Despite continuing ill health, he readily agreed to go through the manuscript. Not only did he go through the manuscript critically but also put forward his thoughtful suggestions to improve the quality and brevity of this publication.

I am also extremely thankful to Dr. Raman Kapur, retired Head of the Division of Crop Improvement, IISR, Lucknow for critically going through the manuscript from the point of view of a practising sugarcane specialist.

I express my gratitude to Dr. A. D. Pathak, the present Director of IISR, Lucknow for providing the necessary permission to go ahead.

I am thankful to my colleagues in the discipline of Entomology for their unflinching support. Special thanks are due to Dr. M. R. Singh and Dr. Arun Baitha for their active participation in the search of Green borer and meaningful discussions.

Finally, I take this opportunity to acknowledge the patience, perseverance and unflinching support that I have received from my wife Kankana during this period when I was completely engrossed in this arduous job. Despite her repulsive feelings about the crawling caterpillars, she has silently tolerated all my misadventures with these creatures at home. In addition, she has actively participated in this endeavour and has critically gone through the manuscript from the point of view of a non-professional. Her suggestions have helped me immensely in improving the clarity, brevity, and readability of this publication.

Lucknow

S. K. Duttamajumder

Preface

The two books that I wrote earlier took me on two different journeys of learning. The first one entitled "*Red rot of sugarcane*" was published by IISR, Lucknow in 2008. It was on my specialized subject '*Plant Pathology*'. The journey with the second book entitled "*Top borer of sugarcane*" needs a bit explanation, as it was a completely different path of journey altogether; a journey on the sister discipline, *Entomology*.

In 2009, I was appointed as the Head, Division of Crop Protection at IISR, Lucknow (2009-2014). My rendezvous with researches on moth borers of sugarcane necessitated to answer a query from ICAR regarding role of nitrogen in the development of infructuous attack of Top borer as an effective means of its management. It was claimed that by inducing faster growth of sugarcane during the peak summer of June, it would be possible to tide over the third brood menace sans pesticide. The idea appears lucrative, as currently management of the third brood of Top borer needs an application of systemic insecticide (like *carbofuran*) during June. In fact, this is the standard practice in sugarcane growing in the subtropical belt. Obviously, the existing knowledge on the biology of Top borer *vis-a-vis* sugarcane crop does not agree with such claim. Dr. R. L. Yadav, the then Director of the Institute requested me to look into the matter critically, afresh and without any prior bias. As an agronomist, he could not fathom the idea that such a faster growth of crown could be induced in summer months and that too with the application of an extra dose of nitrogen (75 kg N). Thus, my journey on the sister discipline Entomology started with the investigation of various facets of Top borer. The second book is the result of an arduous effort of eight years' of study on Top borer.

During the process of learning about the moth borers of sugarcane, it became apparent that there is a dearth of illustrative publications that one can rely upon for the accurate identification of this very important group of insect pests of sugarcane. Mr. B. D. Gupta, who headed the Entomology Section of IISR (1952-1958), published the Technical Bulletin "*Identification of the striped moth borers of sugarcane in India*" in 1958. Since then, many additions as well as refinements have taken place in the accrued knowledge. It was felt that a revision with the current knowledge would be appropriate to cater the present needs. I fervently took up this challenge to develop an identification manual of the important moth borers of sugarcane that even a nonprofessional could use with confidence. The present publication, in addition to the striped borers, encompasses other important moth borers of sugarcane. Moreover, it also contains information on symptoms of infestations, different life stages of the moth borers and key identifying characters for each stage of the life cycle.

Identification of the moth borers of sugarcane is a bit tricky for a field worker or a farmer as these borers are very similar in appearance, display identical symptoms of infestation at shoot stage of the crop and the moths mostly remain out of the common gaze (remain in hiding during the day). It is often axiomed that "*A picture is worth a thousand words*". Obviously, a good colour photograph speaks volumes and often leaves an indelible impression on the mind of the reader that words and explanations cannot do. This publication is an attempt to bridge the longstanding gap of a comprehensive publication detailing the easily identifiable characters of moth borers of sugarcane with suitable and vivid colour photographs of symptoms of infestations as well as different life stages of the pest.

To be honest, it is not a scientific treatise on moth borers of sugarcane written for the use and training of Entomologists. Rather it is intended to serve the more humble but useful purpose of a manual of practical information for the use of the farmers, extension personnel and cane development workers who are interested to learn about the moth borer pests of sugarcane. This publication banks upon the principle of "*Seeing is believing*" and contains photographic depictions of legible larval characters to identify the borer species. More emphasis is given on larval characters as farmers and field workers mostly encounter the larval stage in the field. The larval characters may be confirmed with a simple hand lens (10x). In this regard, one can use the modern gadget at hand. The camera of a smart phone becomes very handy in the field. The picture can be magnified several times without any difficulty.

At the end, it dawned upon me that it would be better to add a chapter on the identification of important bio-agents that are commonly used to contain these borer pests. A clear understanding of the actual identity of these bio-agents is essential for their conservation, augmentation, and gainful inundative releases.

I am fully aware that despite due care, some errors and glitches show their presence in the final print. I sincerely hope that the readers will bear with me in this regard. The failings are solely mine. I fervently hope that the book may be useful to many who are engaged in sugarcane agriculture.

Finally, with all humility, I bow to the revered feet of the Almighty whose invisible inspiration and support has guided me to venture into this arena.

Lucknow — S. K. Duttamajumder

Contents

1

Introduction

The borer pests of crop plants are commonly called '*moth borers*' as they are the larvae or caterpillar stages of moths belonging to the insect order *Lepidoptera* (*Lepid* = scales, *pteron* = wing) – the largest conglomerate of insects possessing scales on their body. The order *Lepidoptera* mainly represents flamboyant butterflies and subdued moths. Butterflies rule the day, whereas the moths rule the night (nocturnal in habit). In general, moths are inconspicuous and smaller in comparison to butterflies. Moths are nocturnal and mostly remain in hiding during the day. Thus, moths mostly remain obscure to the general gaze. Many moths are curiously *phototactic* (attracted to a light source during the night) and fly to reach the light source during night. This *phototactic* behaviour allows these moths to be collected using a light trap. Among the vast majority of moth species that infests plants, larvae of a few species feed on the plants belonging to the family of grasses and these moths are collectively identified as Grass moths or *Crambid* moths (moths with closed wings). The *Crambid* moths, during rest/ sitting put their wings over the abdomen like a thatched roof of a hut.

An insect in the order *Lepidoptera* has four well-demarcated stages in its life cycle viz., *egg* (*ovum*), *larva* (*caterpillar*), *pupa* (*chrysalis*) and *adult* (*imago*). All the moth borers of sugarcane follow the same pattern of life stages. At the larval stage, they feed on the sugarcane plants and inflict damage. The other three stages are non-feeding stages, i.e., *egg, pupa* and *adult* do not harm the sugarcane crop directly. A larva or caterpillar is a voracious eater of the plant tissues, as it has to garner all the nutrition during this period to sustain all the three non-feeding stages of its life cycle. During the growth and development, a larva has to pass through several instars, and during this period it moults several times to attain maturity (*ecdysis* is the process of shedding of the outer older skin and development of new skin underneath of the old one to facilitate growth and development). At the end of the final larval stage (*prepupal* stage), a mature larva moults to completely transform itself into a *pupa*. The egg and the pupa are the two non-feeding stages wherein an insect remains almost defenceless. During the development process within a fertilized eggshell, a larva gradually forms (during the incubation period), and through the process of hatching (*eclosion*) a neonate larva comes out from the eggshell to start a new active feeding life. The neonate larva starts inflicting damage to the host and the damage continues until it stops feeding and gets itself transformed into a *pupa*. In this journey, it passes through several larval instars (usually five) before its final moulting into a *pupa*. A *pupa* is an inactive, intermediate stage in an insect's life, wherein complete metamorphosis takes place. From a crawling caterpillar, a winged adult emerges from the pupal shell. These two life stages are so different that it is difficult to ascertain a direct relationship between them, if observed independently. The developing adult within a pupa is called *pharate adult*. Usually, in moth borers of sugarcane, the pupal period lasts about one to one and a half weeks (depending on the ambient temperature). Before emergence of the moth, a crack develops at the dorsal side in the space between head and thorax of the pupal shell. The moth comes out gradually widening the crack by pushing its head forward. The process of emergence of a moth from the pupal shell is also called *eclosion*. A moth, immediately after emergence, rests for a while near the empty shell to muster strength in the wings and to complete the developmental process. Once a moth gains sufficient strength in its wings, it flies away in search of a mate.

In a given geographical location, important moth borers of sugarcane pass through several generations (*broods*) in a calendar year. Thus, sugarcane crop faces several discrete waves of attack of caterpillars (in each brood) during its stay in the field. In subtropical sugarcane belt, where borers pass the winter months in hibernation, the initial two broods at the beginning of the crop season (February-March) are very distinct. In sugarcane belt of tropical south, due to mild winter, overlapping broods are common (at a given point of time, all the four life stages remain available in the field). Obviously, in overlapping broods, the waves of attack on sugarcane crop appear continuous.

Borer infestations as well as resultant crop damage in sugarcane takes place at different stages of its growth. It is for this very reason that the damage due to borers in sugarcane is cumulative. The cumulative total damage of a sugarcane crop is reflected at the time of harvest i.e., in terms of reduction in the number of millable canes (NMC), yield (tonnage), and quality of the produce (recoverable sugar). As the borers attack at different stages of crop growth, the sugarcane plant also responds differently in different broods. At the early stages of crop growth, an infested plant mostly compensates the loss with increased tillering, whereas at the later stages of crop growth (infestation of later broods), the infested plant fails to compensate and suffers the loss (any infestation after the completion of tillering phase mostly affects the number and quality of millable canes).

Sugarcane is a round the year crop and virtually it remains available in the field throughout the year. For this reason, most of the sugarcane insect pests thrive on the crop round the year and that too without any compulsion of migration to other plants/crops in search of food and shelter (do not leave sugarcane at all). Thus, in sugarcane agro-ecosystem, environment plays the pivotal role in the establishment and subsequent proliferations (population dynamics) of insect pests. In a favourable environment, insect pests quickly attain a population level that is adequate to harm the crop irreparably (infestation cross the economic threshold level, ETL).

It is generally estimated that in agricultural ecosystems, around 1.5 lakh insect species are present and hardly 10% of the insect species have attained the status of crop pests. In any natural ecosystem, a delicate population balance is always maintained between the insect-pests and their natural enemies. In the natural ecosystem, the population dynamics of insect pests mostly follow a *boom and bust cycle*. Initially the population of an insect pest increases and finally the heightened population level is brought down substantially by the increasing multiplication and action of parasites and predators on the insect pest.

A conservative estimate indicates that in a natural ecosystem, most of the insect population is lost during the growth and development of an insect from an egg to an imago. Around 90-95% of insect population is lost (physical, physiological and biological factors) before reaching maturity and to start a new generation by laying eggs. Obviously, insects have evolved to maintain a certain population structure that is adequate for its survival and continuity in the competitive biological world.

It is to be kept in mind that it is only around 5-10% of the insect population (surviving population) that reach the adult stage. Thus, the insect pests have the enormous potential to decimate the crop and in fact, often create havoc in the agricultural ecosystem, if prevailing weather favours their quick multiplication especially in the absence or delayed proliferation of parasites and predators. The parasites and predators that contain the population of these insect pests always maintain a delicate balance in the survival of both the host plant and the insect pests. However, due to the *trophic* relationship (dependence for food), parasites and predators always require more time to catch up in attaining the requisite population structure that can contain the unhindered proliferation of insect pests. Usually, in agro-ecosystem the population balance of insect-pests and their parasites is achieved at a time when much of the crop damage has already taken place. Agriculture is a specialized man-made ecosystem wherein this delicate natural balance has been grossly obliterated. Modern agriculture, due to high density of uniform plant population of a crop variety, regularity of growing a crop in a field, mono-cropping, increased succulence of the plants due to use of fertilizers and irrigation, etc. is more conducive to the insect pests for the build up of a large population. Many a times, the insect pests have a field day; through quick multiplication, they proliferate to a threatening population level and damage a crop with impunity. Traditionally, *mechanical control* measures remained the key management practice to contain the menace of insect pests in various crops including sugarcane. However, with the advent of toxic chemicals (*pesticides*) to combat the insect pests, the situation has eased a little and it has tilted in favour of humankind. In many a cases, insects have overbid the situation with the development of resistance against these toxic chemicals (with a shift in the population structure, many a times insects have attacked the crop with vengeance after a long lull). The use of a large quantity of synthetic pesticides in modern agriculture has put another unwanted residual toxic dimension in the biosphere. These broad-spectrum insecticides not only kill the targeted harmful insects, but at the same time also kill the beneficial insects present in the targeted area. Thus, a major disruption in the natural ecological balance takes place. Moreover, these insecticides are highly toxic to the entire biosphere including human beings. Obviously, as a principle, a restricted and judicious use of these insecticides is always advocated, so long as any other safer chemicals or management methods are available that can sustain the crop productivity sans the use of toxic insecticides.

Among the various agro-ecosystems of field crops, sugarcane agro-ecosystem is a very specialized one and is akin to the ecosystems of perennial shrubs (horticultural crops). Sugarcane is a semi-perennial Graminaceous crop plant (*Poace*) and remains growing in the field round the year. As sugarcane has no dead season, a regular availability of food and shelter to the insects is always ensured. Moreover, the practice of ratooning (raising a stubble crop from the left over stubble after the harvest of the plant/ratoon crop) helps the insect pests further to enjoy the hospitality of sugarcane agro-ecosystem. Obviously, many of the insects that have arrived from the forests (natural ecosystem) in the past to the specialized sugarcane agro-ecosystem did not move out again to their natural habitat. Over the generations, these migrated insect pests have become permanent residents of sugarcane agro-ecosystem. In their long associations with the sugarcane crop, these insect pests have co-evolved with the biological rhythm as well as the nutritional support of the host and adopted the sugarcane agro-ecosystem. Thus, many of the insect pests have evolved to a restricted food habit or host range, i.e. affecting only sugarcane crop and its closely allied genera.

1.1 A glimpse of insect pests' scenario of sugarcane in India

Sugarcane has a fair share of insect pests that thrive on it and every year, on an average these insect pests inflict a loss of around 15-20% in yield, in addition to the loss in quality of the produce. Box (1953) made a world census of the insects that harbour sugarcane crop and listed around 1300 insects that regularly visit sugarcane. However, not all these insects are present in one country or continent; only a part of it comprises the insect fauna of any country. It is accepted that half of these total insects are of new world origin and the rest half are of old world or oriental in nature. In other words, around 650 insects are visiting sugarcane in the old world to which India belongs. So far, in India, around 220 insects have been recorded as the pests of sugarcane (David *et al.* 1986). Obviously, the casual visitors are not included in this list of insect pests. Among the insects of sugarcane that infest the crop in India, the moth borers as a group are the most important causing considerable crop loss regularly. It is variously estimated that, in overall insect pests' scenario of sugarcane, moth borers alone accounts for around 55-60% of the total crop loss (soil insects account for around 5-10%, and various groups share rest 30-35% of the loss). However, at any given location, in favourable situation, individually these insect pests can rein havoc leading to the crop failure.

Out of the vast number of insect pests that subsist on sugarcane crop for their survival and growth, only around 10-15% insect species have attained the status of pests for which any management activity is warranted. Sugarcane possesses a robust compensatory mechanism to withstand the unforeseen biotic and abiotic stresses. Except very few, damage by a vast majority of insect pests is very well tolerated by a sugarcane crop and hardly any perceptible damage or loss takes place. Thus, majority of the insects that subsist on sugarcane are unimportant in the profitable cultivation of a sugarcane crop. These insects are mostly recognized as unimportant or non-damaging insect pests. A few insect pests, in certain situation or in certain locality may affect the sugarcane crop inflicting considerable damage. These types of insect pests are recognised as insect pests of minor importance. Only few insect pests that regularly inflict considerable crop loss in most of the areas and require appropriate management strategies for their containment are considered as insect pests of major importance.

Obviously, only the major insect pests need proper attention from the very beginning to contain them before they cross the economic threshold level (ETL). Farmers need to adopt appropriate steps (mechanical, behavioural, biological and chemical in a judicious blend) to contain them in order to safeguard the sugarcane crop. In India, around a dozen species of moth borers are of major concern as they severely impair the prospect of profitable cultivation of sugarcane.

The insect damage to sugarcane crop starts from the very day when a cane sett is planted in the field. The soil insects immediately try to use the sugarcane sett as their prized food and start devouring the sett with zeal. Among the soil insects, termites play a major role in damaging cane setts in the soil environment. Termites affect the very germination of buds and establishment of the crop by devouring the buds as well as the internal contents of the sett, leaving only the epidermis behind. The other important soil insect is the white grubs. Wherever white grubs are present, they eat away the roots of the sugarcane plant and debilitate the affected clumps leading to premature death.

After germination, a crop stand is established in the field and many insect pests use the young sugarcane plant as their preferred host. The above ground portion of the crop is mostly affected by three groups of insect viz., the borers, the sap-suckers and the tissue eaters. Among the borers, the Green borer, Early shoot borer, Pink borer, Root borer, Stalk borer, and Top borer infest the young sugarcane plants and inflict major damage to the crop during *shoot stage* (a sugarcane crop up to four months of age is called shoot stage) and collectively these borers are called "*shoot borers*". They are also called "*Pre-monsoon borers*" as the shoot period of the crop essentially runs through the pre-monsoon period in the subtropical sugarcane belt of India. The borers like Stalk borer, Internode borer, Gurdaspur borer and Plassey borer are specialized borers of cane stalk and cause extensive damage to the sugarcane crop during the monsoon months and these are also known as "*Monsoon borers*". All these borers infest sugarcane crop at different stages of growth, either alone or in association with one or more species.

The sap suckers like Mealy bugs, Black bugs, White flies, Scale insects, Thrips, Aphids, Hoppers etc., suck the sap either from leaves or stalk and weaken the plants at all the stages of crop growth, but their debilitating effect is more pronounced during the summer months and in post monsoon period. The leafhopper *Pyrilla* is a notorious sucking pest, and occasionally assumes epidemic proportion in certain locations seriously hampering the productivity and quality of the crop. Scale insects are endemic and restricted to certain areas of the country. Wherever these insects are present, they debilitate the crop by sucking the sap from the stalk and affect both quality and productivity. In addition, non-insect pests like Mites also suck the sap from leaves and leaf sheaths. Thrips, Black bugs, White flies, Aphids are of minor importance and in most cases a sugarcane crop is able to withstand the damage.

Among the defoliators, Grasshoppers, Armyworms, Leaf folders and Skippers are prominent. In case of *Hispa*, an adult beetle feeds by scrapping the laminar tissue, whereas at grub stage it leads the life as a leaf miner, and feeds on the parenchymatous tissues by cleverly staying in between the two epidermal layers of the leaf lamina. However, these insect pests are of localized and of minor importance.

Different agro-ecological niches also exercise profound influence on the establishment and subsequent proliferation of certain insect pests in a particular location. In India, sugarcane crop is grown in diverse agro-climatic and edaphic situations. The spectrum of insect pests also varies with the change in agro-ecological niche. The period during which any pest species multiplies profusely is termed as the *period of maximum activity*.

Location or regional specificity is another important factor that determines the establishment and proliferation of a borer species. For example, the Plassey borer is restricted to the eastern part of India whereas, Gurdaspur borer reins in the north-western area of India. On the other hand, Stalk borer is prevalent throughout the subtropical belt, whereas Internode borer, Early shoot borer and Top borer have an all India footprint. Top borer is more problematic in the subtropical belt, whereas Internode borer has an overwhelming presence in the tropical belt. Opportunist *polyphagus* borers, like Pink borer often migrate to sugarcane from other crops whenever these crops share the space with sugarcane or are grown in the vicinity of sugarcane fields. Termites are a pan India problem, whereas the white grubs are localized in some pockets of Bihar, Uttar Pradesh, Maharashtra and Tamil Nadu.

2

Brief history, identification and naming of moth borers of sugarcane in India

"What's in a name? That which we call a rose by any other name would smell as sweet."

That's what William Shakespeare wrote in his famous play, '*Romeo and Juliet*'.

The same sentiment goes with the identification of any crop pest or for that matter anything on earth or solar system. Obviously, the identification is based on the repository of accrued knowledge which is passed on from one generation to the other. Whatever name one uses to identify a crop pest, it will always remain a pest and will continue to inflict losses so long as it uses the crop as its food or shelter or both during its growth and development. The insect does not bother by which name we indentify it. It is the taxonomist's endeavour of naming, grouping and classifying an organism for its straightforwardness and accurate identification in the shortest possible time. A reliable classification and phylogeny are indispensable for correct identification and prediction of various traits (commonality) that are important in the pest management, including the possible exploitation of other organisms (bio-agents) along with host plants.

Frequent misidentifications of sugarcane moth borers in India by different authorities have made the identification a difficult proposition. Different authors have assigned different names for the same moth borer, and accordingly use of different names prevailed in the literature. The correct identification and naming of different moth borers of sugarcane always remained a contentious issue among the entomologists working in India during the early part of the 20th century. At that period, most of the sugarcane moth borers from India were identified and named by Lieut. Col. C. Swinhoe, Sir G. F. Hampson, Mr. G. C. Dudgeon and Mr. T. B. Fletcher.

In the beginning, Mr. E. C. Cotes (Cotes, 1889) who was looking after the Entomology section of Indian Museum, Calcutta collectively placed all the sugarcane borers infesting sugarcane in India as the known American sugarcane borer, *Diatraea saccharalis* F. In that report (Please refer ***Appendix III***), Mr. Cotes erroneously concluded that the communication of Babu Joykissen Mookerjee (1857) to the Agri-Horticultural Society, Calcutta indicating an incidence of disease destroying the Bombay or Red sugarcane in certain districts of Bengal as the infestation of borer. Due to this conclusion of Mr. Cotes, many subsequent authors have taken it as the infestation of *Shoot borer* without going through the original communication of Mr. Mookerjee (please refer ***Appendix I*** for the original communication made by Mr. Mookerjee).

Despite the shortcomings, Mr. Cotes was fully aware that the sugarcane borers present in the Indian subcontinent are different from those found in America. He categorically expressed that

"Further studies are needed to establish the actual identity of sugarcane borers that are present in India".

After the sudden demise of Mr. Lionel de Niceville in December 1901, Mr. E. P. Stebbing of Indian Forest Service stewarded the Indian Museum and made further advancement in the identification of sugarcane borers (Please refer ***Appendix III***). However, it was Mr. H. M. Lefroy, the first Imperial Entomologist, Pusa, Bihar made the beginning of teaching Agricultural Entomology in the newly formed IARI, Pusa and initiated the study of moth borers of sugarcane.

The period of Lionel de Niceville as the Entomologist to the Govt. of India was too short and it was only from January to December, 1901. Further, after his death, the post of Entomologist was transferred from the Indian Museum to the Agricultural Department. Mr. H. M. Lefroy joined as the Imperial Entomologist, Govt. of India in 1903 (filled the position of Mr. Niceville) and was initially posted at Surat, Gujarat for studying the cotton pests. After the establishment of IARI at Pusa, Bihar on 1st April, 1905 he was shifted to Pusa. Due to premature death of his two children, he left India in 1912.

Mr. T. B. Fletcher, succeeded Mr. H. M. Lefroy as the Imperial Entomologist, Pusa, Bihar. Mr. Fletcher addressed this vexed issue of identification and naming of moth borers with the help of his entomological assistants like Mr. C. C. Ghosh. Incidentally, at a later date he initiated Mr. B. D. Gupta in sugarcane entomology to carry out the identification work for the degree of '*Associate IARI*' from Pusa, Bihar. Mr B. D. Gupta made a major stride in the direction of identification of moth borers of sugarcane.

In the Proceedings of the Second Entomological Meeting (1917), held at Pusa Fletcher expressed his view on page 142 as

"With regard to the question of *Diatraea* we are still rather in the dark as to the number of species which occur in India and their identity. Up to a few years ago any borer, found in cane, maize, *juar* and so on, was lumped under the name " Moth Borer " and was supposed to be *Chilo simplex*. On going over the Pusa Collection it struck me that the rather inadequate material lumped together as *Chilo simplex* really comprised more than one species, the specimens bred from cane being on the whole distinct from those reared from maize and *juar*. The difference became of practical importance in 1912 at Coimbatore when the new Sugarcane Station was started there because the surrounding country side was full of fields of *cholam* (*juar*) heavily infested with borers and, if this borer in *cholam* were the same as that attacking cane, the growth of cane on the Sugarcane Station would probably have been seriously affected. I therefore went into the question again with all the material available and came to the conclusion that, generally speaking, the borers in cane and in *cholam* were distinct. This is a conclusion which I have not seen occasion to modify so far, but at present I can only put it forward as a preliminary conclusion. We want to see a good deal more material from all parts of India and I shall be glad if any of you will assist by sending us in long series of specimens in good condition or by sending us parcels of cane, maize or *juar* affected by borers so that we can breed them out for ourselves. It is only by examination of long series in good condition that we shall be in a position to see what species really occur in different localities, by what points they may be separated, and what are the food plants of each."

As indicated earlier, Mr. T. B. Fletcher and his assistant Mr. C. C. Ghosh started working in this direction and presented a detailed report entitled "14. - *Borers in sugarcane, rice etc.*" in the Third Entomological Meeting (1919) held at Pusa, Bihar (page 354-417 + several plates; please refer ***Appendix IV*** for a glimpse of some of the beautiful drawings furnished in this communication). This report provided a detailed account of different moth borers of sugarcane, and indeed, it was the turning point in establishing stability in identification and management of moth borers of sugarcane. In fact, it remained the cornerstone publication in the identification of moth borers of sugarcane in India.

Despite this path breaking effort with excellent drawings, regular misidentification of sugarcane borers remained a major problem with the entomologists of India at large, and it continued for quite a long time. This was due to the quite similar nature of the symptoms produced by most of the borer species at shoot stage, very similar appearance of the larvae and difficulty in spotting the adults (moths) in the field. In addition, the paucity of field identification guide with clear-cut pictures at the disposal of the field entomologists made the situation further difficult. Moreover, inaccessibility of the museum specimens for comparison to establish the actual identity of the borer species (which different authorities named and described differently) added to the confusion. To sort out this problem, Mr. B. D. Gupta started working under the direction of Mr. T. B. Fletcher to develop a key for easy identification of all the stages of the important moth borers of sugarcane with simple characters that can be visualized easily with a simple hand lens. The keys also included the symptoms of infestations. Mr. Gupta came out with a publication in 1940 entitled "*The anatomy, life, and seasonal histories of striped moth-borers of sugarcane in north Bihar and West United Province*" in the *Indian Journal of Agricultural Science*. At that time, he was serving as Cane Entomologist, Sugarcane Research Station, Muzaffarnagar, United Provinces. After this effort, Mr. P. V. Issac, Mr. T. V. Venkatraman and a decade later, Mr. V. D. Puri further advanced the work on the identification of sugarcane moth borers. The Technical Bulletin written by Mr. B. D. Gupta (1958) entitled "*Identification of the striped moth borers of sugarcane in India*" for the identification of the striped borers of sugarcane remained the culmination point of all these efforts. Incidentally, it was the first Technical Bulletin published by the newly established institute, the Indian Institute of Sugarcane Research, Lucknow.

A significant milestone to advance the researches on *Sugarcane Entomology* in India was the establishment of the Indian Institute of Sugarcane Research (IISR), Lucknow, Uttar Pradesh, at the heartland of subtropical sugarcane belt with a pan India mandate. IISR was established on 16th February 1952 by The Indian Central Sugarcane Committee, New Delhi. The Institute started functioning from the day one, as Mr. B. D. Gupta was appointed (Class I, Gazetted) a month earlier (on 4th January 1952) by the Indian Central Sugarcane Committee and subsequently he was relocated at the *Bhadruk farm*, Lucknow on 4th February 1952. He became the first full time Central Sugarcane Entomologist and headed the Entomology Section of IISR. As the Central Sugarcane Entomologist, Mr. Gupta was entrusted with the responsibility to coordinate all the research work related to Sugarcane Entomology in India, in addition to his own research work on sugarcane insect pests as mandated by the IISR.

During the past six decades, a lot of changes have taken place in taxonomy, nomenclature and reassignment of the sugarcane moth borers. The recent advancement of scientific technologies in the taxonomy of insects has made the identification easier. Molecular biology tools have further helped in the refinement in identifying characters and have paved the way for the *DNA barcoding* of different insect pests including the moth borers of sugarcane.

Whatever may be the taxonomic changes taking place in the background, a sugarcane pest will always remains a sugarcane pest so long as sugarcane remains in cultivation and the pest is able to utilize sugarcane for its growth and development. Thus, the symptoms of infestation of a particular insect pest in the cane crop in combination with the actual presence of the pest in the field always help in the accurate identification of the species involved. At the field level a farmer or an extension worker has to be well conversant with the morphological features of borers (larvae) to identify the pest species involved and to respond appropriately to contain them.

The changes that have taken place in last six decades have necessitated a conscious revisit to the subject of identification of moth borers of sugarcane in India. In fact, after Mr. A. P. Kapur (1950), and Mr. B. D. Gupta (1958) and a series of publications of Mr. V. D. Puri in 1950s, it was Mr. G. S. Arora (2000), who made an excellent general effort to review this area of identification of Pyralid moth borers of different crops including sugarcane in the publication "*Studies on some Indian Pyralid species of Economic Importance Part I*". However, it mostly dealt with the adult characters. In the words of Mr. G. S. Arora (2000)

> "... studying the important pests, both major as well as minor, in the family Pyralidae, so as to provide an easy key as an identification guide, not only for the taxonomists but to the agricultural entomologists also to make them aware of the entire pest complex in their field of work."

Taxonomy of insects is largely based on the characters of the adult insects. Obviously, the presence of adults or moths confirms the identity of borer species beyond doubt. In the taxonomy of the members of *Lepidoptera*, the shape, size and colour of the body and wings; shape of the face, presence of *labial palpus* and its size in relation to the compound eye; presence or absence of *ocelli*; venation pattern of the wings; structure of male and female genitalia, presence of *tympanum*, etc. are important for the classification. However, the adult of most of the moth borers of sugarcane (except Top borer) always remain obscure to the normal gaze and thus, farmers rarely encounter adult of these moth borers in the field. Obviously, farmers are not in a position to relate the adult of the moth borers directly to the caterpillar stage, which they mostly encounter in the infested cane. With the discovery of pheromone, getting an adult of the specific moth borer from the infested field has eased a little. Presence of the adults of a particular species of moth borer may be observed in the field with the use of species-specific pheromone traps. These pheromone traps are laced with a chemical, same or similar to what a virgin female moth releases in the environment to draw attention of the male moths. Obviously, this trap attracts the male moths and thus, only male moths of a species are detected.

The current effort is rather restricted only to sugarcane moth borers. This effort is to provide a fresh look into the above subject and that too more from the standpoint of a cane farmer, a sugarcane development worker, and a practising sugarcane entomologist, who are actually responsible for management of borers in sugarcane crop. Pictures depicting the symptoms of infestation in detail by different moth borers are provided in this treatment, as symptoms alone are generally the guiding principle in establishing the identity of the pest involved in the crop damage. This treatment is centred on the identification of the moth borer species with the help of a simple hand lens (10x).

To make the presentation simple and easily comprehendible, the common entomological jargons have been avoided (left out to retain the simplicity), like the details of *chaetotaxy*, explanations of appendages on head, thoracic segments etc. This publication puts forward the motto "*Seeing is believing and learning by doing*". It embodies photographic essay of important moth borers of sugarcane. Many a times even an accurate drawing or sketch fails to impart a lasting impression on the reader that a good colour photograph does easily. A telling colour photograph explains many things at a glance, without the need of much explanation. A range of colour photographs, provided in this publication will help even a nonprofessional to identify the borer species accurately without much difficulty.

Some line drawings from the publication of Fletcher and Ghosh (1919) have been appended (please refer ***Appendix IV***) to illustrate the brevity of the corner stone publication on Sugarcane Entomology. In the present endeavour appropriate photographs of different stages in the life cycle of moth borers have been taken to highlight different characters that have a direct bearing on identification and classification.

Most of the photographs furnished here are of live insects (to retain their natural lustre, shape and size). For taking photographs, a number of larvae of each species was collected from the field and maintained in the laboratory for pupal transformation as well as for the final transformation to an adult (moth). A simple technique using sugarcane stalk was developed and used for rearing the larvae. Sugarcane stalk of Co 1148 and CoJ 64 (stalks without any pith) were chosen for this purpose. In the internode portion, a hole was made with *IISR inoculator* (a plunger type instrument that is commonly used to make holes in the standing cane for red rot inoculation). A moth borer larva was placed inside the hole made in the internode of the cane. The opening of the hole was blocked with a portion of the extricated stalk tissue and finally the edges were lightly sealed with adhesive tape to avoid moisture loss. Stalks with two to three crown leaves were kept in standing position and the cut end of the stalk was placed in water kept in a big bucket to retain the turgidity of the stalk tissues. The pre-pupal larva invariably prepared an exit hole in the stalk to facilitate the emergence of the moth. Once the exit hole became apparent, the portion of the stalk carrying the larva was carefully cut and placed with appropriate orientation in a two feet long cylindrical glass specimen jar. Moist sand was kept at the bottom of the jar to maintain the humidity. The mouth of the glass jar was covered with a muslin cloth to prevent the escape of the moth on emergence. Each morning the exit holes were observed for the emergence of the moth. Once an adult was observed in the jar, it was collected for photography and further study.

3

An overview of sugarcane moth borers in India

The borers are the larvae of the moths that actually damage the sugarcane crop. A larva enters into the cane by making a small hole in the epidermal tissues (leaf, leaf sheath and stalk) and makes its way inside by making a feeding tunnel in the tissues. A larva mostly stays inside the protecting cover of the cane tissues except occasional movement outside in search of new feeding location. Pupation mostly takes place inside the feeding tunnel and the adult emerges through the exit hole. Due to the drab straw colour of most of the moths (adult of Top borer is white) they remain almost camouflaged with the dry sugarcane leaves. When these moths sit on a cane stalk, they quickly align their body almost parallel with the stalk (vertical alignment) and thus, become inconspicuous to the normal gaze. Most of moth borers that are important pests of sugarcane in India are grouped into the super family *Pyraloidea* with the family *Pyralidae/ Crambidae* (the snouted moths/grass moths) and *Noctuidae* (the owlet moths, nocturnal in habit).

In India, thirteen species of moth borers have been recognised as the pests of sugarcane. Out of these, only nine species are considered important with substantial economic significance. These nine species belong to six genera of Lepidoptera (five of *Pyralids*), viz., *Acigona (1), Chilo (4), Emmalocera (1), Raphimetopus (1), Scirpophaga (1)*, and (one of *Noctuids*) *Sesamia (1)*.

The four species viz., *Acigona tauromma* Kapur, two purple borers *Phragmatoecia purpurea* Fletcher and *Phragmatoecia terebriefer* Fletcher and one six-striped borer *Procometis trochala* Meyrick are mostly found in the wild species of *Saccharum* and its allied genera and may occasionally infest sugarcane crop. These four moth borers are rarely found in the commercial plantations of sugarcane.

The Maize or Sorghum borer (*Chilo partellus*) occasionally infests sugarcane in an insignificant manner. As apprehended by Fletcher (1917), the Sorghum borer is not a regular and important borer pest of sugarcane (please refer Chapter 2 for the passage he outlined). As on today, for all practical purpose, *Chilo partellus* is not considered a pest of sugarcane. Despite this fact, a chapter on *Chilo partellus* is included here to retain the continuity of earlier publications and to avoid any confusion that often arises in the identification and differentiation of larvae of Maize and Plassey borer.

All these nine important moth borers of sugarcane are not equally important. These borers differ with respect to their occurrence, distribution, and intensity of crop damage. Moreover, these borers have also adapted to

row is called *uniserial*, two rows arrangement is called *biserial* and arrangement of more than two rows is called *multiserial*. Dorsally, on the 10th abdominal segment there is also a *sclerotized plate* called the *anal plate* or *anal shield*.

The occurrence and distribution of the setae and punctures present on the larval body are of major taxonomic significance. The arrangement of these structures on the larval body is called *chaetotaxy*. (*Chaetotaxy* is not dealt here as in field identification this bears very little or no significance).

3.1.3 Pupa (chrysalis)

Pupa is the inactive and almost defenceless state of these moth borers. In the pupal state internal organs and structures of the larva are completely reorganized (*metamorphosis*) to produce a winged adult. The head, thorax and abdomen of the pupa can be readily differentiated. The head features a pair of prominent compound eyes, well-defined antennae and other structures. The thorax has three distinct segments. Among these, *mesothorax* is the largest and prominent segment. The prominent forewings supported by the *mesothorax* largely conceal the hind wings. Usually the *prothoracic shield* forms one arch each at both the distal ends to make room for the *mesothoracic spiracle pair*. *Metathorax* is relatively smaller and supports the hind wings. Abdomen consists of ten segments; the last three segments always remain fused together and bear the external genital structures or markings. *Spiracle* pair of the first abdominal segment is not visible. On the 2nd and 3rd abdominal segments, *spiracles* often show characteristic *arch formation*. The wing pad area extends up to the 4th abdominal segment and thus, up to that segment the pupal body remains stiff and a pupa is unable to sway these segments. *Spiracle* pair present on the 8th abdominal segment is poorly developed and non-functional. The 5th, 6th and the 7th abdominal segments usually remain free and a pupa can sway these segments. Many a times these three segments possess certain growth or processes, recognised as *ornamentations*. These pupal ornamentations vary widely with the borer species and thus, play an important role in the quick identification of a species. Usually the *ornamentations* are in the form of spine like projections (Stalk borer, Plassey borer) or processes (Internode borer) or ridges (Early shoot borer). These epidermal growths on the pupal body help in the movement of a pupa in distress. The pupae of moth borers of sugarcane are typically *obtected* type i.e., developing wings, legs, mandibles, maxillae and antenna are strongly cemented to the surface of the body. However, in case of Top borer such adhesion is not so firm, terminal portion of legs, especially in male pupa remain free from the body surface.

Identification of the gender of a larva is difficult, whereas the gender of a pupa is easily discernible. Female pupa of a moth borer is larger than the male pupa of that species. In a female pupa, the 8th abdominal segment supports the genital marking and in male, the genital marking is present on the 9th segment. The 10th or the anal segment bears anal markings and various types of ornamentations (*cremaster*) and these ornamentations often become very characteristic for a species.

3.1.4 Adult (imago)

The winged phase of the moth borers helps in their quick dispersal from one location to another. Insect taxonomy is largely based on the morphology of an adult or winged phase of the insect. The body of the adults of sugarcane moth borers can be divided into three well-defined parts i.e., *head*, *thorax* and *abdomen*. The head is generally rounded with some modifications of the face. These modifications of face help in the identification of the species (Gurdaspur borer and Maize borer have pointed projections that can be felt on a subtle touch). The compound eyes are more prominent on the head and the diameter of the compound eye is taken as the reference parameter for measuring the length of *labial palpus.* Behind the compound eyes, antennae are present. Antennae are mostly *filiform* in female moths. However, in male moths there are variations among the borer species (*filiform* in Top borer whereas, *pectinate* in Root borer). Behind the *scape* of the antenna and just above the compound eye, a dark coloured round structure called *ocellus* is present (*ocellus* is characteristically absent in the Internode borer).

The thorax consists of three segments i.e., *pro, meso* and *metathorax* and one pair of legs are present in each thoracic segments. *Mesothroax* supports the pair of forewings and the *metathorax* supports the hind wings.

3.1.5 Wings and forewing venation

The wings of *Lepidoptera* bear scales. The wings are essentially membranous sacs with both the upper wall and lower wall pressed together. The forewing has three sides like that of a triangle. The leading edge of the outstretched wing is called *costa;* the outer margin joining the *costa* is called *termen* and the posterior margin is called *dorsum.* The tip of the outstretched wing forming the angle between the *costa* and the *termen* is called *apex.* The *apex* is the distal portion of the wing. Six main veins are recognized in the wing, viz., *costa* (C), *subcosta* (Sc), *radius* (R), *media* (M) *cubitus* (Cu) and *anal veins* (A). The closed area formed between the stem of CuA and the stem of R is called *discal cell* or simply the *cell.* The coastal vein runs along the leading edge of the wings and provides the necessary strength to the front and other veins follow the order as listed above.

In Comstock and Needham system of designation of veins of the wings, branches of *radius* in the forewing are designated as R_1, and Rs (*radial sector*). The branches of *radial sector* (Rs) are further termed as R_2, R_3, R_4 and R_5. When the R_3 and R_4 develop as forking of one *radial sector,* these are called stalked. R_5 originates from the cell arch. The branches of *media* are designated as M_1, M_2 and M_3. The media veins appear on the cross vein of the *discal cell.* Branching of *cubitus* is designated as Cu1a and Cu1b. In addition to these veins, in Top borer there is a *post cubitus vein* (CuP). It is present in the area between *cubitus* and *anal veins* and it starts from midway and ends at *termen.* In general, three *anal veins* are recognised but in most cases, only two veins are present and out of these, only one *anal vein* (A) is well developed. Due to the specific position, it takes on the forewing, the main anal vein is designated as A_2 in this publication.

In *Chilo* moths, vein R_3 and R_4 are always stalked. There are variations in the development of *subcosta* (Sc) and R_1. These two veins may anastomose or remain free (the anastomosed or connate nature of this two veins lead to the erection of the genus *Chilotraea* by Kapur (1950).

3.1.6 Abdomen

The abdomen is made up of ten segments; the last three are always fused together to form the external genitalia. In an abdominal segment, the dorsal plate is called *tergum* and the ventral plate is called *sternum.* These two plates are joined at the side of the body by the *pleural membranes.* Functional *spiracles* are present on the *pleural membranes* of the first seven segments. A pair of *tympanal* organ is present at the base of the abdomen and remains in hiding (covered) under the *coxa* of hind legs. The colour of the abdomen varies depending on the development of the scales and their colour. The external genitalia of a female moth i.e., *papillae anales* or ovipositor are densely clothed with sensory setae.

3.2 General biology of the sugarcane moth borers

In the subtropical north India, sugarcane has to withstand extremes of temperature during both summer and winter and the variation in ambient temperatures regulate the life cycle of these borers (duration of brood, number of broods and period of hibernation). In this extreme ambient environment, all the moth borers also follow a similar pattern; hibernate during the cold winter months and resume their activity in spring (around February) when the ambient temperature becomes favourable. However, Gurdaspur borer delays it emergence and waits till the summer temperature continues.

In general, an adult (moth) is seldom noticed in the infested fields of sugarcane (except that of Top borer which can be noticed easily due to its silvery white colour, especially during morning and afternoon hours). Moths are nocturnal and fly during the night. During daytime, moths mostly remain in hiding among the dry or old cane leaves and in the soil cracks. As these moths are relatively smaller and light-brown or straw coloured (very similar to the dead and dried sugarcane leaves), spotting of a moth in the field is extremely difficult, even for an expert.

3.2.1 Egg laying habit, hatching and neonate larva

Egg laying habit of these moth borers is also very similar. Generally, these moths lay eggs to form a well-arranged imbricate row (serially overlapping like tiles on the roof) of egg mass, except the Root borer, which lays egg singly or in a scattered manner. The Noctuid moth *Sesamia* pushes the eggs within the leaf sheath and collectively the eggs form linear to sheet arrangements but never overlap. Similarly, the egg masses of these moth borers mostly remain naked i.e., without any protective cover (except the egg masses of *Scirpophaga*, which are covered with anal hairs).

In general, before the commencement of egg laying, a female moth selects a suitable spot, aligns itself parallel to the mid rib and then lays an egg. Once an egg is laid, the moth carefully attaches the egg with the leaf surface by pressing it down (flattening it) against the leaf with the ovipositor. In this way the female moth ensures that all the egg are securely attached to the leaf surface. Hereafter, the moth moves down a bit, so that another egg can be deposited in a manner in which the new egg covers around ½ to ⅔rd of the egg previously deposited. By a pendulum like sideways movement of the ovipositor, the moth lays two to three parallel rows of eggs and forms an egg mass that usually contains 20-30 eggs in a cluster.

Generally, the eggs of moth borers need around 7-10 days of incubation to hatch. Hatching usually takes place during the morning hours. All the eggs in an egg mass do not hatch at one go; rather these eggs hatch in batches. Hatching of an egg-mass mostly completes in an hour or so.

3.2.2 Larval growth and behaviour

The larval behaviour differs with the different borer species. In some, the neonate larvae are *gregarious* (Gurdaspur borer and Plassey borer) or *partially gregarious* or *pseudogregarious* (Internode borer, Stalk borer) and in others, larvae are more individualistic (Top borer, Root borer). Some borers are not satisfied at feeding in one location or plant and migrate to another location of the same shoot/stalk or to the adjoining shoot/stalk and thus, a larva of these borers damage a number of shoots or canes before pupation (Internode borer, Early shoot borer, Gurdaspur borer and Plassey borer). There are others that always remain restricted to a cane or shoot (Top borer). Some prefer high temperature and dry weather (Early shoot borer, Green borer), some tolerate cold weather (Stalk borer) and some others prefer high humidity and medium ambient temperature and proliferate well during the rainy season (Stalk borer, Plassey borer and Gurdaspur borer). Usually one species of borer larva(e) is found in the affected cane. In many instances, two species of larva share the same cane. It is also not uncommon to find three or more borer species infesting the same cane (Top borer, Internode borer, Stalk borer and Root borer).

The period from hatching until the first moult of the larva is called the *first instar*. The larval state between the first and second moult is called *second instar* and thus, with growth of larva the number of instars advances until the larva moults to the pupal stage. The number of instars a larva usually passes through during its growth and development is usually around five. However, larval stage is very flexible and amenable to intermediate change in the duration. A borer banks upon this stage for the need based adjustments in the duration of larval stage. Depending on the prevailing ambient environment (especially the temperature), the number of larval instars may range from four (in high temperature) to seven (in cold environment) in a brood. These moth borers mostly tide over the cold winter months in larval stage. Obviously, at this phase a larva has the longest span.

Generally, the egg period is around 1 to $1\frac{1}{2}$ week and so is the pupal period. An adult (moth) usually lives up to 4 to 5 days. The larval size (length) usually ranges from 25-35 mm and this larval growth is attained in around 30-35 days. Thus, a larva usually attains a mean growth rate of around 0.8 to 1 mm/day.

3.2.3 Moulting or *ecdysis*

Moulting or *ecdysis* (shedding of old skin or cuticle) remains an important activity of the borer larva for sustaining growth and development. During *ecdysis* a larva discards the old skin and head capsule by recasting a larger new skin and head capsule underneath of the old. The developing head capsule appears white to light creamy. Before the onset of the moulting process, a larva stops feeding for a few hours and appears sluggish or motionless. At certain point of such rest, the old dorsal skin at the back of the head (in front of the *prothoracic* shield) ruptures to make way for the new skin to get exposed to the air. Hereafter, through a series of contractions and wriggling movement the old skin is pushed backward over the body. The newly formed head underneath dislodges the old head capsule. The increase in size of the head capsule in each instar follows a definite *geometric progression* and thus, it is possible to establish the actual instar a larva is passing through by the measurement of head capsule. In this process of *ecdysis*, dislodging of the old head capsule takes place separately. The whole process of *ecdysis* takes about 15-60 min. In the case of Internode borer, the *ecdysis* needs about 15-20 min, whereas in Top borer *ecdysis* needs around an hour to complete. In the *ecdysis* of first instar larva, the head capsule remains intact (does not separate into two halves). During the last moulting, transformation of a larva into a pupa takes place. In this *ecdysis*, the two halves of the head capsule are partially separated and remain attached to the rest of the skin; the thoracic legs also go away with the *exuvium* (all the remains are dislodged at the same place i.e. at the anal end of a pupa).

3.2.4 Pupal stage

Location or place of pupation varies depending on the species and it again depends on the brood a moth borer is passing through. Regarding location, most moth borers stay inside the cane during hibernation and pupate within the feeding tunnel (except Green borer, which pupates in soil). Pupation may takes place within the cane stalk or outside the cane stalk. Whenever, pupation takes place within the cane stalk (feeding tunnel), a mature larva (prepupal) makes an exit hole (larger enough for the safe exit of the moth) on the stalk tissue and seals the hole with the host and silk. Whenever, a larva pupates outside the host tissues, it weaves a strong silken barrier around it. Before entering into pupation phase, a mature larva passes through a near sedentary stage, called *prepupal stage*. A *prepupal larva* weaves a specialized resting chamber with silk called the *pupal chamber*. A larva also prepares an exit area for safe emergence of an adult from the pupal chamber. It may be noted that by cleverly residing in the feeding tunnel during pupation a borer gets the requisite protection for the pupa from the attack of parasites and predators. Just before transformation into a pupa, a prepupal larval body contracts longitudinally, especially at the thorax, and suddenly a crack appears at the head capsule to mark the beginning of the process of shedding of the skin. In this moulting process, the head capsule as well as all the larval legs accompany the *exuvium*.

The developing adult within a pupa is called *pharate adult*. When development of the *pharate adult* is complete, a crack appears just behind the head region of the pupa to facilitate the emergence of the moth. The emerging moth pushes forward its head widening the crack both vertically and horizontally, and finally the winged adult emerges outside. The scales present on the body act as a dry lubricant and help the safe exit of the moth from the pupal shell and its passage through the feeding tunnel for the final escape through the exit hole. Immediately on emergence, the wings of the moth remain close to the body and the newly emerged moth rests for a while to allow the wings to expand, to garner strength and complete the development process. Finally, the moth flies away in search of a mate. Through laying of fertilized eggs, moths start a new generation. In the first brood, the brood that appears after winter inactivity or hibernation, a greater uniformity is often observed in the emergence of moths (duration of peak emergence is short). However, in subsequent broods, the duration of peak emergence as well as the brood period is variable, and as a result overlapping of broods is common.

3.3 Tiding over the unfavourable ambient environment

Sugarcane growing areas of subtropical north India experience two extremes of temperature; very high in summer (maximum temperature > 40°C) and very low in winter temperature (minimum temperature < 5°C). Both these extremes of temperatures are not favourable for the growth and development of most of the moth borers. Almost all the moth borers of sugarcane withstand the summer temperature in active state and respond to the situation by reducing the brood period or generation time. (Gurdaspur borer appears only after the period of peak summer temperature is over. Stalk borer tides over the high temperature period with very limited activity and responds with an increase in the brood period. In the Green borer, the period of inactivity starts from the rainy season and continues to the following winter till it appears in next February as an adult moth). Moreover, in the summer months, dry summer winds drastically reduce the chance of survival of neonate larvae due to quick desiccation. This situation arrests the quick proliferation of most of the borer pests during the summer (Early shoot borer predominates during the dry summer months). The monsoon borers (Stalk, Internode, Gurdaspur and Plassey borer) are more comfortable during the monsoon months (high humidity and moderate temperature).

In the cold winter months of subtropical India, moth borers of sugarcane feel uncomfortable with the lower ambient temperature and majority of them tide over the situation in quiescent or inactive larval stage (Stalk borer larva withstands colder temperature well and remains active during the winter months except very severe cold days). The hibernation in some of these moth borer species is not a complete one. A hibernating larva quickly responds to any warm days in between, becomes active, and munches some of the surrounding plant tissues wherein it has taken shelter. With the onset of favourable ambient temperature, it makes an exit hole and proceeds for pupation.

3.4 Symptoms of infestations

Borer infestation on sugarcane may cause several types of damages viz.,

(1) Formation of *dead heart* at shoot stage (up to four months crop age), and death of mother shoot and tillers.

(2) Formation of dead crown /*dead heart* in older plants (when cane formation has set in).

(3) Complete yellowing or partial yellowing of the crown leaves.

(4) Sprouting of buds of the stalk, especially below the affected portion. Often rooting takes place from the adjoining nodes. Shortening of internodes also takes place.

(5) Formation of *bunchy top* in mature cane due to growth of side shoots just below the damaged crown region. In immature cane, such bud proliferation takes place at soil level and this side shoots often act as tillers.

(6) Damage of internal content of stalk tissue and resultant feeding cavity remains either hollow or filled with frass. Due to secondary infection by microorganisms, rotting of the internal content also takes place.

(7) Damaged or broken-over stalks, weakening of the basal portion of the cane leading to wilting or lodging.

(8) Loss in weight and extractable juice of cane.

(9) Loss in sucrose content, quality of juice and in sugar recovery.

(10) Difficulty in milling due to shortening of internodes as well as hardening of nodes and

(11) Deterioration of the quality of seed cane (germination of the cane is impaired due to injury to bud and cane stalk).

The injury known as '*dead heart*' is recognized by the drying up of only the inner central leaves of the crown (mostly unfolded) in contrast to the healthy and green outer leaves of the crown (in other words, heart of the crown dies and further growth of the plant stops). The infestation in the shoot stage when cane formation is yet to start, due to the eating up of the underground portion of the stalk or base of the immature crown leaves of a young shoot, a '*dead heart*' is formed. Depending on the number of leaves involved in the formation of a *dead heart,* it is grouped as either small (dying of up to two central leaves of the crown) or large (dying of more than two central leaves of the crown. A small *dead heart* is usually formed due to the infestation of Top borer, whereas formation of a large *dead heart* usually takes place due to the infestation of Pink and Root borer. Majority of *dead heart* formation takes place during the shoot stage. However, formation of *dead heart* continues further in later stages of crop growth. In case of Top borer, *dead heart* is formed in every brood, irrespective of the growth and maturity of the cane whereas, in case of Root borer *dead heart* is formed only in the shoot stage when cane girth is not adequate to sustain the borer damage; when cane girth increases, no *dead heart* is formed. Plassey borer also form dead heart at shoot stage. At later stage (Grand growth phase) the infestations of both Plassey borer and Gurdaspur borer result in drying of the entire crown.

Dead heart formation at the initial stage of the shoot growth often boosts tillering due to the breaking of apical dominance (like in Top borer) but in certain cases, where stalk portion has been damaged badly (as in case of Early shoot borer in dry summer weather), the affected shoot often dies. Due to this variable response of cane plant, it is difficult to determine losses from the formation of *dead heart* in the young sugarcane plants.

Borer damage is not easily recognized in mature/old plants (except in case of Top borer) as in the young, unless the infestation is too serious to reckon with. Some borers (like Plassey borer), tunnel up and down the stalk, gnaw their way out through the hard rind and re-enter at another place. Thus, these species of borer cause maximum damage to the immature portion of the stalk and as a result the crown leaves turn yellow and eventually die. Usually, the leaf sheaths need to be stripped off to detect the entry and exit holes (Stalk borer, Internode borer). The frass often pushed out by a larva from the feeding tunnel helps in the field identification (Internode borer, Plassey borer). In older plants, when a larva attacks the immature stalk near the apical dome (like the Internode borer), the severed portion dies out (may form a *dead heart* or a few leaves of the crown die). As a result of the death of apical growing point, the apical dominance is broken and buds near the growing point may sprout resulting in a '*bunchy top*' appearance of the crown. The borers that tunnel near the epidermal region of the stalk or make tunnel encircling the stalk eventually render the stalk very weak. Upper part of the cane plant or the crown portion of such damaged stalks often fails to carry the load and breaks away. The crown portion topples down, completely breaks up and touches the ground or remains hanging from the stalk (like in case of damage caused by the Internode borer and Gurdaspur borer).

3.5 Grouping of borers

For an easy understanding of the borer complex of sugarcane, different borers are grouped according to the time (season) and nature of infestation of the crop, physical appearance of the borer larva and their endemicity at certain location. All these nine important moth borers are categorized according to the body colour of the larva and the number of coloured stripes present on the body. The striped borer may be taken as the partially coloured borer. Pink borer and Green borer are the two coloured borers whereas, Top borer and Root borer are the two white to cream coloured borers. Early shoot borer and Stalk borer are the two five-striped borers. Internode borer, Plassey borer, and Gurdaspur are the three four-striped borers.

3.5.1 The Shoot borers

In the growth and development of a sugarcane crop, first four months of its life from germination is called shoot stage. Accordingly, the borers that infest cane plants in the formative phase are collectively called '*shoot borers*.' In subtropical India, spring planting of sugarcane (February-March) is commonly practised and in this period, sugarcane setts need four to six weeks to complete the germination phase. Thus, the shoot stage of the crop coincides with the summer months (April, May and June). These borers are also called '*pre-monsoon*' borers as their infestations increase considerably before the onset of monsoon.

At shoot stage infestation following borers are often encountered.,

(i) Early shoot borer, *Chilo infuscatellus* Snellen
(ii) Green borer, *Raphimetopus ablutella* (Zeller)
(iii) Pink borer, *Sesamia inferens* (Walker)
(iv) Root borer, *Emmalocera depressella* (Swinhoe)
(v) Top borer, *Scirpophaga excerptalis* (Walker)
(vi) Plassey borer, *Chilo tumidicostalis* (Hampson).
(vii) Stalk borer, *Chilo auricilius* Dudgeon
(viii) Internode borer, *Chilo sacchariphagus indicus* (Kapur)

Among these, the first mentioned four borers predominate at the shoot stage of the crop and Early shoot borer takes the lion share. Due to this, *Chilo infuscatellus* is often recognised as the '*shoot borer*'.

The Pink borer is not a regular pests of sugarcane but at the time of maturity of wheat crop, it often migrates from wheat to the nearby sugarcane field and causes *dead hearts* at shoot stage. In addition, during this period, Green borer, Stalk borer, Internode borer and Plassey borer also infest the crop and produce similar *dead heart* symptom. However, the intensity of infestation of these borers remain much less than the early shoot borer. At the shoot stage (mostly in ratoon sprouts), Stalk borer infestation is mostly observed during March and its infestation sharply declines with the rise in ambient summer temperature in the late April, May and June.

At shoot stage, these borers follow a similar pattern of infestation. The neonate larva feeds singly or in groups on the tender leaf and or leaf sheaths for four to six days. Hereafter, these larvae disperse to find a suitable shoot to feed. On reaching an uninfested shoot, a larva bores a hole in the *psuedostem* (the upright portion of a young shoot above the apical dome formed by the tight leaf sheaths), and enters into the shoot. The larva feeds on the soft immature central crown leaves and ultimately destroys the growing point including immature stalk portion below. The central whorl of leaves dries up and forms the characteristic large *dead heart.*

Larvae of these shoot borers are migratory in habit and a larva damages a number of young shoots before pupation. To catch a larva, an affected shoot bearing freshly formed *dead heart* has to be split open carefully. The larvae of these borers are very active, on extrication from the shoot quickly move away, and try to hide under a cover.

All these shoot borers also damage the crop as *cane borers* (borer attacking cane stalk during grand growth phase and later). However, their incidence and damage potential diminish substantially in the grand growth phase of the crop. For this reason, these borers always draw limited attention in comparison to the other specialized borers (Stalk, Internode, Gurdaspur and Plassey) that damage the stalk with much aggression.

3.5.2 Larva with and without body colour

Depending on the body colour of a larva these nine important sugarcane moth borers are further grouped into following four categories:

(a) ***The fully coloured borers***

Green borer and Pink borer

(b) ***The partially coloured borers (striped borers)***

Early shoot borer, Stalk borer, Internode borer, Plassey borer, and Gurdaspur borer.

The striped borers are further subdivided into two categories viz.,

(i) *Four striped borers*

Internode borer, Plassey borer and Gurdaspur borer

(ii) *Five striped borers*

Early shoot borer and Stalk borer

(c) ***The colourless or white/cream coloured borers***

Top borer and Root borer.

Usually larvae bearing stripes are called '*striped borers*' and are treated separately. The colour of the stripes (pinkish violet) is not a permanent feature present on the larva and usually this pinkish-violet colour fades away during hibernation and also in a prepupal larva. For this reason, I have termed these striped borers as '*partially coloured*'.

3.5.3 The two borers with white/creamy larva (Top and Root)

The field identification of two moth borers of sugarcane with white/cream coloured larva is not difficult due to their physical appearance, behaviour of movement on extrication from the host tissue, feeding habit and the damage symptoms they result in. Top borer always attacks the crown portion of the cane plant (for this reason it got the name), whereas the Root borer always infests the underground portion of the stalk and reside near the root (but not in the root). Top borer infestation displays characteristic entry tunnels on the mid-rib of crown leaves, characteristic *dead heart* and *shot holes* as well as *drooping leaf* symptom.

The difficulty sometime arises during the summer months when the sugarcane shoots remain small and there is very limited stalk formation. This situation forces both the borers to utilize the same food i.e., the immature stalk tissue of the shoot. However, a close look at the borer larva, its physical appearance and its behaviour of movement immediately helps to differentiate the two.

Top borer larva is sluggish and fragile, whereas the larva of Root borer is relatively more active and stout. Top borer larva is creamy white, while the larva of Root borer is shiny milky white. Root borer larva remains white throughout the larval period, whereas colour of a Top borer larva changes with maturity and a fully mature larva is creamy yellowish in colour. Body of a Top borer larva is mostly cylindrical throughout and just abruptly tapers at both ends and possess relatively smaller head capsule in comparison to its body. Root borer larva is wedge shaped (conical), much broader at the thoracic region, gradually tapers towards the anal end and the head capsule is relatively larger in comparison to its body. Root borer larva has a relatively larger, well-developed light yellowish *prothoracic shield*. *Spiracles* as well as body hairs are prominent in Root borer. Whereas, in Top borer, the larva is smooth bodied (except neonate larva) with inconspicuous *spiracles*. Crochets on the *prolegs* are poorly developed in Top borer as compared to that of the Root borer. In both crochets or hooks on the abdominal *prolegs* are arranged in a complete circle (*elliptic* in Top borer and *pear shape*d in Root borer). These two borers also differ in the number of hooks on the *calspers* (*anal prolegs*). Root borer bears around 11 crochets whereas Top borer has around 20-22 crochets.

3.5.4 The two borers with coloured larva (Pink and Green)

The two coloured borers (body of the borers are fully coloured) are essentially pre-monsoon borers. These two borers proliferate during the summer months and inflict maximum damage to a sugarcane crop at shoot stage. Field identification of these two borers is not difficult. Pink borer is flesh coloured at dorsal and subdorsal areas and creamy to light yellowish at the ventral side i.e. below the *tracheal trunk*.

The Green borer is dorsally brown-coppery and ventrally green. Typically, *prothorax* remains green. In contrast to the Pink borer, the Green borer has a very restricted host range. Green borer mainly attacks the young shoots of sugarcane during February-May.

3.5.5 The three four-striped borers

The commonly called '*striped borers*' may be considered as partially coloured borers as indicated earlier. The pinkish-violet colour of the stripes varies depending on the season and growth stage of the larva. The colour of the stripes usually fades away during hibernation and in larva proceeding for pupation. At this stage, these borers display variable shades of colour or appear colourless but they never become milky or creamy white like that of Root borer and Top borer.

The four striped borers have one pair subdorsal and one pair lateral stripes on the body. The three four-striped cane borers viz., Internode borer, Plassey borer and Gurdaspur borer are categorised as *monsoon borers*, as these borers usually attain a sizeable population and cause accelerated damage to the sugarcane crop during the rainy season. The infestation of these borers and consequent damage to the canes continues up to October-November until the prevailing ambient temperature (colder winter temperature) forces these borers to retreat into the phase of inactivity (*hibernation*). Internode borer and Plassey borer make their appearance early in the season, as the temperature in the eastern region of India becomes favourable much earlier, whereas Gurdaspur borer that prevails in the western region of the country continues to hibernate during the summer months and appear only after receiving the cue from pre-monsoon showers.

The Plassey borer and Gurdaspur borer have two distinct phases, i.e.,

(i) An initial *gregarious phase*, wherein most of the larvae of one egg mass follow the leader larva and enter into the stalk and share the same feeding tunnel, and

(ii) A *solitary phase* or dispersal phase, wherein the larvae sharing the same feeding tunnel start migrating to the uninfested stalk area of the same or different stalks.

3.5.6 The two five striped borers

Presence of the median stripe is very characteristic and in dorsal view three stripes remain always visible. These two well-recognised five striped borers i.e., Early shoot borer, and Stalk borer, actually possess seven stripes on their body. In addition to the one pair of subdorsal, one pair lateral and a median stripe, just corresponding to the alimentary canal underneath, a larva also bears another pair of ephemeral stripe well below the *tracheal trunk* and runs through the tubercles. Five stripes are very prominent, continuous, and found on the larval body most of the time. Generally, ephemeral pair appears discontinuous, only the tubercles and adjoining area retain the colour of the stripe. Colour fades away very fast after moulting, and in a mature larva, it leaves only a faint impression. The colour is relatively more intense (prominent) in the Early shoot borer in comparison to that of Stalk borer. These stripes are easily noticed when the larva is viewed from the anal end. Most of the time, these two ephemeral lateral stripes are difficult to make out with a hand lens and thus, these two stripes are not taken as the stable identifying character.

The nature of median stripe differs in these two borers. In Stalk borer, the median stripe terminates at the two conjoined larger dorsal tubercles (D_1) on the 8th abdominal segment whereas, in Early shoot borer the median stripe continues further and reach the 9th abdominal segment. Similarly, in Stalk borer the large dorsal tubercles (D_1) are square like, relatively larger, and the distance between the two large dorsal tubercles (D_1) in a segment is narrow. Due to this, in Stalk borer the median stripe touches these large tubercles (D_1). In Early shoot borer, the large dorsal tubercles (D_1) are relatively smaller and the distance between the two large dorsal tubercles (D_1) in a segment is more and thus median stripe runs free (does not touch the dorsal tubercles).

In the forewings of these two five striped borers, the veins Sc and R_1 anastomose to form the combined vein Sc+R_1 and this was considered an important character by Kapur(1950) for the erection of the genus *Chilotraea* in the fold of *Chilo*. In *Chilo auricilius* these two veins anastomose strongly and the combined vein covers a good distance before bifurcation to separate again as Sc and R_1. In *Chilo infuscatellus*, the R_1 tangentially touches the Sc and only a partial anastomosis takes place (*tracheal* connection from Sc often reaches R_1).

The immediate distinction between the two larvae of these moth borers is derived from the arrangement of hooks on the abdominal *prolegs*. In case of Early shoot borer, the arrangement of crochets forms an *incomplete circle*, whereas the crochets form a *complete circle* in case of Stalk borer. The pupae are quite distinct in their respective appearances. Early shoot borer has wavy ridges as the ornamentations on 5th, 6th and 7th abdominal segments, whereas Stalk borer has small spines as ornamentations on these three segments. Early shoot borer proliferates during the summer months, whereas the Stalk borer proliferates during the monsoon season and continues its activity in the winter months. *Chilo infuscatellus* mainly damages the crop at shoot stage, whereas the *Chilo auricilius* mainly damages the crop after sufficient stalk formation.

3.6 Adults in relation to taxonomy and identification

The adults of stalk boring *Crambinae* are characterized by rather uniform drab colour of the forewings and by the reduction of veins in the forewings. In the female genitalia, the *papillae anales* are laced with sensitive hairs and always coalescent like the placement of two bedroom slippers side by side; this arrangement leaves a small space between the two halves. The *ocelli* are often reduced or absent. With a few exceptions, the vein R_5 in the forewing is free (Bleszynski, 1969). In India, most of the important moth borers of sugarcane belong to the Genus *Chilo* Zincken, whereas in USA the moth borers belong to the genus *Diatraea* Guilding. In the genus *Chilo, ocellus* is generally present, whereas in *Diatraea, ocellus* is characteristically absent.

Bleszynski (1970) described the genus *Chilo* as:

> *"The genus Chilo was erected by Zincken in 1817, for a heterogeneous cluster of species from which Duponchel, 1836 selected Tinea phragmitella Hubner as the type species".*

The subsequent described genera i.e., *Proceras* Bojer, *Borer* Guenee, *Diphryx* Grote, *Hypiesta* Hampson, *Nephalia* Turner, *Silveria* Dyar and *Chilotraea* Kapur were considered as junior synonyms by Bleszynski (1970).

Ocellus (*stemma*) in most of the *Chilo* species is well developed; *chaetosemata* are moderate, *labial palpus* porrect, usually at least three times as long as the diameter of the compound eye; it is shorter in male. Face is round or conical, sometimes with a strong ventral ridge formed by *clypeus*; antenna in male is serrated.

In the forewing, R_1 is free, approximate to or coincident with Sc. R_2 is always free. R_3 and R_4 are always stalked and R_5 is always free. Forewing of most species is yellowish to brown. Many species bear metallic scales on the forewings; distal dots are often present but median dot is absent in most of the species.

The wing venation of *Scirpophaga excerptalis* is quite similar to that of *Chilo* moths but the presence of post *cubitus* vein (CuP) helps its easy differentiation. Similarly, in the Noctuid *Sesamia*, formation of "O" due to bifurcation and anastomosis of radial veins helps its easy identification.

List of plates on an overview of sugarcane moth borers in India

Descriptions of plates

Plate 3.1. Egg stages of Top borer, Root borer, Internode borer, Stalk borer, Pink borer and Early shoot borer.

Top borer, *Scirpophaga excerptalis*

Root borer, *Emmalocera depressella*

Internode borer, *Chilo sacchariphagus indicus*

Plassey borer, *Chilo tumidicostalis*

Gurdaspur borer, *Acigona steniellus*

Maize borer, *Chilo partellus*

Stalk borer, *Chilo auricilius*

Early shoot borer, *Chilo infuscatellus*

Pink borer, *Sesamia inferens*

Plate 3.2. Adults of different sugarcane moth borers

A1. Egg mass, dome shaped and covered with yellow-orange anal hairs

A2-A6. First instar larva,Second instar larva, Third instar larva, Fourth instar larva, Fifth instar larva

A7. Newly formed female pupa (light yellowish)

A8. Shiny white female moth sitting/ resting on the sugarcane leaf.

Plate 3.3. Typical life cycle of a moth borer of sugarcane (Top borer)

Plate 3.4. Internode borer larva in the process of *ecdysis*

Plate 3.5. A larva (Internode borer) in the process of *ecdysis*. Immediately after *ecdysis* head capsule remains white except the mandibular area, which appears dark coloured. Larval body is basically white and bears four prominent stripes. *Spiracles* are prominent but tubercles take body colour.

Plate 3.6. Prepupal Top borer larva just on the verge of *ecdysis* to a pupa. The process of *ecdysis* to form a pupa from a larva.

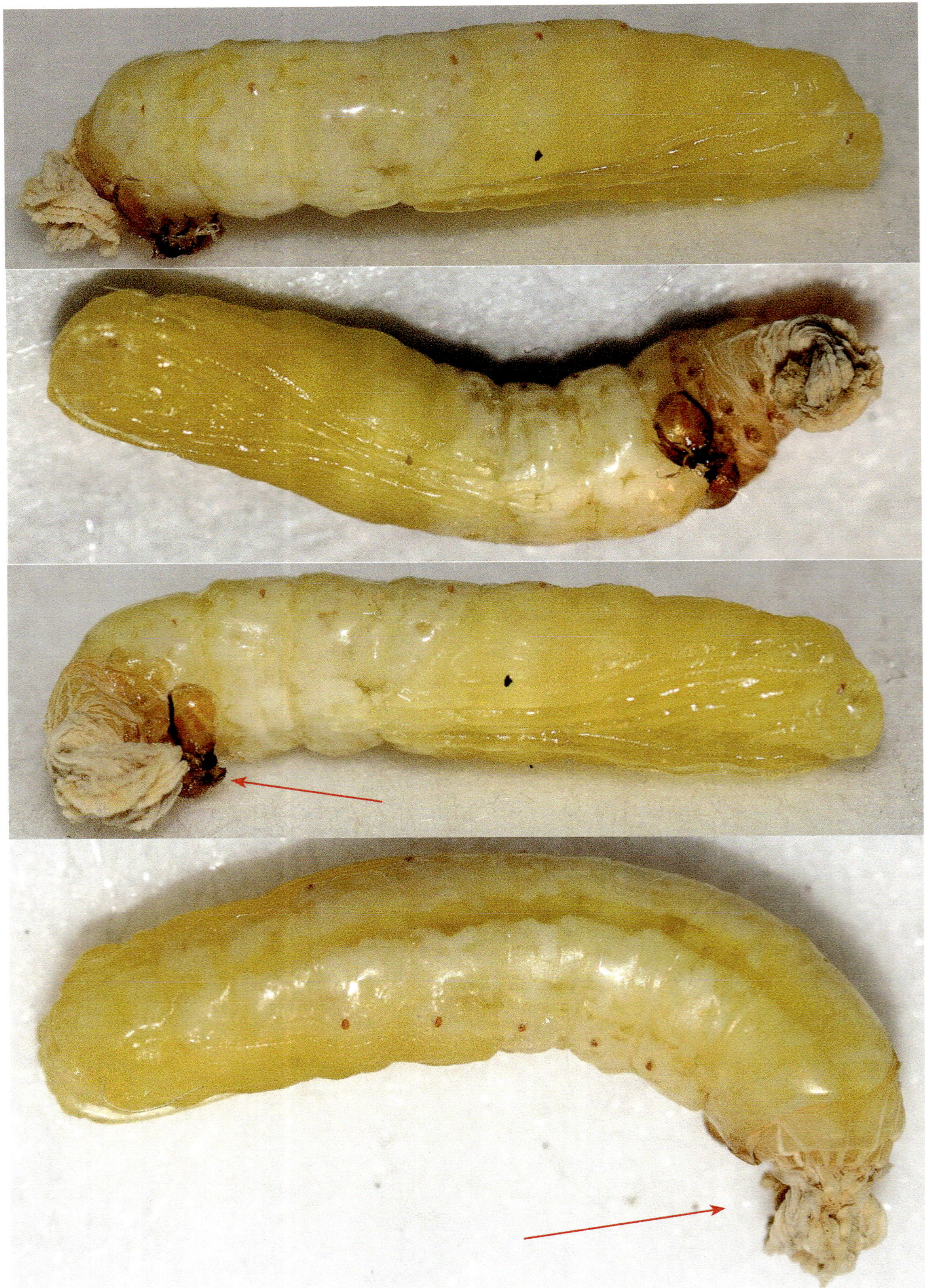

Plate 3.7. Top borer larva in the process of *ecdysis* to form a pupa. Freshly formed pupa is yellowish white. *Spiracles* are prominent on the body.

4

Top borer, *Scirpophaga excerptalis* (Walker)

Top borer, *Scirpophaga excerptalis* (Walker) is a major pest of sugarcane in India, especially in the subtropical belt where it is recognised as the number one entomological problem in the profitable cultivation of sugarcane. The silvery white moth often observed in the sugarcane agro-ecosystem is easily identifiable in contrast to other moth borers of sugarcane. As per the available records, the moth was first described by Walker in 1863 from an insect collection of Mr. Steven from north India (maintained at the British Museum (Natural History). He named this species of moth borer as *Chilo excerptalis* Walker. In spite of its striking appearance, various authorities have identified and named this moth borer differently. The genus *Scirpophaga* was erected by Treitschke in 1832. In 1863, Zeller included Top borer as *Scirpophaga auriflua*. Hampson (1895 & 1896) placed the Top borer in the genus *Scirpophaga* as *Scirpophaga excerptalis* (Walker) Hampson. In 1960, Common erected the genus *Tryporyza* and included Top borer of sugarcane as one of the constituent species. Of late, the Top borer of sugarcane has been retained in the fold of the genus *Scirpophaga*. The current accepted name of Top borer of sugarcane is *Scirpophaga excerptalis* (Walker) as settled by Madam Angoon Lewvanich (1981), who revised the old world species of *Scirpophaga*. She categorically explained that *Scirpophaga nivella* is a pest of rice, not a pest of sugarcane.

4.1 Currently accepted name

Scirpophaga excerptalis (Walker)

4.2 Synonyms

= *Chilo excerptalis* Walker

= *Scirpophaga auriflua* Zeller

= *Scirpophaga intacta* Snellen

= *Scirpophaga chrysorrhoa* Zeller

= *Scirpophaga monostigma* Zeller

= *Scirpophaga butyrota* Meyrick

= *Tryporyza nivella* (Fabricius)

4.3 Symptoms of infestation and brief biology

A neonate larva (freshly hatched first instar larva) of Top borer wanders on the crown leaves for a while in search of a suitable place to make an entry in the crown. It bores a tiny hole in the immature mid-rib of a rolled crown leaf from the ventral side (underside). At this period, only the ventral side of the mid-rib of immature leaf remains exposed to direct sunlight. After successful entry in the mid-rib a borer larva makes its way downward through the mid-rib by making a long feeding tunnel (just below the upper epidermal layer of mid-rib) to reach the deeper area of the inner crown leaves. The feeding tunnel made in the mid-rib forms a very characteristic *white streak* symptom, which later changes to red. During the mid-rib journey, a Top borer larva cuts several small holes almost in a regular interval on the dorsal epidermal layer of the mid-rib. Opportunistic microbes use these tiny openings to infect the mid-rib tunnel and as a result reddening of the mid-rib tunnel takes place. The presence of white streak in the mid-rib of the crown leaves confirms the infestation of Top borer. The vertical travelling period of larva in the mid-rib is around 24-48 hours. During this period a first instar larva covers a distance of around 16 to 48 cm within the mid-rib to reach the heart of the crown. On reaching there, a first instar larva comes out of the mid-rib tunnel and laterally bores inward to the centre of the immature spindle leaves. During the process of horizontal feeding movement of a first instar larva-from the outer leaf mid-rib to the inner leaf mid-rib, *first order of shot holes* (*horizontal shot holes*) are formed on emerging rolled lamina. These shot holes are smaller, always present at one level and usually vary from three to five in number. On reaching the inner core, a larva again starts moving downwards feeding through the young developing spindle leaves. Hereafter, a larva feeds vertically down through the young developing spindle leaves of the crown and as a result, a large number of much bigger shot holes, the *second order of shot holes* or *vertical shot holes* are formed. These shot holes appear much later, only when the affected leaves get unfurled with the subsequent growth of the plant. However, these vertical holes may also be observed in the immature leaves before unfurling, if the mature leaves are carefully separated from the crown exposing inner developing leaves. Keeping a steady space with the growth of the crown, the larva feeds on the young rolled leaves above the growing point for about three to four weeks. After attaining certain maturity, the larva carefully nibbles the tip of apical dome (growing point) and goes further down by devouring the immature cane tissues. Much before eating away the growing point, it completely devours the central portion of the immature crown leaves and this results in the formation of a small, slender *dead heart,* usually comprising of only one immature leaf. Due to the immaturity of the leaf, on drying it takes a light reddish straw colour. Another characteristic of infestation of Top borer is the formation of '*drooping leaf*' symptom. Due to occasional damage of mid-rib, due to feeding of the larva, the damaged mid-rib often breaks away or gets twisted from that damaged point as it fails to bear the load of distal area of the leaf. The distal end of the leaf remains hanging from the affected area of the mid-rib. No other moth borer of sugarcane produces such a symptom.

If the crown portion of the cane is cut and split open at this stage, a creamy white larva (of the Top borer) may be found at the growing point or in the tunnel at the top immature internodes of the cane. The fully-grown larva (prepupal) cuts an exit-hole, mostly at the node of the cane stalk and covers the exit hole with host tissues and silk. After preparing the exit-hole, a larva moves backward and weaves several protective discs with silk. Each disk covers the entire width of the tunnel. Finally, the larva settles down behind these discs in a specially woven silken chamber within the tunnel. It pupates within the silken chamber. In a week or ten days, a silvery white moth emerges from the pupal shell. The moth comes out from the tunnel by pushing through the silken discs and dislodging the barrier of exit hole. During the period of peak emergence, especially during first and second generations (broods), moths may be observed resting on the leaves during early morning, and in late afternoon or evening hours. Mating takes place during night and after mating, a female moth lay eggs in clusters (egg mass) and covers the eggs with yellowish orange anal hairs. Usually, the eggs are laid on the underside (parallel to mid-rib) of an upright or near upright leaf of the crown.

In a comparatively grown up crop (in grand growth period usually the infestation of 4th brood takes place), the borer attack arrests further apical growth of the crown and induces the top crown buds to sprout and put forth side shoots giving rise to what is commonly called the "*bunchy top*". There is considerable deterioration in juice quality in canes with *bunchy top*. Majority of the bunchy top formation takes place due to the infestation of 4th brood. After the attack of 5th brood, winter sets in and the prevalent cold weather restricts the growth of the apical side shoots. The growth of the side shoots (*bunchy top*) accelerates when the ambient environment becomes favourable in February. By this time, harvesting of most of the sugarcane crop is over. Thus, bunchy tops formed due to the infestation of 5th brood exert much less impact on the quality of the cane. Moreover, during harvesting of the cane, this sprouted portion of the cane is also removed from the stalk.

4.4 Egg and neonate larva

In contrast to the egg masses of other moth borers of sugarcane, the egg mass of Top borer is not naked; it is covered with several layers of yellowish orange anal hairs. Due to the hair cover, the egg mass appears as dome shaped and remains easily visible on the green background of sugarcane leaf. Once the hair cover is removed, three parallel overlapping rows (imbricate rows) of eggs strongly glued to the leaf surface are observed. The egg is dull white or water-soaked faint yellowish white (like desi ghee) in colour and an individual egg is a bit fleshy, not scaly like in other borers. The length of the egg mass varies widely but the width hardly changes, and it measures around 3 mm. After hatching from the egg, a neonate larva comes out making its way through the anal hair cover. Neonate larvae stay on the crown leaf only for a few hours and are often found hanging from the leaf spinning silken threads. To the naked eye, it appears like small black crawlers moving on the leaf. The neonate larva is a hairy caterpillar with light brown abdomen. The head and the thoracic segments are dark brown to black in colour. A definite *white band* girdling the first abdominal segment immediately follows the black thoracic segment. This *white band* is very characteristic of the neonate larva of Top borer. The presence of this white band helps in its quick identification in the field. During its journey through the mid-rib a first instar larva retains all the body hairs and also grows in size considerably (5-6 mm). It moults only after reaching the central portion of the immature crown leaves. During this journey to central core of the immature crown leaves, a larva has to face a lot of competition from the fellow larvae and a large number of neonate larvae generally perish before reaching the target.

The neonate larva of Top borer is a very active, light brown hairy caterpillar and measures around 2.5 mm (2.22 to 3.00 mm). The larva has a characteristic *white band* just after the dark thoracic segments.

4.5 Larva

A neonate larva enters the crown portion of the cane through leaf mid-rib and passes the rest of the larval period within the crown tissue. It never ventures out of the crown like larvae of many other moth borers. Thus, at any given point of time, after the successful entry and establishment in the immature inner core of the crown, only one of the II, III, IV, or V instar larva of top borer can be observed in an infested crown. As indicated earlier, from the standpoint of a sugarcane farmer it is very important to fully understand the larval characters of all the borer pests to identify them easily and accurately in the field. Top borer is a member of *Schoenobiinae* and a larva of this group may be differentiated by the presence of a *thoracic sac* in front of each thoracic leg pair. A larval body can be divided into three well distinguished parts viz., *head, thorax* and *abdomen*. The head bears prominent mandibular mouth parts, numerous setae and *ocelli* (*stemmata*).

A mature top borer larva is very sluggish and takes long time to move away on extrication from the immature portion of the cane stalk. It is distinctly creamy yellow in colour, with occasional translucent longitudinal dorsal area allowing visibility to the dark alimentary canal. It is largely cylindrical, relatively thicker and abruptly tapers towards both head and anal end. Skin is smooth and has numerous folds. Except the setae present on head capsule, prothoracic shield and thoracic legs, all other setae are very small and indistinct. Head is relatively small in comparison to the body. In a mature larva, head capsule is yellowish-brown with off-white network of stripes and has prominent 'V' notch at the *vertex*. A portion of the lateral terminal area of the head capsule is dark. The *epicranial suture* area is yellowish white and it prominently continues further in the prothoracic shield. Prothoracic shield is light yellowish in colour with dark patches near the terminal area.

The larval head or *cranium* is a hard and highly sclerotized structure with a 'V' shaped mid-dorsal indentation (*epicranial notch*) or vertical triangle at the front. The two mandibles have five teeth each; the lower four pair are sharp and pointed. On each side of the head, six single-lens visual organs, popularly known as *ocellus* (actually this structure is called *stemma*, plural *stemmata*) are present. These six *ocelli* follow a 1+3+2 arrangement. In most of the larval instars, these *ocelli* appear as dark raised structures. The visibility of the dark *ocelli* on the developing white head capsule in a larva indicates that the larva is in the process of moulting. The ocelli appear clearer at the last larval instar when the larva readies to transform itself into a pupa.

The abdomen of a larva also consists of ten segments but the last three are always fused together to form the anal segments. The first eight abdominal segments bear breathe hole or *spiracle*. In the second and third instar larvae, all the abdominal *spiracles* are easily visible as an elongated hole surrounded by a white area (air-filled trachea). In a mature larva (fifth instar), the *spiracles* are inconspicuous on the body folds. These *spiracles* appear as elongated light orange-yellow streak, just one shade darker area on the creamy yellow body. The *spiracle* on the 8th segment is present laterally and almost in line with the other abdominal *spiracles*.

The three pairs of thoracic legs are five-segmented structure and bear numerous setae and *sensilla*. The tarsal end of the leg ends in a stout claw. As described earlier, one characteristic sac like structure is present in front of the each leg pair (towards the head). In addition to the thoracic legs, a larva has five pairs of abdominal pseudolegs or *prolegs*. These are not true legs. These are contractile structures bearing hooks at the terminal area to steadfast the larva during movement. These *prolegs* are shed with the skin during moulting. A pair of *prolegs* is present each on the 3rd, 4th, 5th, 6th and the 10th abdominal segments. In a mature larva of Top borer, the *prolegs* are inconspicuous on the large larval body. Crochets or hooks on the four abdominal *prolegs* are poorly developed. These hooks follow *uniserial* and *uniordinal* arrangements. Hooks are arranged in a complete elongated elliptical circle (ellipse, to be exact; it appears as if composed of two half circles or first brackets). However, *claspers* or prolegs of anal segment bear around 20-22 hooks each and the hooks cover only half of a circle. (In comparison, Root borer has around 11 hooks on each *clasper* and covers only $1/3^{rd}$ of the imaginary circle).

A Top borer larva usually passes through five instars in the life cycle and attains maximum length at the end of fourth instar or at the beginning of fifth instar stage. The length of a freshly moulted fifth instar larva varies from 32-38 mm. However, a prepupal larva gradually reduces its length and a female prepupal larva measures around 25 mm just before pupation. It is difficult to distinguish between fifth and sixth instar larvae. The larva is very sensitive to moisture loss and needs a moist environment for survival on extrication from the cane top.

4.6 Pupa

An immature and freshly developed pupa of Top borer is faint/pale yellowish in colour and with maturity the colour of a female pupa gradually changes to white due to the formation of white scales on the body whereas, in male the colour becomes ochreous yellow to off-white. Male pupa is smaller in size (14-18 × 2.5-3.0 mm) and weighs around 80-90 mg while the female pupa is quite larger (19-23 × 3-4 mm) and weighs around 130-140 mg. The pupal chamber formed in the resting tunnel made in the cane stalk is slightly larger than the pupal size and the pupal chamber usually measures around 25-30 mm. In around 7-10 days, the pupal stage is completed.

Like the larva, the head, thorax and abdomen of a pupa is also readily distinguishable. Like the larva, the abdomen of a pupa also consists of ten segments; the last three are fused together. The fused segments impart a sharp conical shape to the anal end of a male pupa. Pupa of Top borer is smooth and does not have any distinctive ornamentation (spines or ridges) on 5th, 6th and 7th segment. The anal end in the immature female pupa remains yellowish and as it gradually matures, the anal segments initially turn distinctly white, but with further maturity, this white anal area gradually changes colour and finally becomes orange red due to the development of anal hairs underneath. No such distinctive colour changes takes place in male pupa except the colour of the abdomen that gradually changes to pale ochre to whitish yellow.

The pupa of most of the moth borers of sugarcane are typically *obtected* type i.e., developing wings, legs, palpi, maxillae and antennae are tightly cemented to the surface of the body. In all these moth borers hind leg pair usually remain more or less in line with the wing pad area. However, in a female pupa of Top borer, the hind leg pair reaches up to the 5th abdominal segment, and in male pupa, the hind leg pair extends even up to the 7th abdominal segment. The wing pad area in a male pupa is longer in comparison to its body and extends to the 5th segment, whereas in a female pupa it ends at the 4th segment. Thus, the wing pad area and hind legs impart an appearance of collared shirt with a long necktie. A pupa of Top borer cannot be called *obtected* type in the true sense, as all the appendages are not completely soldered to the body surface; the tips usually remain free. The distal portion of the hind legs of a male pupa remains completely free and become visible to the naked eye when a male pupa sways the abdominal segments. Determination of gender at the larval stage is difficult whereas, the gender of the pupa is easily discernible due to specific developments of genital structures at the anal end in addition to the colour and size of the pupa.

A pupa of Top borer is very delicate, slender and cylindrical. The body wall (cuticle) of the pupa is very thin. Due to the translucent nature of the body wall, the development within the pupal shell (development of *pharate* adult) becomes visible. Eyes are prominent; initially appear as light yellowish area, with passage of time a brown streak appears in the centre and gradually the eye area turns brown in colour. In general, a male pupa has larger eyes (comparative to body size) than that of a female pupa. With maturity, the brown eye areas become black and finally developing white scales cover the black eyes. Thus, in a fully mature pupa the eyes appear as white before the emergence of the moth. Probably these coverings of scales protect the eyes during the journey of the moth in the host tunnel as the moth pushes the barriers (a number of silken discs and the cover of the exit hole) with the head to make its way to reach the exit-hole.

The *mesothoracic spiracles* are prominent, arch type with prominent hollow inside. As outlined earlier, there are seven visible *spiracles* present on the abdominal portion of a pupa of moth borer. *Spiracles* are visible from 2nd to 7th abdominal segments. The *spiracle* pair, present on the 8th abdominal segment is rudimentary and non-functional. This pair is faintly visible in the immature pupa but in a mature pupa, this *spiracle* pair is hardly traceable. In the pupa of a Top borer, abdominal *spiracles* always stand elevated from the body surface. *Spiracles* are slit like (elliptical) with prominent thick yellow-brownish or orange coloured rim.

Genital marking in the male pupa appears as a small slit placed on the 9th segment. It appears as a depression with two prominent conical lobes on either side. In the female, the marking is present on the 8th segment as two slightly raised areas on either side touching the 7th segment. *Cremaster* (prolongation of anal segment carrying spines) is typically absent from the pupal body.

4.7 Moth (Adult)

In contrast to the other borer moths of sugarcane, moths of Top borer are easily visible in the field due to their silvery white colour and sitting habit on the crown leaves of sugarcane. Moths of Top borer are mostly observed sitting on the leaves during the morning hours and in the late afternoon and evening hours, especially during the first two broods (February-March and April-May). Both the wings of the moths are milky white in both sexes. Occasionally, one centrally placed black spot may be found on both the forewings.

The body of the moth has three distinct region i.e. *head, thorax* and *abdomen*. The head is a hard sclerotized structure. The most conspicuous appendages of the head are the two large black compound eyes, the two *filiform* antennae and a pair each of *labial palpi* and *maxillary palpi*. One *ocellus* is present on each side adjacent to the compound eye and posterior to the antenna. The *proboscis*, composed of two lateral tubes is very much reduced and bears small scales at the base.

The *labial palpi* and *maxillary palpi* are clothed with white scales. Occasionally, a few fuscous scales are observed at the distal segment. The *labial palpus* is porrect and about 1.5 times of the diameter of compound eye, sometimes slightly longer but never more than twice the diameter of the eye. It has three segments and the second segment is the largest. It occasionally takes a slight upward curvature. The *maxillary palpus* is small, consists of four short segments and their combined length does not exceed half of the length of *labial palpus*.

The antenna is *filiform* in both sexes. The *filiform* antenna can be divided into three parts. The basal segment or *scape* is attached with the head. In turn, a *pedicel* attaches with the *scape* at one end and a long *flagellum* containing several segments on the other end. Antenna in a freshly emerged moth remains clothed with white scales; the dorsal side appears white while the ventral side appears black. However, these white scales often get dislodged with time, exposing the black colour of *flagellum* underneath. The *flagellum* consists of 39-47 segments and appears finely serrated. In male, the serrated nature of antenna is more pronounced and the number of segments is more. Each segment of the *flagellum* has several *setae* and *sensilla*. The antenna of a male moth is highly ciliated in comparison to that of a female moth. The apex of the *flagellum* ends in a *sensory cone* or a *fork*.

The thorax bears three pair of legs and two pair of wings. Each thoracic segment ventrally bears one pair of jointed legs. The hind legs or *metathoracic* legs are much longer, reach the anal end, and probably help in dislodging and placement of anal hairs on the egg mass. The *mesothorax* supports the forewing pair and the hindwing pair is attached on the *metathorax*. Hindwings are more fragile than the forewings. Occasionally, one centrally placed black spot is present (black scales develop at the base of M_2, M_3 and Cu_1 covering the cubital peak or arch of the cell). Rarely, another minor black spot (below the larger black spot, towards the radial peak) is also found. Black spot is mainly found in the male moths. Another variation in spotting of the fore wings is also found. The centrally placed spot in the fore wings, instead of black appears as a *creamy yellowish white* patch in the white background. One interesting point with regard to spotting on the fore wings is that whenever a black spot was observed, the moth was a male, and whenever a *creamy white spot* was observed, the moth was a female.

The abdomen is slender and has ten segments; the last three are always fused together. Functional *spiracles* are present on the pleural membranes of the first seven segments. A pair of *tympanal* organ is present at the base of the abdomen and remains in hiding (covered) under the *coxa* of hind legs. The colour of the abdomen varies depending on the development of the scales and their colour. In male, the abdomen is white to pale yellow ochreous and relatively narrow. At the central portion (*meson*) of the distal end of the 7th segment (*sternite*) a flattened scale tuft that extends and covers the 8th segment is present. In a freshly emerged female moth, the abdomen is white on upper side, pale ochreous on underside. The female moth has a typical orange red anal tuft. It is to be noted that all the hairs of anal tuft are not orange red; only the outer layers are orange red and impart the distinctive appearance; the inner layers are ochreous/buff yellow. The orange red hairs are not used to cover the egg mass, rather it is the central ochreous/buff yellow hairs that are used for the purpose.

Papillae anales or the ovipositor is densely clothed with sensory setae. Part of the ovipositor mostly remains visible. Ovipositor appears like a betel leaf or heart shaped. The two halves of *Papillae anales* are folded in the ventral side with large space in between. The dorsal side form a define cone and without any fold.

4.8 Forewing venation

Among the moth borers of sugarcane, the wings of Top borer are relatively larger (Internode borer has also larger wings). In the forewing, the *costal* vein(C) is not very prominent but the *subcostal* vein (Sc) is prominent. The *radius* (R), *media* (M) and *cubitus* (Cu) veins are almost similar in thickness. The central veins viz., *radius* and *media* are variously branched and provide strength to the wings at the centre. The *radial* and *cubital* veins are connected with a wavy bimodal *cross vein* to form the head of the central cell (*discal cell*) of the forewing. This *cross vein* supports the development of one *radial vein* (R_5) and three *Media(M) veins* (M_1, M_2 and M_3).

The *subcosta* receives a branch of the *radius* and gets anastomosed to form the combined vein $Sc+R_1$. In all, the *radius* vein gives rise to five prominent branches (clockwise). The first branch, R_1 joins the *subcosta* (Sc) and the combined vein usually continues to the costal area. Occasionally, in some female wings, the combined vein $Sc+R_1$ bifurcates at the terminal area. The R_2 remains free throughout and it ends at the costal area. R_3 and R_4 is on a long stalk. R_3 is smaller than R_4 and both the veins end at the coastal margin. R_4 meets just a notch above the leading corner (*apex*) of the wing.

From the radial peak of the cell (it connects both *radius* and *media* and is adjacent to *radius*), apparently two branches arise. The first branch is part of *radius* and termed R_5 and it ends at *termen*. The second branch, due to its proximity to *radius* and thickness apparently appears as the part of Rs but it is actually the first branch of *media*. This vein is termed M_1 and it is connected with the base of the *media* with a long wavy connection. A cross-bridge from base of M_1 to the base of R_5 increases its strength. The *media* has three branches (M_1, M_2 and M_3). The anastomosis of *media* and *cubitus* results in the development of Cu1a and Cu1b. The presence of CuP vein (*post cubitus*) on the forewing distinguishes this species to be a member of *Schoenobiinae*. The CuP vein originates mid-way from the wing membrane and ends in *termen* but it only runs around one third of the wing length. CuP is present in the space between *anal veins* and the *cubitus*. The *Chilo* moth borers do not possess this vein. Among the anal veins, only A_2 is well developed and prominent. A_3 is very feeble and rudimentary and in most cases misses the detection.

A close examination of forewings (stained appropriately) under the higher magnification of a microscope revealed that the veins of both forewings and hindwings end with a definite brown spine like seta at the edge (*termen*). Moreover, mostly two circular thick walled structures with sensory roots (*mechanoreceptor*) all around are present just before periphery (*termen*) of the wing where a vein terminates. These *mechanoreceptors* precede the seta. Such type of round structures (*mechanoreceptor*) are also present (scattered in the veins) and these are mostly observed in the veins attached to the *discal cell*.

In addition, another curious fact that emerged during these microscopic observations was that the first bimodal peak of the cell from which R_5 arises, many a times is supported by a bridge or a channel from the middle portion of the radius (normally in most of the forewings it remained obscure). The supporting branch usually develops from Rs and the location is variable i.e. appear either before or after the branching of R_1 and connects near the base of R_5 (this type of variation was also not described earlier for this species).

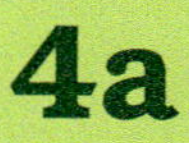

List of plates on Top borer, *Scirpophaga excerptalis* (Walker)

Description of plates

Plate 4.1. Symptoms of Top borer infestation. Typical slender *dead heart* formation. *Dead heart* is relatively smaller, usually of one rolled leaf. Formation of vertical shot holes on the leaf, drooping of leaves due to damage of the mid- rib.

Plate 4.2. Symptoms of Top borer infestation. Typical *dead hear*t formation in the infestation of third brood. *Dead heart* is relatively smaller and reddish orange in colour.

Plate 4.3. Tunnelling in the mid rib. Initially the feeding tunnel of the neonate larva remains white. Later on, due to microbial infection red colour develops. The neonate larva makes small holes in the tunnel at regular intervals probably to have passage of air.

Plate 4.4. Formation of horizontal shot holes due to the entry of first instar larva in the rolled crown leaves (A). Formation of vertical shot holes due to the feeding of 2nd and 3rd instar larvae (B). No shot hole formation (C &D).

Plate 4.5. Formation of vertical shot holes. Drooping of leaves due to the feeding damage of the mid-rib.

B

A

Plate 4.6. Infestation of third and fourth brood. Fourth brood infestation results in bunchy top formations. The side shoots grow considerably before the onset of winter and may contribute lanky millable cane (impair quality).

Plate 4.7. Bunchy top formation due to the infestation of fifth brood of Top borer. Due to winter, the growth of the side shoots remains limited (A &B). Side shoots resume quick growth when ambient temperature become favourable in February (C).

Plate 4.8. Neonate larvae of Top borer. Emergence of a larva from an egg mass. Neonate larvae of Top borer has black head and thoracic region. First abdominal segment bears a white band. Larva is very hairy with dark grey coloured body.

Plate 4.9. Neonate larvae of Top borer. Head and thoracic region of a neonate larva of Top borer is black. First abdominal segment bears a white band. Larva is very hairy.

Plate 4.10. Larvae of Top borer in different instars - note the change in shape, size and colour.

Plate 4.11. Feeding of 3rd instar larva inside the rolled crown leaf and formation of vertical shot holes. Finally the larva eats away the central immature portion of the crown leaves and totally snaps the upper portion.

Plate 4.12. A larva initially feeds on the rolled crown leaves, then devour the apical dome and finally feeds in the stalk tissue. Gradually by feeding, a larva moves inside of the cane. Eventually it makes an exit hole.

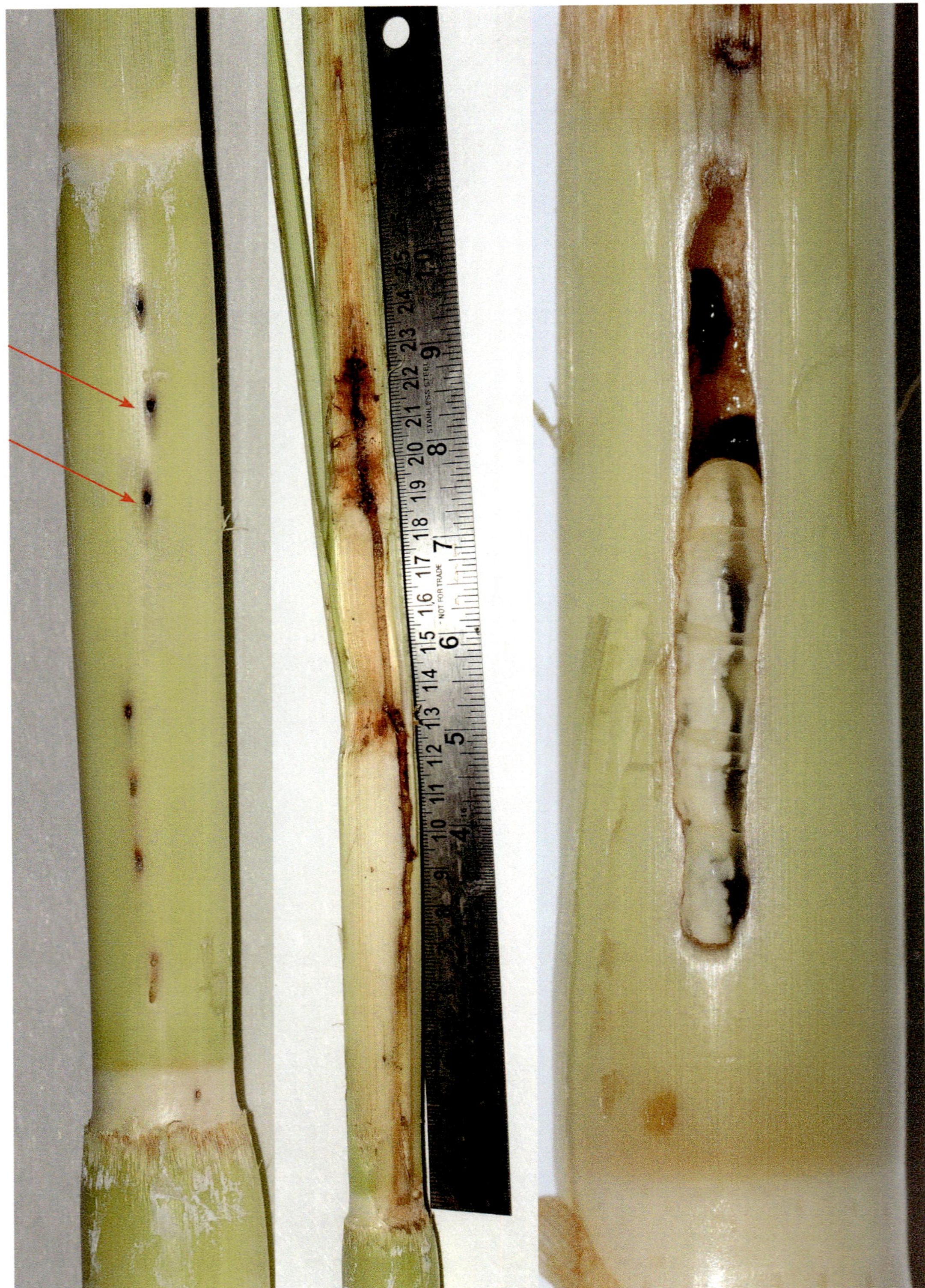

Plate 4.13. Tunnelling of the larva inside the immature stalk portion of the cane. Larva first devours up the apical dome and arrests further apical growth of the crown. Sometimes (by mistake) a larva feeds on the epidermal tissues and exposes itself to the attack of the larval parasites.

Plate 4.14. Larva of Top borer tunnelling inside the immature stalk. A prepupal larva makes an exit hole preferably at the nodal region for the safe exit of the moth.

Plate 4.15. Larvae of Top borer found inside the immature stalk tissues. Fifth instar larvae, the colour of the head capsule and body is creamy yellowish. Central alimentary canal visible through the translucent body.

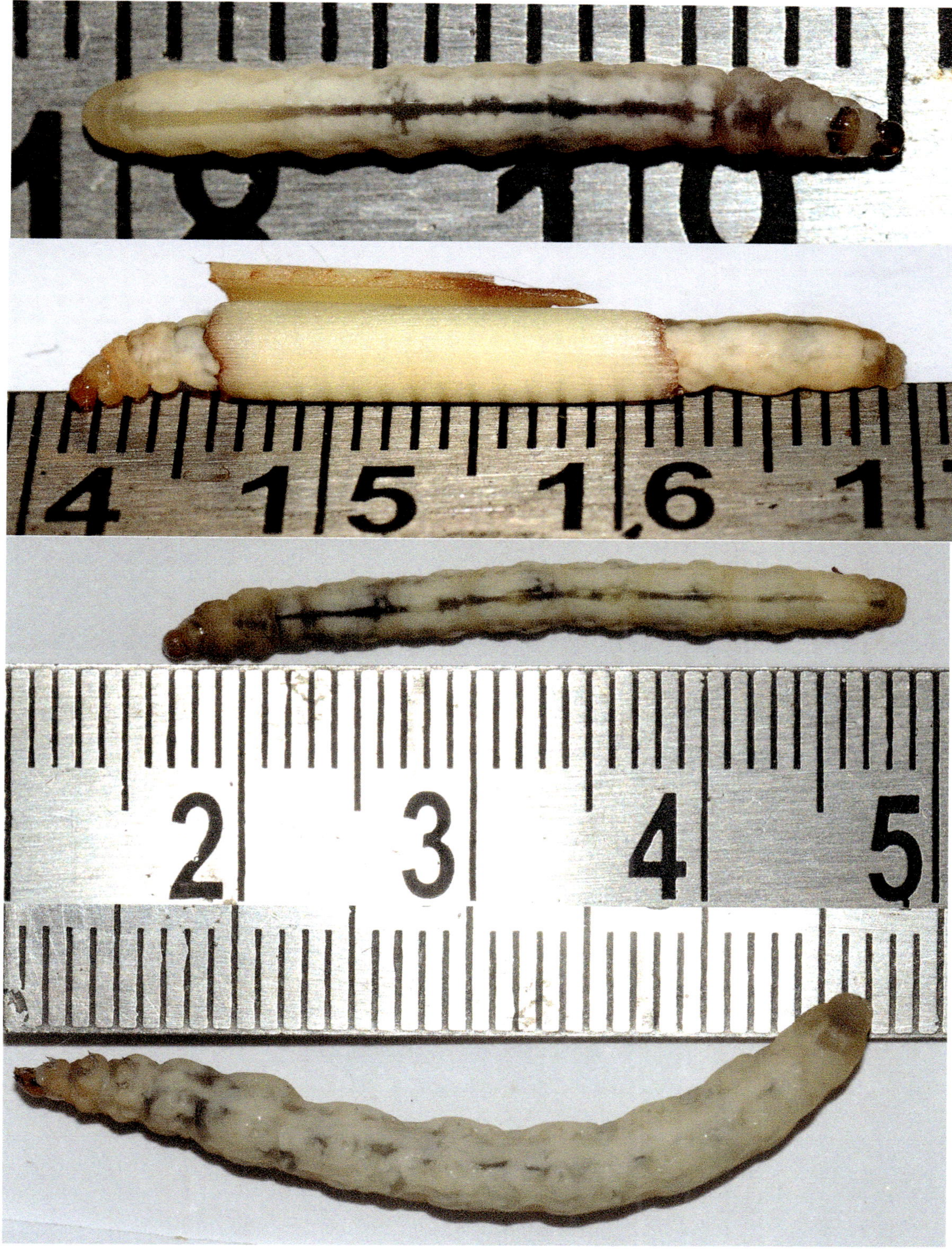

Plate 4.16. Size of the larvae of Top borer (3rd and 4th instar). Maximum length observed during the fourth instar. Length may reach 39 mm (as in the last picture).

Plate 4.17. Mature larvae (fifth instar) of Top borer.

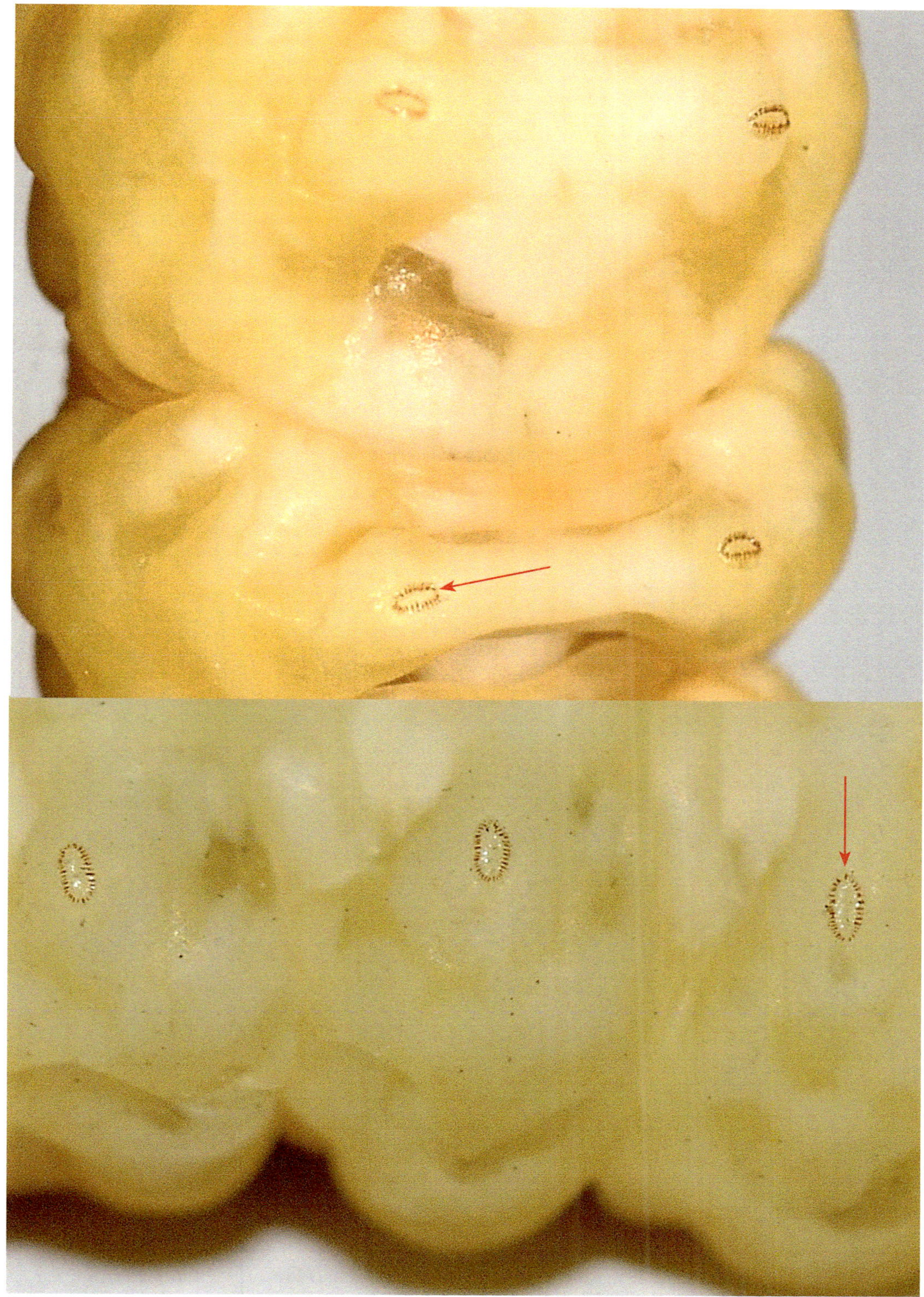

Plate 4.18. Mature larvae of Top borer. Abdominal Prolegs are hardly visible. Crochets are tiny, *uniserial* and *uniordinal* and arranged in an elongated circle like the two small brackets ().

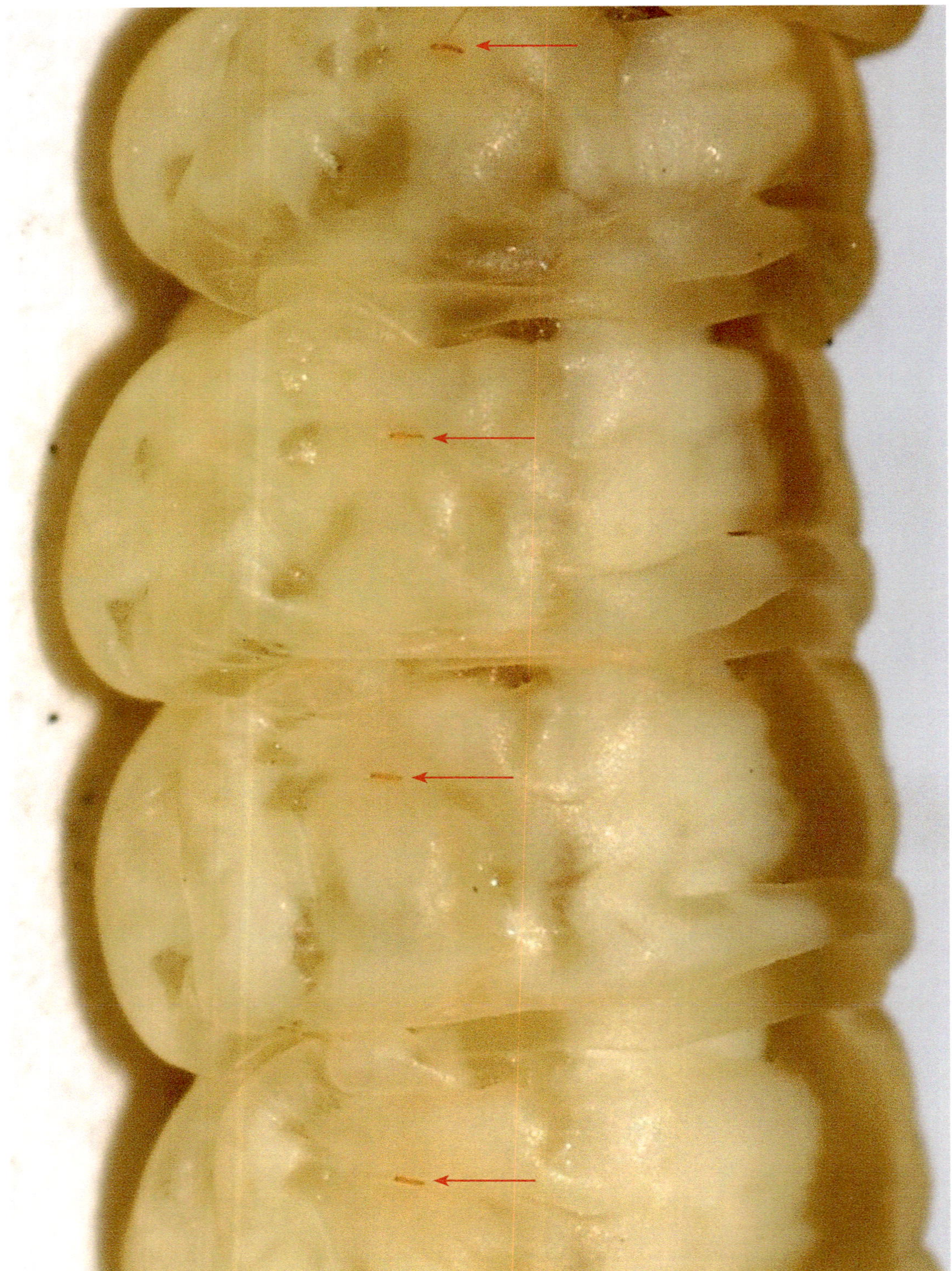

Plate 4.21. Fifth instar larva: *Tracheal trunk* is not visible. *Spiracles* are slit like, orange coloured and inconspicuous on the body. Body is smooth i.e. without any prominent hairs.

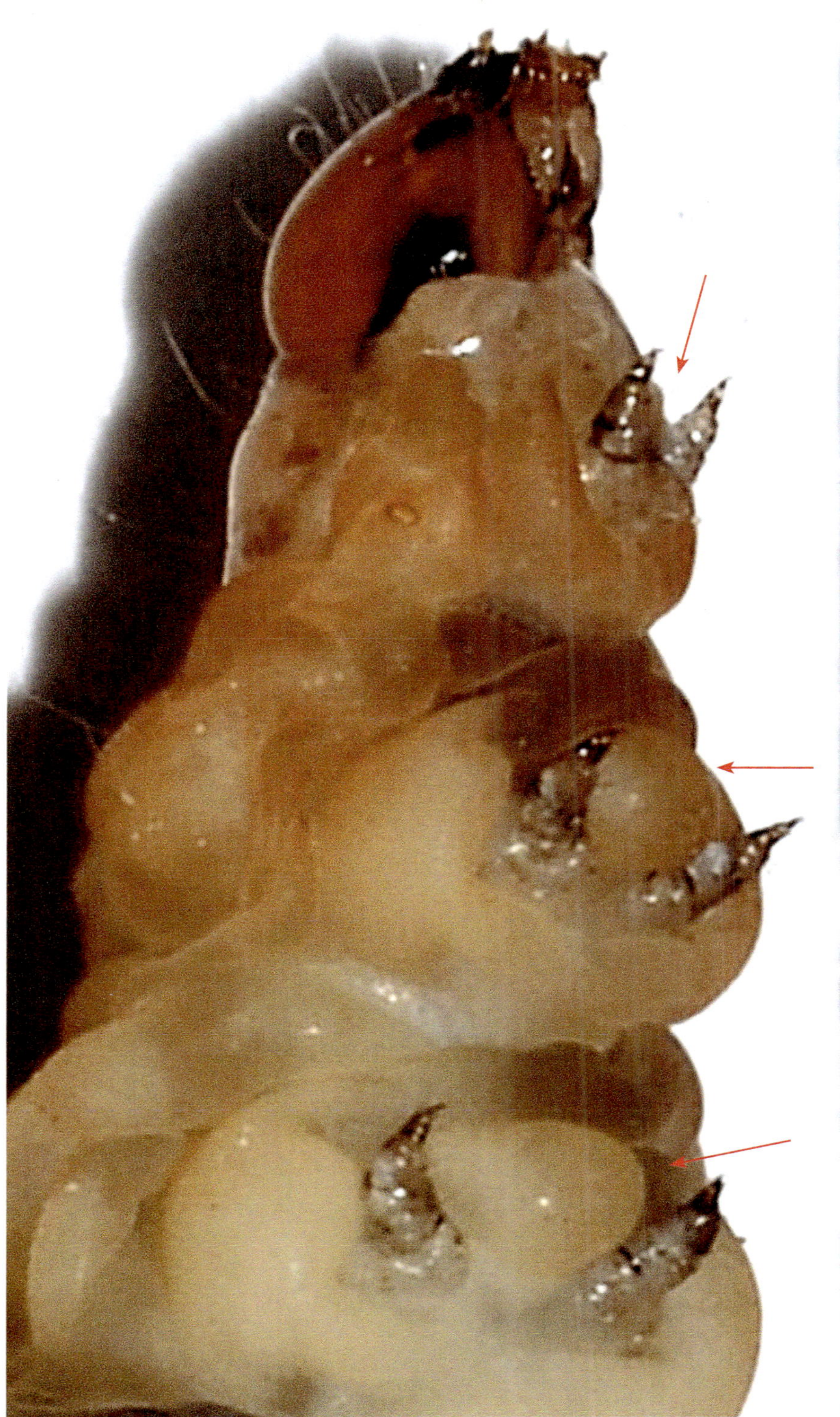

Plate 4.22. Presence of *sac like structure* in front of each of the thoracic pair of legs. This is very characteristic for this group of moth borers.

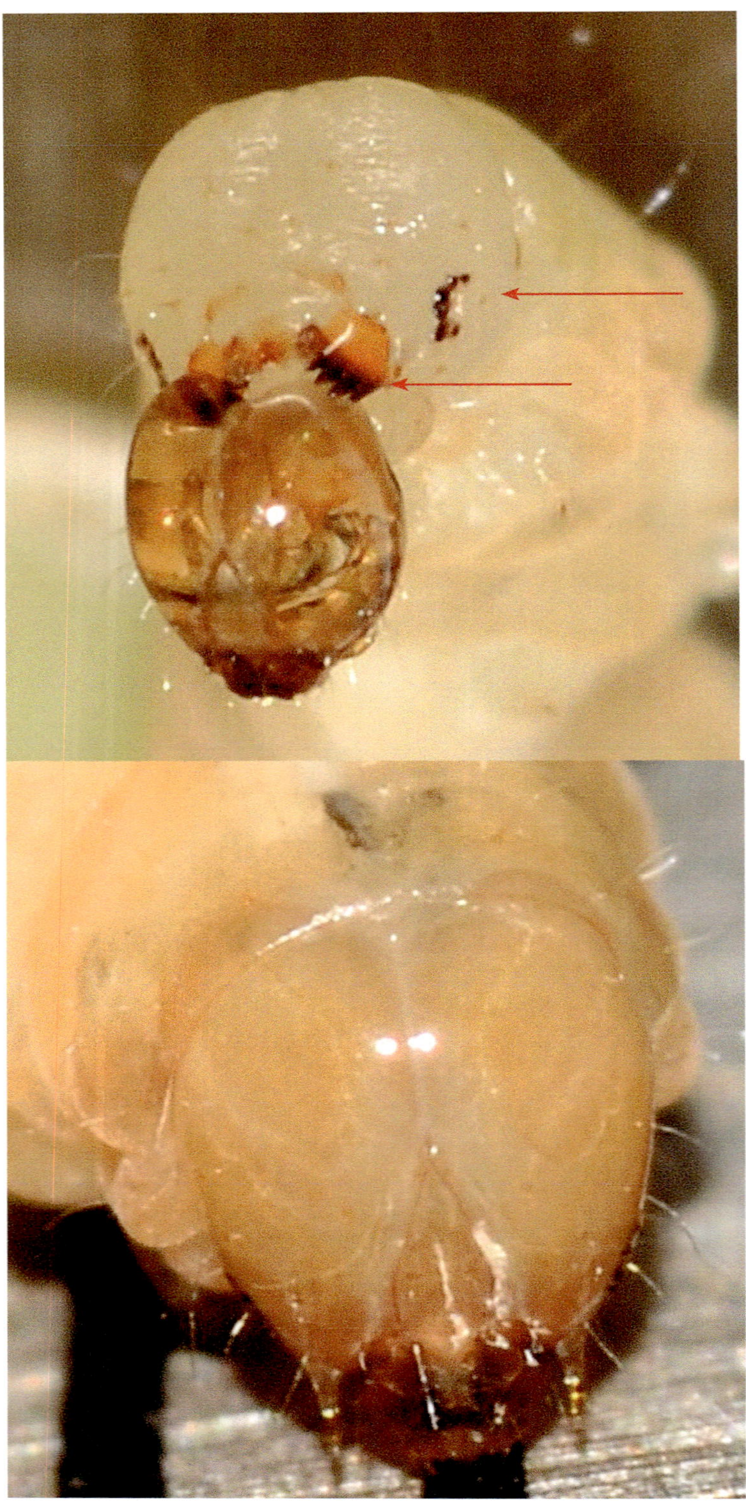

Plate 4.23. Immediately after moulting head capsule remains white except the mandibular area. Mandibles have five teeth each, four are dark and sharp.

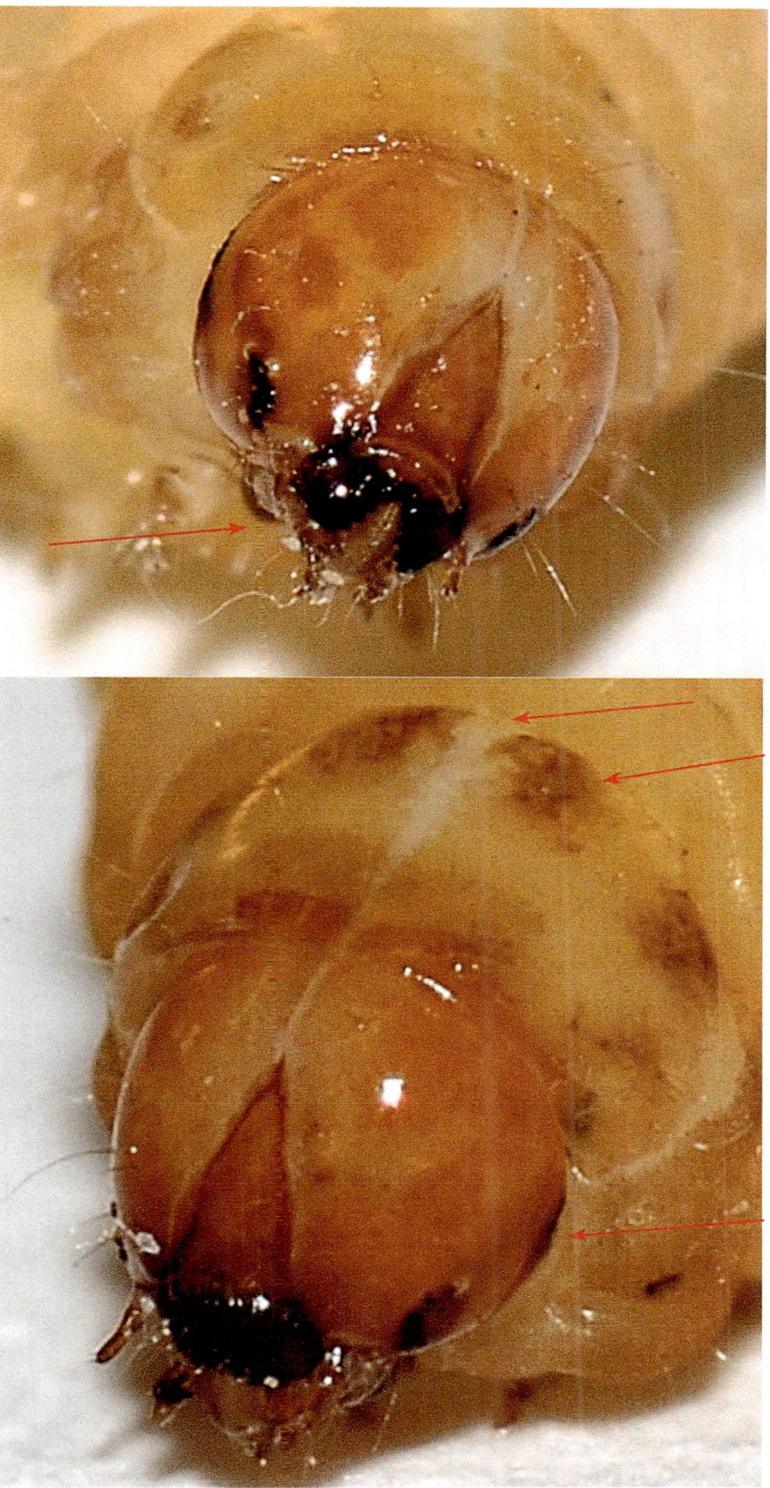

Plate 4.24. Head of the larva. *Prothoracic* shield is well developed and bear white median line. Depending on the instar a larva is passing through, colour and dark patches present on the *prothoracic* shield changes.

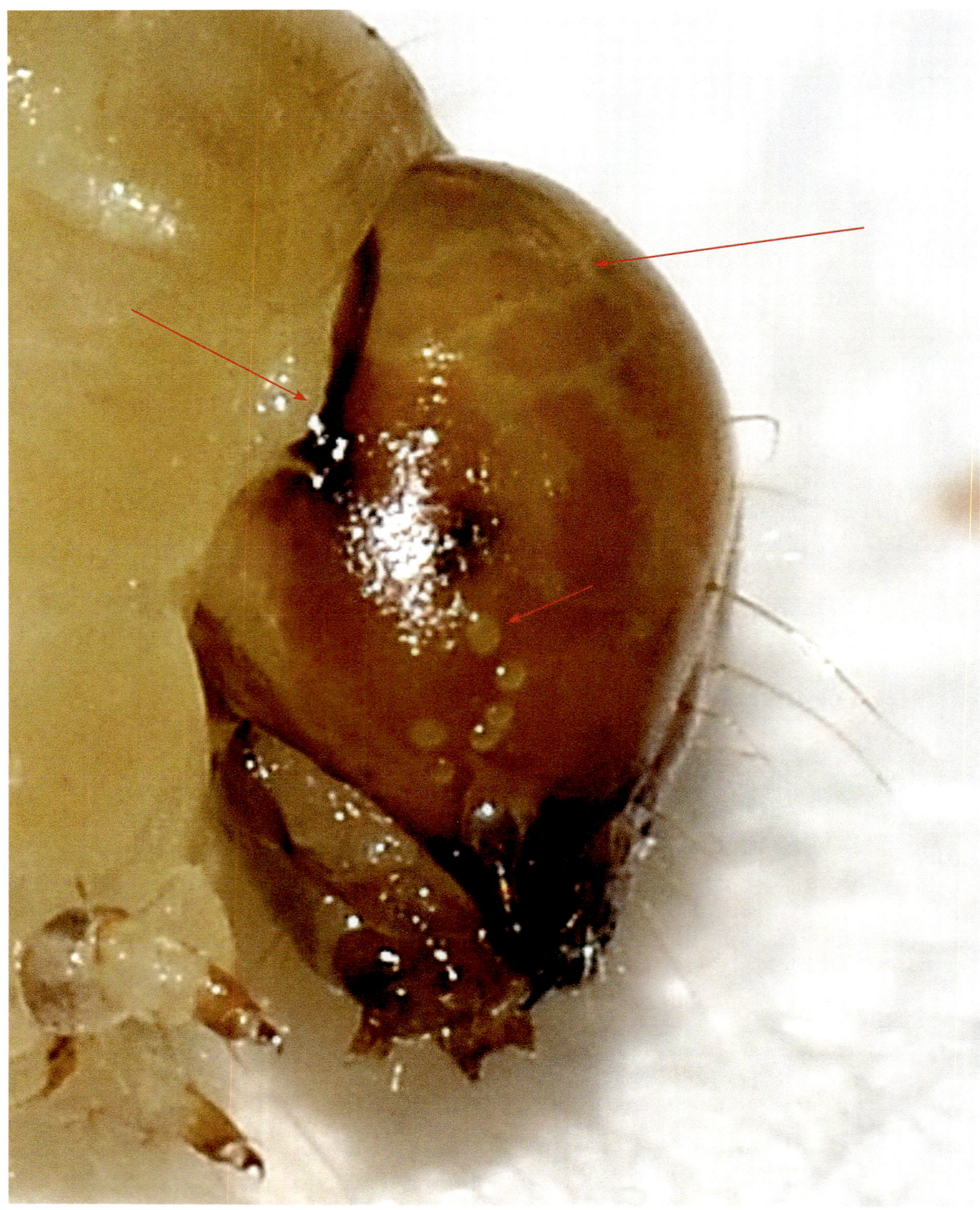

Plate 4.25. Head of the larva (close up, side view). Arrangement of six *ocelli* in groups. Formation of colour stripes on the head capsule.

Plate 4.26. Immature pupae of Top borer are yellowish white in colour with prominent spiracles on the body. Female pupa is larger and broader, whereas male pupa is smaller and narrower.

Plate 4.27. Immature female pupa of Top borer. A freshly formed pupa is yellowish white in colour with prominent orange rimmed *spiracles* on the body. Hind leg pair extends beyond the wing pad area and covers the 5th abdominal segment.

Plate 4.28. Mature female pupa of Top borer (dorsal and ventral view). The 8th and 9th abdominal segments are distinctly orange coloured. Formation of *papillae anales* in the *pharate* adult is visible.

Plate 4.29. Mature male pupa of Top borer (dorsal and ventral views). Anal end is conical and yellowish orange in colour. Dorsal side yellowish orange and ventral side yellowish white in colour. Terminal area of the hind legs are free from body and fully cover the 7th abdominal segment.

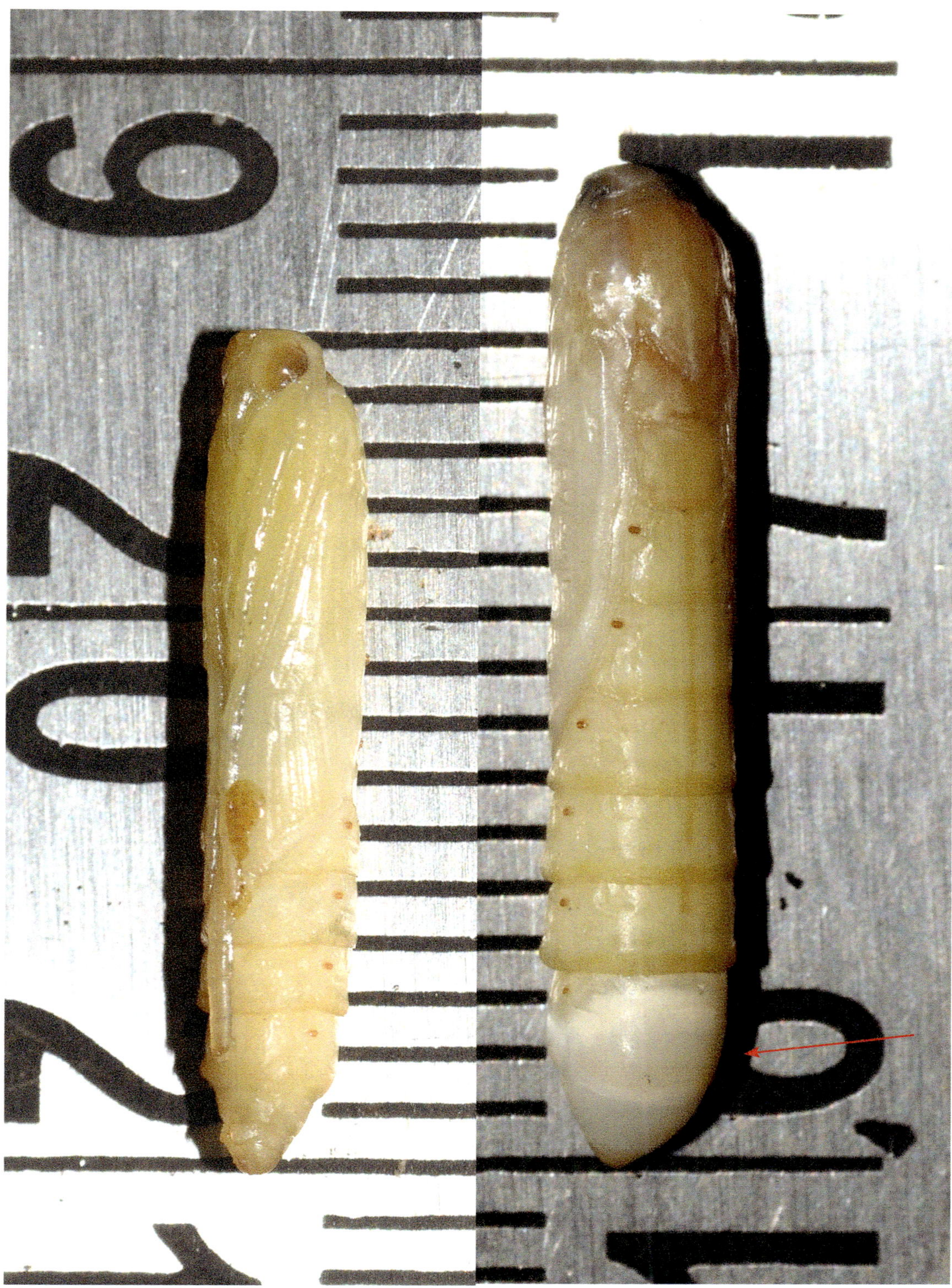

Plate 4.30. Top borer pupae on a centimetre scale. A male pupa measures around 15 mm and a female pupa around 18 mm. In the process of development, dorsal area of the anal end (from 7th segment) of a female pupa turns distinctly white.

Plate 4.31. Female pupae of Top borer. Size of the pupa gradually decreases with maturity and the colour of the pupa gradually changes due to the internal development.

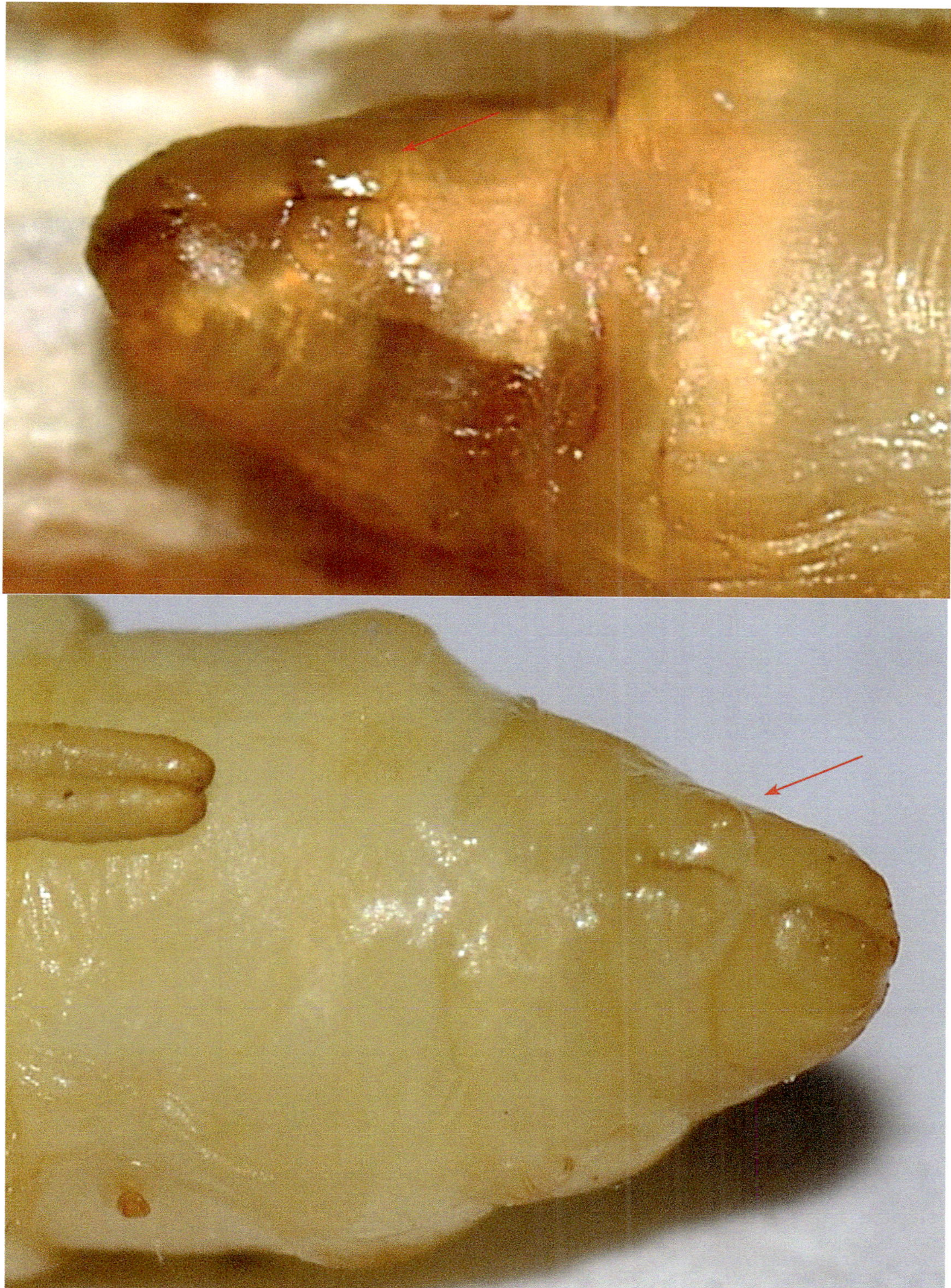

Plate 4.32. Genital area of the male pupa is marked with two broadly raised pointed lobes with a median depression on the 9th segment. Anal area with a median depression.

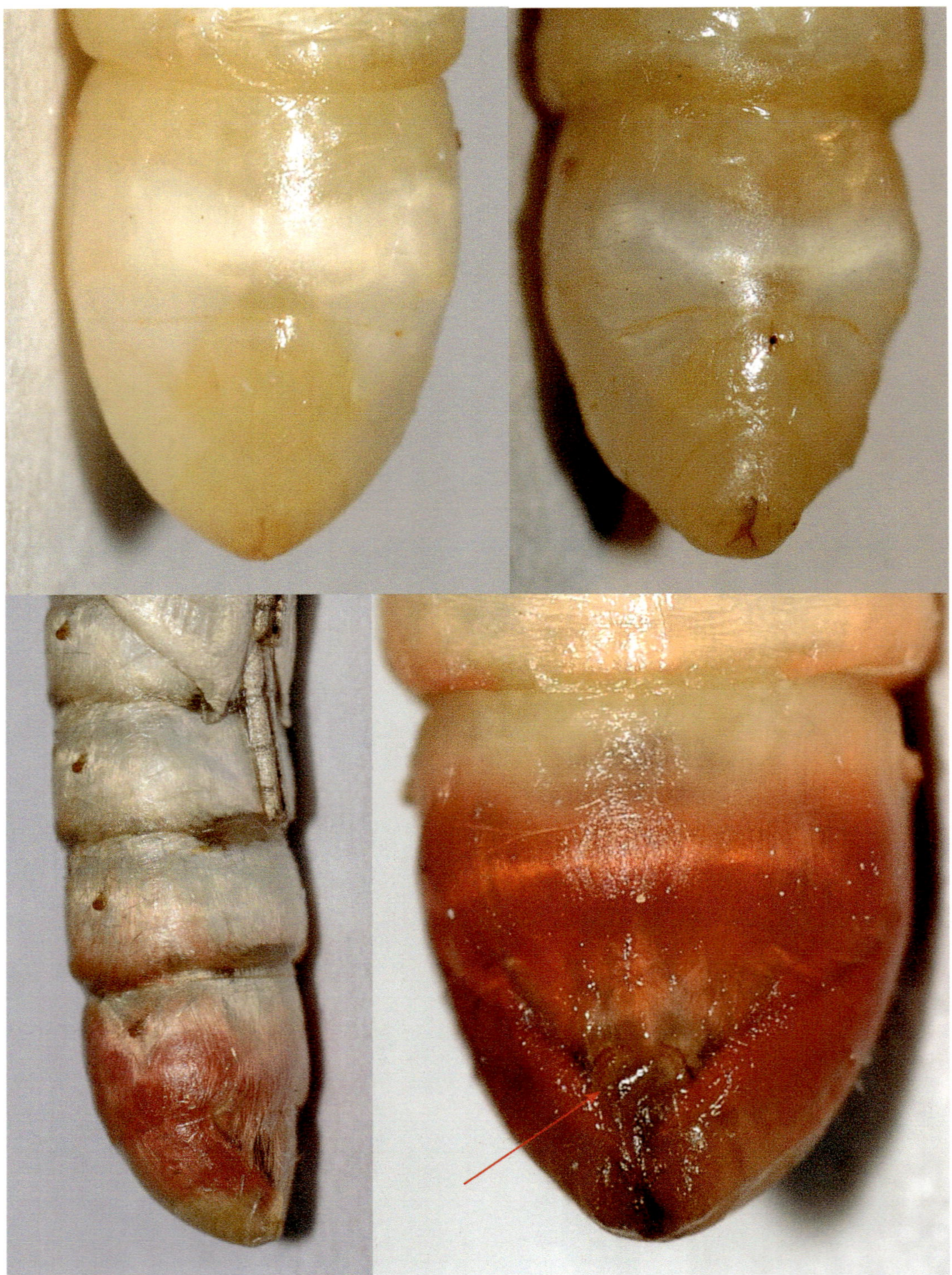

Plate 4.33. Genital area of a female pupa is marked with a small depression on the 8th segment. In the dorsal side, 7th and 8th segment turns white. The area gradually changes due to the development of *pharate* adult. This area changes to *papillae anales* and orange colour hairs.

Plate 4.34. Head and thoracic region of pupa (side and ventral views). *Metathoracic shield* is well developed. *Mesothoracic spiracle* is prominent and arch type. With development, pupal body appears white. Later, black eyes are also get covered with white scales. Development of serrated antennae of the pharate adult is clearly visible.

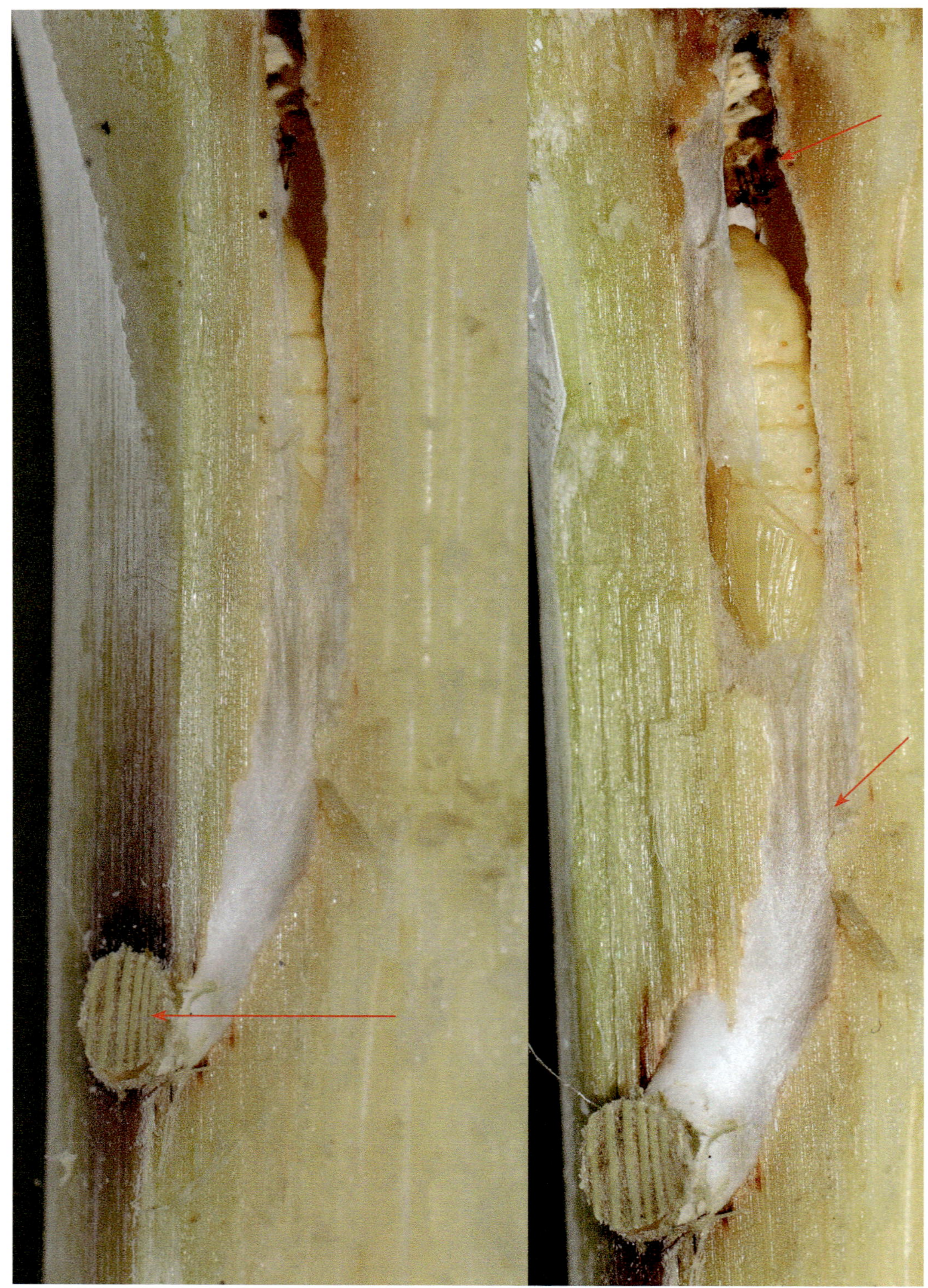

Plate 4.35. The exit hole is round and covered with the host tissue. A prepupal larva prepares a nice white silken chamber to rest within it.

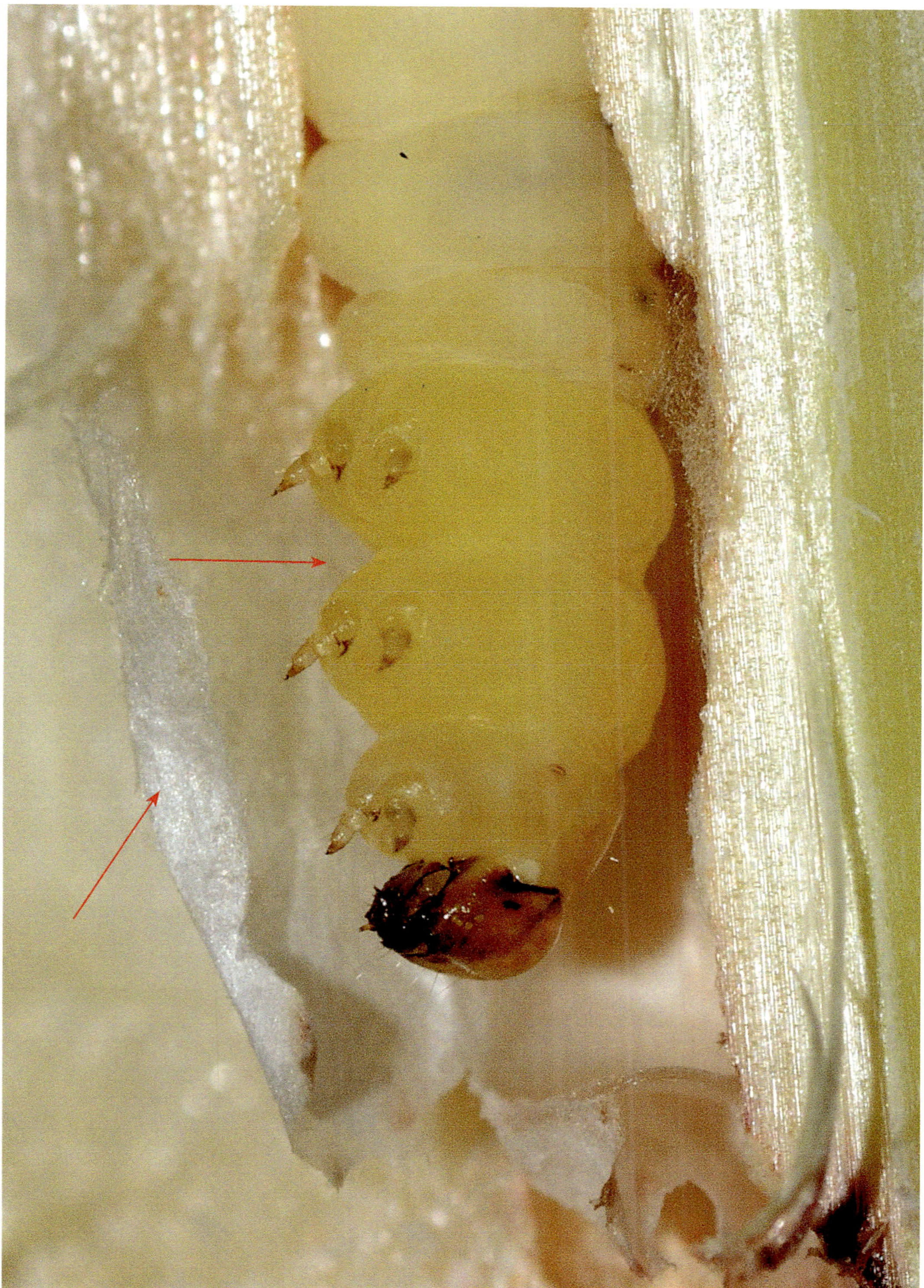

Plate 4.36. A prepupal larva prepares a silken chamber before going for pupation. In prepupal larva thoracic region turns lemon yellow in colour.

Plate 4.37. A pupa rests in the silken chamber. A prepual larva prepares several silken discs in the tunnel to prevent the ingress of parasites. Here, 14 silken disks are visible behind the exit hole.

Plate 4.38. Female moth of Top borer is in the process of emergence leaving the pupal exuvium behind in the tunnel.

Plate 4.39. Female moth of Top borer showing milky white forewings, orange coloured hairs at the anal end. Simple *filiform* antennae, *porrect labial palpi,* black eyes and a tuft of scales behind the eyes extending to the thoracic region.

Plate 4.40. Sitting habit of female moth of Top borer. Showing *crambid* nature. Milky white forewings, orange coloured hairs at the anal end and visible *papillae anales*. Simple *filiform* antennae, *porrect labial palpi*, black eyes and tuft of scales at the head extending to the thoracic region.

Plate 4.41. Sitting habit of female moth of Top borer. Showing *crambid* nature. Milky white forewings, and orange coloured hairs at the anal end. Simple *filiform* antennae, *porrect labial palpi*, black eyes and tuft of scales at the head extending to the thoracic region.

Plate 4.42. Male moths of Top borer. Male moths are identical to the female, except its size and orange coloured anal area. In some moths, black spots are present on the forewings.

Plate 4.43. Size of male and female moths of Top borer on a centimetre scale.

Plate 4.44. Ventral view of a female moth of Top borer.

Plate 4.45. Ventral view of a male moth of Top borer.

Plate 4.46. Head of the Top borer moth (female). Showing porrect, three segmented *labial palpi* with darker terminal segment. *Proboscis* and large black compound eyes and pattern of scales behind the eyes

Plate 4.47. Head of the Top borer moth. Showing serrated and ciliated dark coloured antennae, *porrect labial palpi* and black eyes. Dorsal side of the antennae remain covered with white scales.

Plate 4.48. In the anal area of the female moth, long orange coloured hairs predominate at the dorsal side. On the ventral side hairs are short. *Papillae anales* are beetle leaf like and folded from both the sides. In male, from the *meson* area of ventral side, a tuft of flat scales arise from the 7th segment (*sternite*) and cross the 8th segment.

Plate 4.49. Forewing of female moth (stained and unstained). Vein *post-cubitus* (CuP) is present. Sc anastomoses with R_1 and ends as one vein. R_2 is well developed and relatively nearer to stalk of R_3 and R_4. R_4 is almost twice of the R_3 and ends at costal area just a notch above the apex. R_5, M_1, M_2 and M_3 are independent and appear equidistant from each other. A_1 is not visible, A_2 is prominent and A_3 is rudimentary.

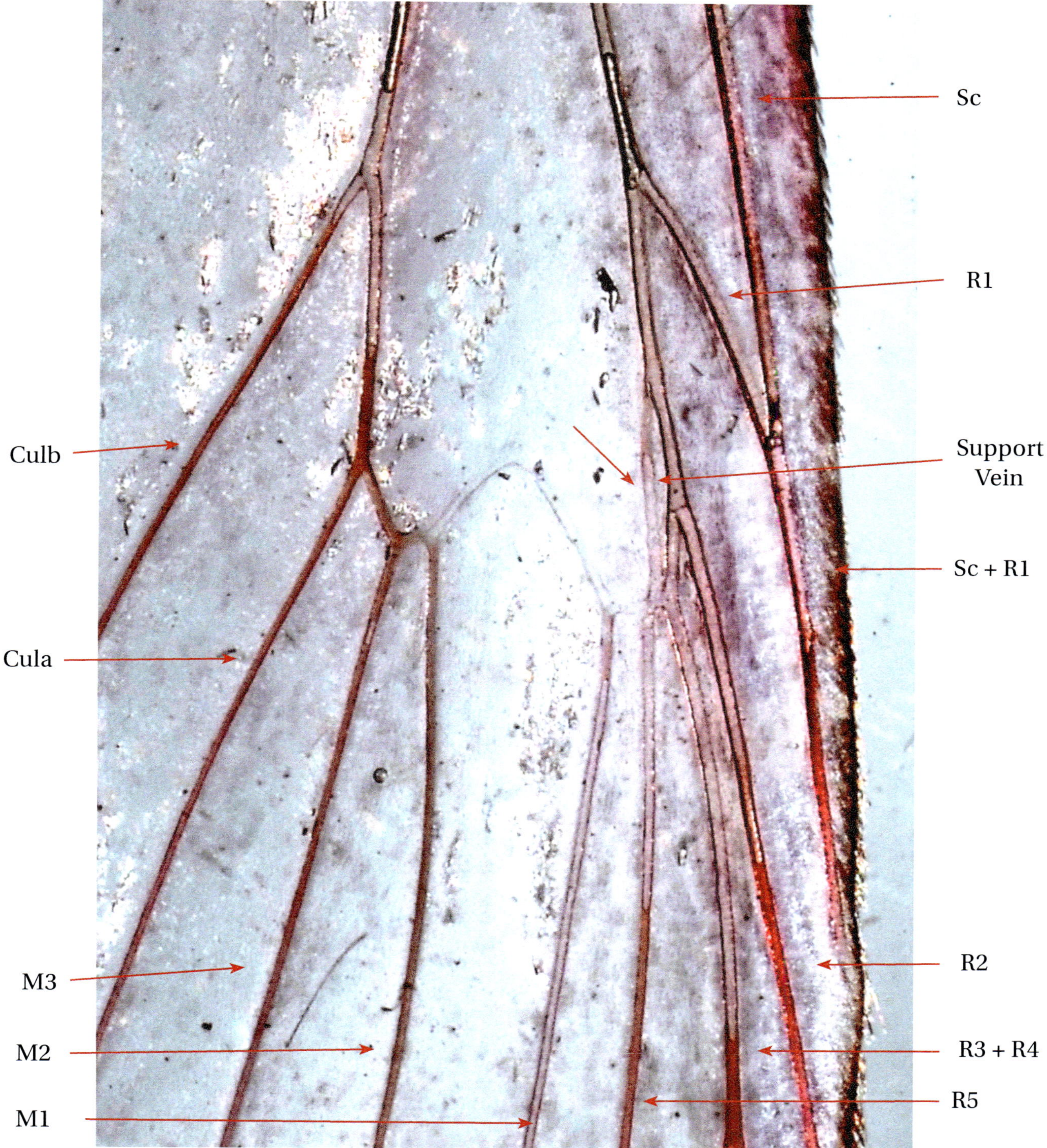

Plate 4.50. Discal region (close up). Showing ending of combine vein Sc+R_1. Bases of R_1 and Cu1b occupy the same plane at the discal cell. Support vein of R_5, radial angle remains higher than the cubital angle.

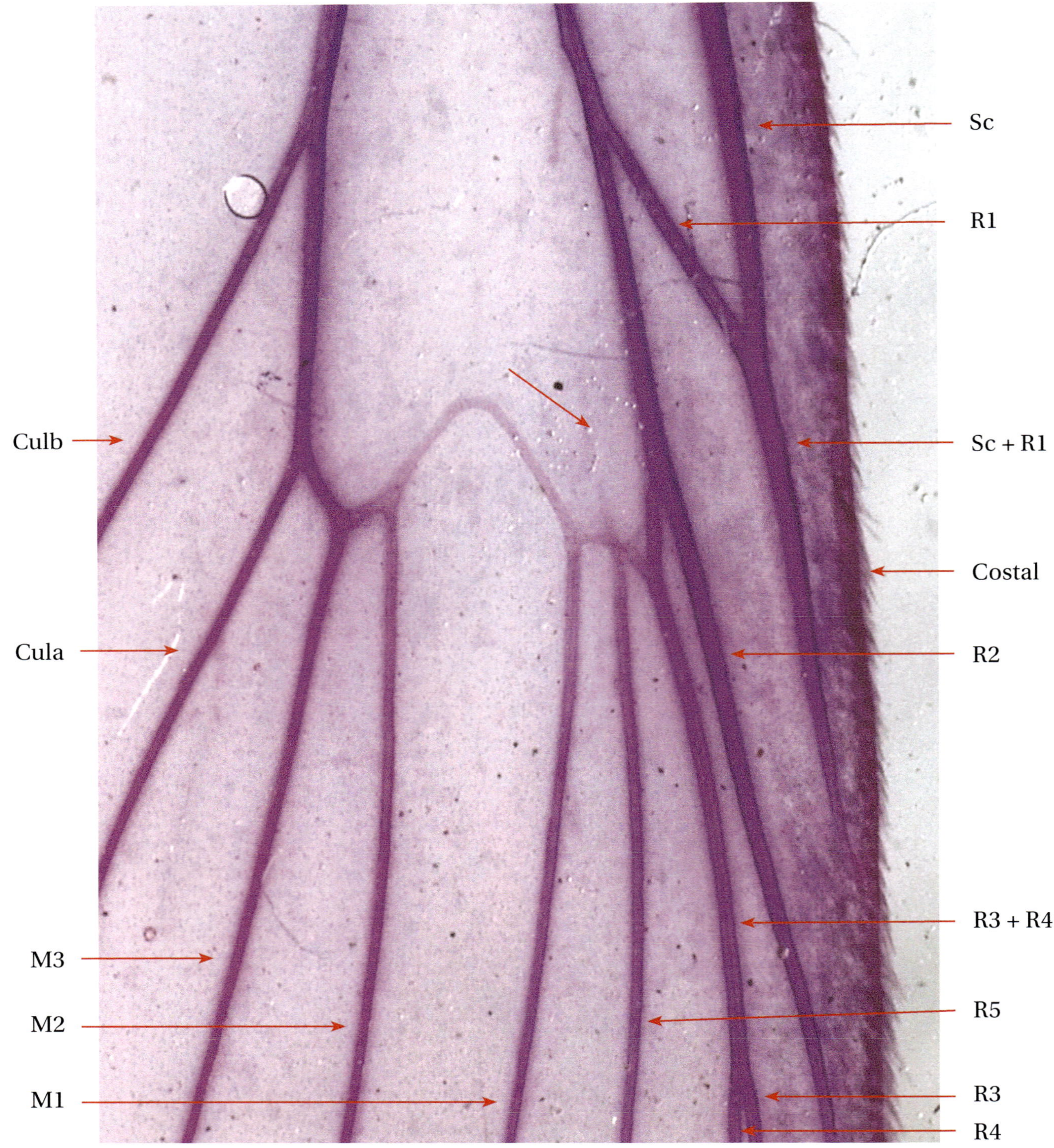

Plate 4.51. Close up of discal region showing ending of combine vein Sc+R_1. Bases of R_1 and Cu1b occupy the same plane at the discal cell. Traces of the support vein of R_5 and radial angle remains higher than the cubital angle (variations).

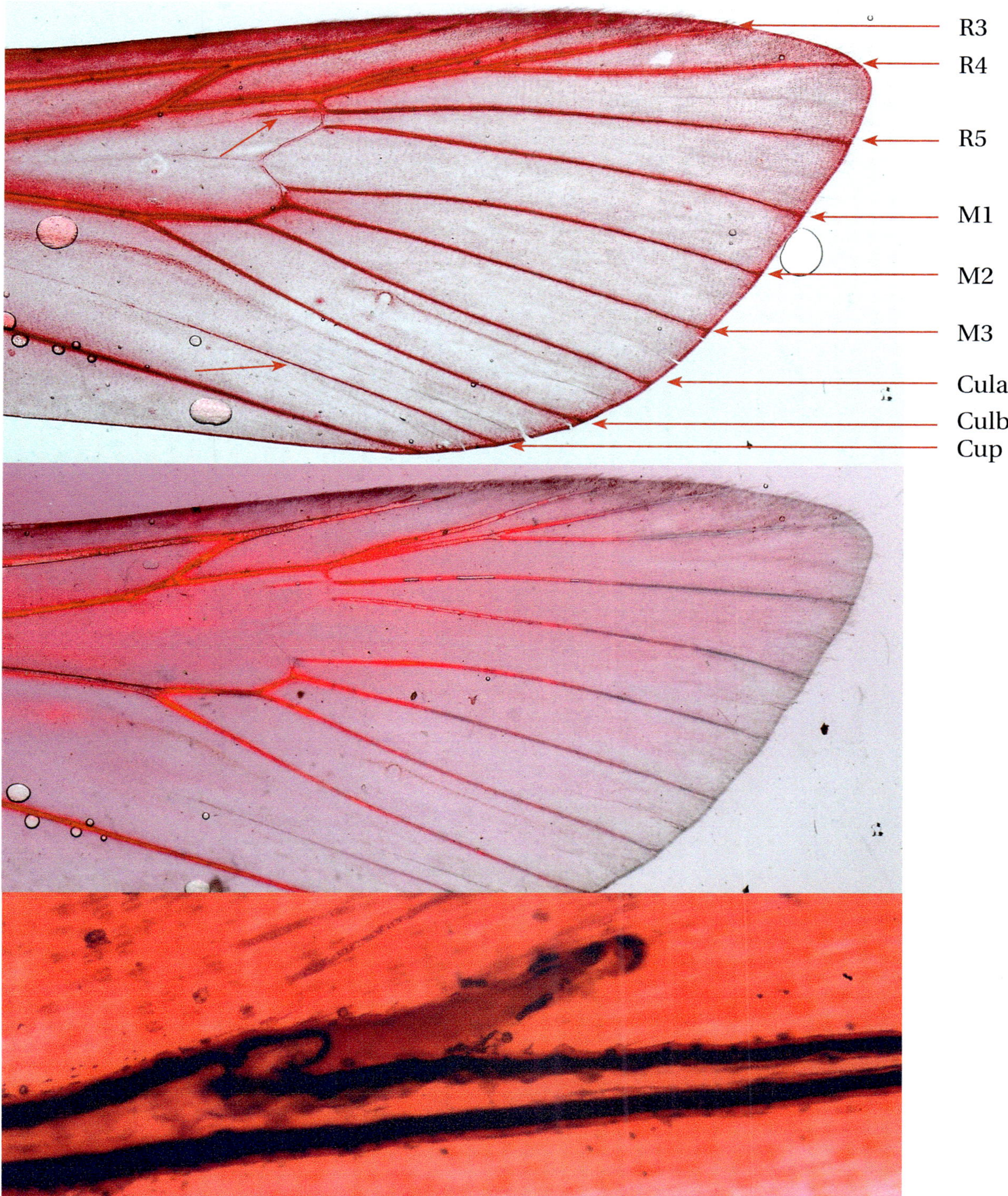

Plate 4.52. Close up of discal region showing ending of combine vein Sc+R_1. On way to the costal margin occasionally the combined vein throws a rudimentary branch (Sc).

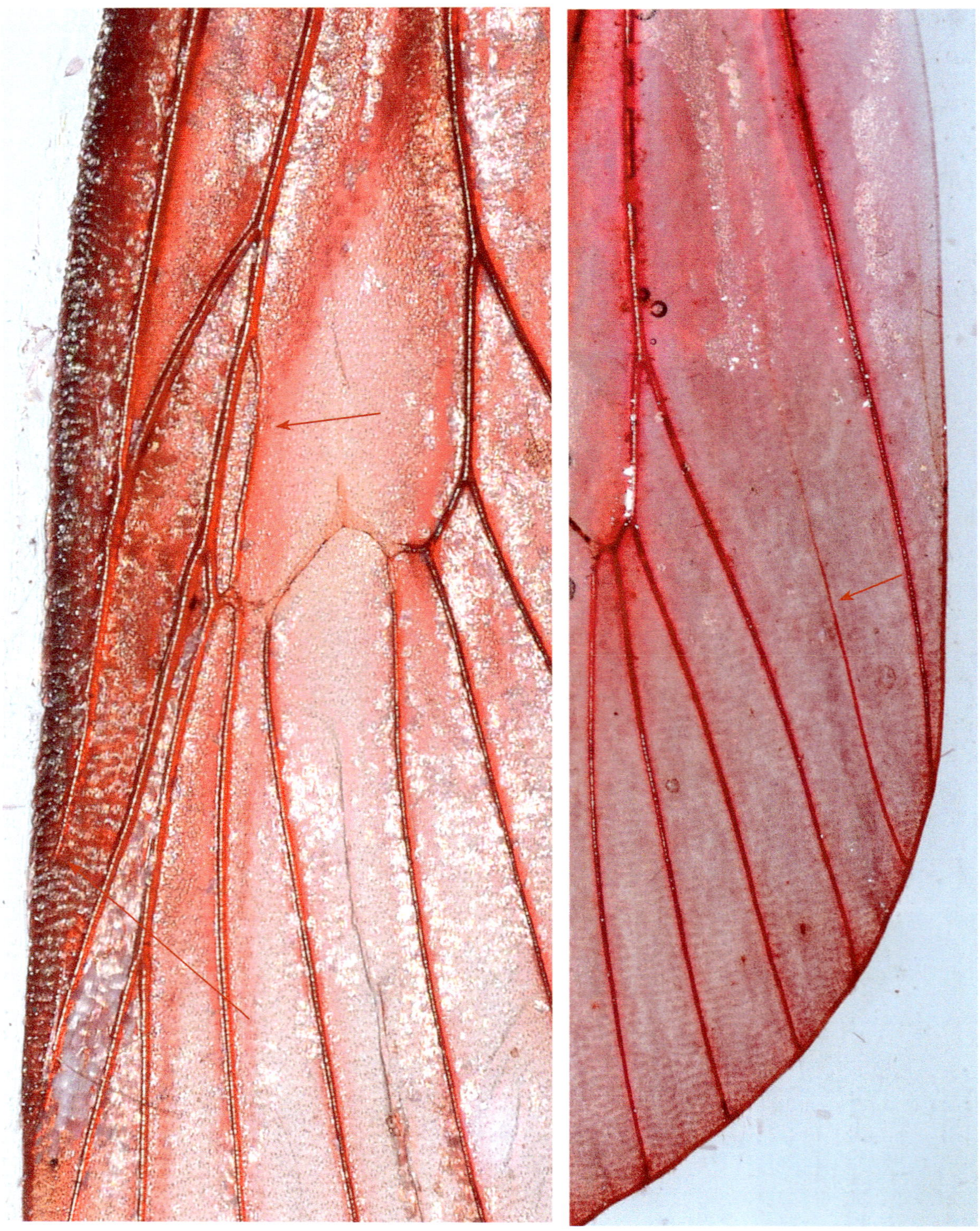

Plate 4.53. Close up of discal region showing ending of combine vein Sc+R_1 (no bifurcation). Radial angle remains higher than the *cubital* angle. *Post- cubitus* vein is present between Cu1b and A_2 and it is the characteristics of this moth borer.

Plate 4.54. Egg mass of Top borer. An egg mass is typical dome shaped and covered with light orange yellowish anal hairs. Egg mass mostly observed on the underside of a leaf near and parallel to the mid-rib.

Plate 4.55. Egg mass of Top borer. Typical dome shaped elongated egg mass covered with light orange yellowish anal hairs. Egg mass mostly observed on the underside of a leaf near and parallel to the mid-rib.

Plate 4.56. Close up of egg masses of Top borer. Typical dome shaped egg mass covered with yellowish orange coloured anal hairs. Size of an individual egg mass varies widely and may reach 19 mm or more. Eggs are laid in an imbricate manner. Size and arrangement of eggs after the removal of hairs. Eggs are a bit fleshy in comparison to that of *Chilo* moth borers.

5

Root borer, *Emmalocera depressella* (Swinhoe)

The Root borer was first described by Swinhoe (1885) as *Melissoblaptes depressellus* in the family *Galleriidae* from a specimen collected from Poona, Maharashtra, India. The description as appeared on page 876 of the Proceedings of the Zoological Society, London is furnished below:

GALLERIIDÆ.

175. MELISSOBLAPTES DEPRESSELLUS, n. sp. (Plate LVII. fig. 5.)

Poona, October.

Antennæ and palpi greyish yellow; thorax grey, with a brown band in front; abdomen whitish. Fore wings greyish yellow, irrorated with grey, with the irrorations intensified into pale streaks on the costa, in the centre and on the hinder margin, also some grey streaks on the veins, on the marginal border. Hind wings pure white, unmarked.

Expanse of wings 1$\frac{2}{10}$ inch.

A decade later, Hampson (1896) named the Root borer as *Polyocha depressella* and placed it in the subfamily *Anerastiinae.*

4312. **Polyocha depressella,** *Swinh. P. Z. S.* 1885, p. 876, pl. 57. fig. 5; *C. & S.* no. 4733.

Polyocha variegatella, *Rag. Nouv. Gen.* p. 39; *id. Mon. Phyc.* pl. xxxvi, fig. 16; *C. & S.* no. 4618.

♀. Head pale pink; thorax brown; abdomen ochreous. Fore wing pink, with a white costal fascia slightly suffused with pink and narrowing to base and apex; the area below the costal fascia broadly suffused with brown; the veins of outer area streaked with brown. Hind wing white with a slight yellow tinge.

Hab. Dharmsála; Poona; Nilgiris; Pulo Laut. *Exp.* 22–36 millim.

Dudgeon (1905) while studying the moths collected from north Bihar by Mr. Mackenzie and Mr. Lefroy, gave another name, *Polyocha saccharella* to the Root borer. He identified this species as a new species in the subfamily *Anerastiinae*, as he observed some characters that were quite different from the earlier descriptions.

ANERASTIINÆ.

POLYOCHA SACCHARELLA, n. sp.

Section II, B. Differs from Sect. II, Hmpsn. in Moths of India, Vol. IV, in vein 10 of the forewing not being stalked with 8 and 9. Antennæ of the male with a large tuft of bicoloured scales in the sinus at the base of the shaft and uniseriate rather short branches. Palpi obliquely upturned in the male, rostriform in the female.

♂. Head, thorax, abdomen and forewing uniform pale ochreous brown more or less suffused with fuscous in and beyond the cell. Hind-wing white.

♀. Differs from male in wanting the fuscous suffusion except slightly at the apex of the forewing.

Habitat :—Burogah, Bengal. *Exp.* 32-34 mill.

Types sent to British Museum.

Several specimens of this species were sent me by Mr. Maxwell-Lefroy reared by Mr. Mackenzie. It is said to be destructive to sugarcane in Behar.

In 1917, the genus *Polyocha* was transferred to *Papua* and this moth borer was redesignated as *Papua depressella*. However, in 1918, Hampson reassigned this species in the genus *Emmalocera* as *Emmalocera depressella* (Swinhoe). Since then this name is in use for the Root borer of sugarcane.

(7) †EMMALOCERA DEPRESSELLA.

Melissoblaptes depressella Swinh. P. Z. S. 1885, p. 876, pl. 57. f. 5; Hmpsn. Moths Ind. iv. p. 63; id. Rom. Mém. viii. p. 324, pl. 36. f. 12.

Polyocha saccharella Dudgeon, J. Bomb. Nat. Hist. Soc. xvi. p. 405 (1905).

ADEN; PUNJAB; BENGAL; BOMBAY. The larva feed on the roots of sugar-cane.

Of late, different authors are trying to revive the old name *Polyocha depressella*. If at all, the name has to be changed and if it has to be placed in the genus *Polyocha*, then it will be appropriate to identify this species as *Polyocha saccharella* Dudgeon, as the description provided by Mr Dudgeon is more appropriate to this species.

5.1 Currently accepted name

Emmalocera depressella (Swinhoe)

5.2 Synonyms

= *Melissoblaptes depressellus* Swinhoe

= *Polyocha depressella* Hampson

= *Polyocha saccharella* Dudgeon

= *Papua depressella* Swinhoe

5.3 Symptoms of infestation and brief biology

The name "*Root borer*" to this species of sugarcane moth borer is definitely a misnomer. The larva of this moth borer species does not attack or damage the roots of sugarcane plant. Larva mostly damages the underground portion of the cane stalk. However, as no other borer remains confined in the underground portion of the cane stalk, the name got acceptance to indicate its habit of staying near the root zone. This is the only species among the cane borers that goes down well below the soil surface (2 to 5 cm) to initiate an attack. Root borer larva enters the sugarcane plant mainly by making a minute bore-hole through the underground portion of the stalk. Like in any biological system, aberrations are also there, and one may encounter the entry of Root borer even above the soil surface, but it hardly goes few inches above the soil level. As the borer larva goes down the soil surface to enter into the cane plant, crop grown on a sandy to sandy loam soil gets higher infestation than the crop grown in clay soil. Similarly, the borer has a higher preference for early cane forming varieties (not necessary the early maturing one). Like most of the borers Root borer also prefers the ratoon sprouts and autumn (October) planted crop for the infestation of first brood (emerges after hibernation) in February-March. In subsequent broods, moths venture out to the nearby plant crop.

The Root borer larva damages the cane stalk below the soil surface and thus, it necessitates a careful digging of the infested clump to ascertain the actual presence of a larva inside the cane stalk. In the infestation at shoot stage, when cane formation is yet to gain momentum, a *dead heart* is formed. The *dead heart* is large (covers many leaves of the crown) and it is formed as a result of devouring the immature stalk tissues by the larva from below the growing point. As the growing point of crown remains intact (it does not damage the growing point like Early shoot borer or Pink borer), there is no foul smell from the lower end of the *dead heart*. Therefore, pulling out the *dead heart* is quite difficult or needs much force. In a very young shoot, on pulling, the immature stalk breaks away from the damaged dark-rotten area. In most of the cases of such pulled *dead heart*, it does not contain the root borer larva; the larva stays behind in the damaged stalk tissues. When the larva comes out with the *dead heart*, a milky white larva with prominent yellowish-brown head, staying at the centre of the black tissues (dead crown tissues) is observed. The formation of *dead heart* indicates killing of the apical buds and arrest of the growth of the cane. In many cases, the affected shoots die prematurely. Thus, infestation of Root borer at shoot stage affects the crop stand. As the growth of the crop advances and cane formation takes place, (at grand growth and later) a damaged cane stalk sustains the growth of the above ground portion. With the increase in the girth of the stalk, only a partial damage takes place and the undamaged portion of the stalk maintains the growth of the above ground portion and in such situation no *dead heart* is formed. Obviously due to the partial damage at the base of the stalk, the affected cane plant shows some debility. Partial yellowing of the crown, one-sided yellowing of leaves to the total yellowing of the crown takes place.

If in the affected shoots, underground buds remain undamaged, these affected buds quickly respond to tide over the situation. These buds sprout for survival and eventually act as effective tillers. Thus, the affected shoot manages to compensate the loss. This type of situation is often encountered due to the infestation of first brood of the Root borer that appears during February-March. The second brood is delayed considerably and appears after passing the dry summer months of April, May and a week or so of June (in June-July). It appears, as if Root borer waits for the signal of a pre-monsoon shower to resume its activity. The major infestation of root borer in a sugarcane crop takes place during the grand growth phase i.e., during July-August-September. With the onset of winter, larvae of this moth borer also proceed for hibernation staying within the feeding tunnel of the cane.

Root borer, as it stays in the underground portion of the cane plant, needs a special silken tunnel for the escape of the moth in the aerial environment (above the soil surface). Depending on the situation, a prepupal larva prepares a long silken tunnel lined with tiny bits of host tissue attached all around at the outer surface. The opening of the tunnel obviously always remains above the soil surface. Sometimes in shoot stage, a Root borer larva moves up through the central portion of the *dead heart* and reaches well above the soil surface. After assessing its position above the soil surface (staying inside the shoot) it opens up an exit hole in the *pseudostem*, and weaves a silken tunnel extending outside the host tissue. However, in the latter period of the crop growth, a Root borer larva stays at the underground portion of the stalk. At the clump base, a prepupal larva usually prepares a long silken tunnel extending above the soil surface for safe exit of the moth. In later broods, if the damaged stalk area remains above the soil surface, a Root borer larva may not produce the silken tunnel, it may just close the opening of the tunnel with silk.

A Root borer larva is not a voracious consumer of cane tissues like the larvae of other moth borers of sugarcane. In a thick or mature stalk, a substantial amount of the stalk tissue always remains undamaged around the feeding area. An affected cane survives well by utilising the water and nutrients supply through the undamaged portion. However, due to the debility of plants, many a times affected plant may turn pale yellow or show partial yellowing and stunted growth. In the infestations of the later broods, the affected plants may or may not show any stunting of growth. Variations in symptoms and growth of the affected plants largely depend on the stage of the crop when infestation is taking place.

The Root borer larva is very independent in nature as the eggs are mostly laid singly and in a scattered manner. Obviously, a neonate larva of Root borer has to defend itself from the very beginning of its life. In general, only one larva is found in the young infested cane shoot during summer months. However, later on in the season, a cane stalk attains its full thickness (sufficient diameter), more than one larva may be found feeding side by side in different tunnels in the same stalk. A Root borer larva usually does not push the frass outside (as it does not have the chance to do so in the underground, as adjoining soil tightly adheres to the stalk) like others borers that feed on the stalk tissues above ground. The frass of Root borer is rather dry type and the larva stays with it in the feeding tunnel. Initially at shoot stage, the tunnel is usually made at the edge of the stalk and then it runs across the stem (in an irregular fashion). The tunnel may also go round the stalk in a semicircular manner. In the shoot stage, the feeding tunnel always turns black.

In infestation during summer months, sugarcane plant usually compensates the loss with accentuated tillering (in shoot stage infestation when characteristic *dead heart*s are formed). During the grand growth phase, (July, August, and September) major damage of the canes takes place from the infestation of Root borer. A Root borer larva is not migratory in habit like that of Early shoot borer or Pink borer, but if situation demands, it migrates to an adjoining shoot or cane stalk. Thus, a Root borer usually larva damages one or two shoots in completing its life cycle. Thus, there is no sudden flare up of *dead heart* formation in case of Root borer The canes at the grand growth phase withstand the Root borer infestation well; consequently there is no *dead heart* formation and the infested plants show a near normal growth. However, due to damage to the underground portion of the stalk, growth and development of the infested plant gets a setback. This is reflected in the consequent reduction in yield. In contrast to the damage by other borers, there is hardly any impairment in juice quality in case of the infestation of Root borer. In heavy infestation, spanning different broods, the cumulative yield loss is substantial. Usually with the onset of winter (during November) a Root borer larva enters into hibernation phase.

A direct association of Root borer with *wilt disease* of sugarcane is a misconception and is not based on the actual experimental findings. This misconception has been pushed forward in a number of publications. In these publications, a direct correlation between the incidence of wilt and Root borer was indicated. An assumption was made that the Root borer larva provided the needed opportunity or a conduit for the wilt pathogens to make inroads in the cane. In statistical terms, these are the occurrence of two independent events sometimes sharing the same space and time. However, these events differ in their respective and characteristic ways. In no publication, it was convincingly established that an underground artificial damage (similar to that of Root borer damage) would predispose the cane plant to wilt pathogen(s). In a field, incidence of Root borer was observed in both the wilt affected and healthy clumps of a wilt susceptible variety, Co 7717. Similarly, nobody could establish the brood of the Root borer that predisposes the sugarcane plant to wilt infection. Moreover, the dark and dry frass of root borer usually remain free from apparent fungal growth. In a wilt tolerant variety like Co 1148, there is no development of wilt symptom despite the high incidence of Root borer.

5.4 Egg and neonate larva

Eggs are generally laid singly or rather loosely in a cluster on the stalk and less frequently on leaves (mid-ribs). Eggs are even laid at the clump base. It is estimated that a female moth lays around 300 eggs. The eggs are elongated and creamy white in colour when freshly laid and about 0.8 to 1.0 mm in length and width is around 0.65 to 0.70 mm. Individual eggs remains firmly attached to the host surface and appears a bit dome shaped. In general, in a sugarcane field, these eggs mostly remain undetectable or untraceable even by an expert eye. The neonate larva is about 2 mm long. It has a yellow brown head and strong mandibles. A Root borer larva generally crawls down towards the base of the shoot from where it can make an entry. Usually a larva goes further down along with the leaf sheath/stalk and reaches around 4-5 cm below the soil surface. It usually takes the advantage of the soil cracks or uses the little gap that usually forms around the stalk due to swaying of the upper ground portion with normal breeze/wind. Staying below the soil surface, a larva makes an entry hole in the sugarcane tissue and feeds on the inside tissues of the cane. Occasionally, it remains outside the stalk for a longer time and feeds the peripheral stalk tissues by scraping.

5.5 Larva

A full-grown larva of Root borer, at rest measures around 25-30 mm in length. It has a strikingly shiny creamy white or milky white body with a yellowish-brown head and *prothoracic shield*. The body often appears somewhat wrinkled transversely, but the body retains a good amount of lustre. The larva of Root borer has prominent body hairs and this helps its quick identification. The hairs may be easily observed with a hand lens. The thoracic region of a Root borer larva is broader and the larval body gradually tapers towards the anal end. This imparts a distinctive conical appearance to the larva (especially in the dorsal view). Characteristically the *prothoracic shield* is well developed and relatively larger than the head. The colour of the *prothoracic shield* is much lighter; usually light yellow or water soaked yellow in colour. The head capsule of the Root borer larva is relatively larger in relation to the body. The colour of the head capsule varies from light yellowish to light reddish orange. The mandibular area remains distinctly darker. The head casule shows formation of intricate patterns of yellowish white bands. The *epicranial* suture appears yellowish white centrally and larger yellowish-white areas extended to both the sides. Occasionally the *epicranial* suture takes dark colouration. The *epicranial* notch (triangular head capsule free area) at the *vertex* is very prominent. There are six *ocelli* present at the lateral side and these ocelli follow a 1+4+1 pattern. In a freshly moulted larva, the head capsule remains creamy white, except the mandibular area, which remains light to dark brown in colour.

In the larva, *spiracles* or breathe holes are well formed and always remain visible. These appear as distinct orange-brown rimmed oval to round structures at the lateral side. The orange colour of the rim appears prominent in the milky white background of the larval body. *Tracheal trunk* remains mostly in hiding in the white body. However, in a stretched larva, *tracheal trunk* becomes visible as a distinct white band running at the lateral side supporting the orange oval *spiracles*. In Root borer larva, the position of *spiracle* on the 8th segment appears more lateral (like the Green borer).

Crochets on the four abdominal *prolegs* are *biordinal* and are arranged in a complete *obovate*/pear shaped circle (outline of a pear). The narrower area of the pear shaped circle is towards the interior (*meson*) of the body. On each of the *anal proleg* (*clasper*) only 7-8 crochets are present and the crochets are *uniordinal* and covers around 1/4th of the imaginary circle on the *proleg* base.

A larva also bears faint tubercle marks on the dorsal surface of the body, but tubercles are not developed or prominent. On extrication from the host tissue, a Root borer larva is rather sluggish in movement in comparison to the larva of other *Chilo* moth borers but it always remains more active than the larva of Top borer. On disturbance and provocation, a mature Root borer larva tries to defend itself by attacking with its tough mandibles. Probably, this aggressive behaviour is needed to ward off any ingress of termites in the damaged cane tissues.

As the larva feeds on the underground portion of the sugarcane stalk, it needs a passage for the safe exit of the adult, much above the ground surface. For exit of the moth, a prepupal larva, weaves a long silken tunnel in the soil extending to the soil surface. Pupation usually takes place inside the cane; however, occasionally a mature larva comes out from the feeding tunnel of the cane to the adjoining soil for pupation. In that situation, pupation takes place in the silken tunnel and the pupa remains at or near the soil surface.

5.6 Pupa

Like larva, pupa of Root borer is also much broader at the thoracic region and gradually tapers towards the anal end. This imparts a typical conical appearance to the pupa. The pupa of Root borer may be easily distinguished from the pupa of other cane moth borers by the presence of a *hump* like protuberance on the dorsal side of the anal end (*cremaster*). Presence of numerous small pits (depressions) on both the dorsal and the ventral surfaces of the pupa provide it the distinctive identity. The pits on the pupal body may be observed with a hand lens. (Small pits are also present on the pupa of Pink borer, *Sesamia inferens* but the pupa of *Sesamia* is stout-cylindrical with broad spine like projections at the *cremaster*).

A freshly developed pupa of Root borer has light yellowish wing-pad area and whitish body, except the hump like protuberance at the anal end, which appears light to dark brown. However, the tip of the hump appears brown from the very beginning of its development. With the passage of time, the hump becomes very dark. Like pupa of other moth borers, colour of the pupa of Root borer also becomes dark with maturity and finally a pupa turns yellowish brown.

The head of the pupa is smooth. The median ridge on the thoracic shields is prominent. *Mesothoracic spiracle* is relatively smaller. *Spiracles* on 2nd and 3rd abdominal segments show prominent arch formation. There are no ornamentations present on the 5th, 6th, and 7th abdominal segments. Except for the presence of pits, the abdominal surface is smooth. However, there is marked gradation of colouration; lower and upper sides of each segment are dark coloured. At the ventral side of the pupa, leftover marks of *prolegs* of larva are prominent. In a male pupa, genital opening is present in the 9th abdominal segment and it is a prominently raised structure with slit like median depression. In female pupa, the genital area is less prominent with shallow slit like depression present on the 8th abdominal segment. The anal opening in both sexes is prominent and deep slit like.

5.7 Moth (Adult)

The moth of Root borer is also quite different from the other moth borers of sugarcane. In contrast to the drab straw-coloured moths of other cane borers, Root borer moth has a pinkish tinge all over the forewings and the body, and appears more lustrous. Sporadic occurrences of small light black to greyish scales are observed on the forewings. The greyish scales are more prominent from the radial angle to the apex, and these scales help its quick identification along with other characters. Head is pale-pink and the face is round. *Labial palpi* are stout, conical and porrect in female moths, whereas in male moths, *labial palpi* are conical and projected upward. The length of the *labial palpus* is about twice the diameter of the compound eye. *Ocelli* are prominent and present behind the antennae and just above the compound eyes. In female moths, *papillae anales* or the ovipositor is densely clothed with sensory setae. The ovipositor is elongated and relatively narrower (ventral side is slight wider). The two halves of the *papillae anales* are with prominent space in between. The ovipositor mostly remain hiding behind the scales (not easily visible like in other moth borers of sugarcane).

In addition to the variations in size and genitalia, *sexual dimorphism* is more pronounced in the antennae. Female moths are larger in size (around 15-17 mm), whereas male moths are smaller (13-14 mm). Among the moth borers of sugarcane, Root borer has the longest antenna (8-10 mm). A female moth has long and *filiform* antennae, with broader scape and smaller pedicel. Whereas the antennae of male moths are typical *pectinate* (comb shaped) type with a distinctive ladle shaped bottom (*scape*). In addition, the male antenna has an inward sickle like curvature (forming the *sinus*) near the base (*pedicel*) and the concave pedicel is covered with numerous scales (rows of scales arise from both dorsal and ventral side in the sinus area). In the distal area of the sinus (near the end of pedicel) a tuft of black scales are present. In male, an antenna has more than 60 segments, and in female, there are about 50 segments. The terminal segment of an antenna ends in a cone. Root borer moths also feign death under any perception of threat like that of Top borer.

5.8 Forewing venation

Among the moth borers of sugarcane, the shape of the forewings of Root borer and Green borer is rather elongated (not triangular). In a straightened forewing, the *radial angle* of *discal cell* is almost in line with the *apex*. In the forewing, the costal area is prominent, like that of Plassey borer. *Subcostal* vein (Sc) is also prominent and much thicker at the base. R_1 is free. R_2 is also free and both the veins end at *costal area*. R_3 and R_4 are stalked. R_3 is smaller than R_4 and both end at the *costal area*. In Root borer, traditionally it has been taken that the vein R_5 is absent in the forewing. Usually, only one vein is found at the location that is normally occupied by R_5 in the forewing (originate at or near the cell angle and terminates at the *termen* just near the *apex*) of other *Chilo* moth borers of sugarcane.

However, in Root borer due to convention, the vein present at the location of R_5 is recognised as M_1. A closer look at the *tracheal* development clearly indicates that this vein is actually R_5 not M_1 and it will not be appropriate to call it M_1. In reality, M_1 is not present in the forewings of Root borer. This needs to be corrected.

During this study it became apparent that in good percentage of the moths of Root borer, this so-called vein M_1 (actually R_5) is bifurcated near the *termen* i.e., a bifurcation with a long stalk. This type of bifurcation opens up the possibility of designating the bifurcated veins as R_5 and M_1. In the situation of bifurcation, the first vein of the fork from the costal side be recognised as R_5 and the second vein of the fork as M_1. The two median veins, M_2 and M_3 are also stalked. The stalk is short and it originates from the *cubital* peak of the *discal* cell. The stalk bifurcates to form the M_2 and M_3. This stalk formation is a very striking feature in the wing venation of Root borer.

It appears that the stalk formation (or bifurcation of a vein) of the veins is very characteristic of this species. In some cases, there are three instances of stalk formation in the veins; two bifurcations are present in *radials* (Rs) and one bifurcation in the *median* veins. Probably, the Root borer as is known today is a complex of two subspecies. One sub-species has two instances of forking and the other subspecies has three instances of forking of veins in the forewings.

Cu1a and Cu1b are present. Two anal veins are also present; A_1 is absent, A_2 is very prominent and A_3 is feeble, but it is long enough to reach the *dorsum*. At the terminal area of each vein, two *mechanoreceptors* with well-distributed roots are also present.

List of plates on Root borer, *Emmalocera depressella* (Swinhoe)

Descriptions of plates

Plate 5.1. Symptoms of infestation of Root borer. Formation of large *dead hear*t at shoot stage. *Dead hear*t is not easily pulled out and doest not emit foul smell. At shoot stage a larva mostly attacks the immature portion of stalk.

Plate 5.2. Symptoms of infestation of Root borer. Formation of large *dead heart* at shoot stage. A larva feeds at the underground portion of the young cane shoot. Small entry hole remains blocked with frass. Becomes visible after removal of the outer leaf sheaths.

Plate 5.3. Symptoms of infestation of Root borer. Initially a larva feeds at the soft epidermal tissues of the stalk and results a pattern of semi spiral nature of feeding tunnel. Finally larva feeds inside the tissue and mostly in a cross wise fashion.

Plate 5.4. Symptoms of infestation of Root borer. A neonate larva makes a fine hole in the stalk well below the soil surface. The dry frass remains inside the feeding tunnel. Occasionally the larva may attack at the crown portion of the shoot and totally damage the shoot.

Plate 5.5. Symptoms of infestation of Root borer. When the cane develops sufficient in thickness, the larva feeds only a portion of the cane tissue and rest of the stalk tissue remain unaffected and continues to carry water and nutrients.

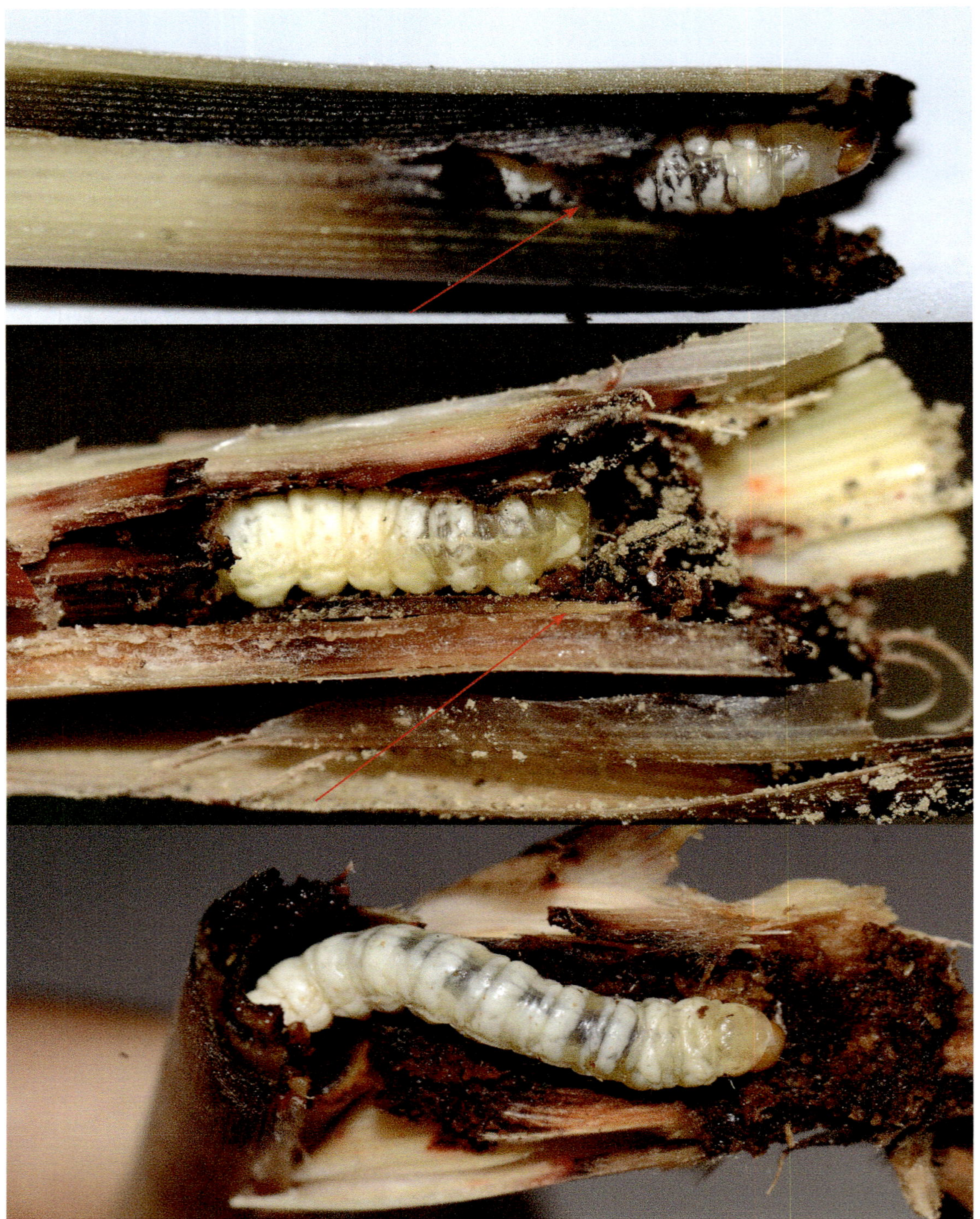

Plate 5.6. Symptoms of infestation of Root borer. On pulling with force, the crown portion often gets broken from the damaged area and comes out. Mostly a larva stays inside the damaged stalk portion of the shoot. Occasionally, when the larva stays inside the immature central leaf portion, it comes out with the *dead heart*.

Plate 5.7. Larvae of Root borer. (Dorsal view) *Prothoracic shield* is very large, thoracic region is broader and a larva gradually tapers towards the anal end.

Plate 5.8. Larvae of Root borer (Side view). Body of a lava tapers towards the anal end. Head capsule is light yellowish orange in colour. Body hairs and *spiracles* are visible on the white body.

Plate 5.9. Larvae of root borer. Larvae at different instars showing variations in colour.

Plate 5.10. A larva of Root borer. (Dorsal view) Larva is glistening white to creamy white in colour. Setae are prominent on the body. Larval body has several skin folds and the body tapers towards the anal end. *Prothoracic shield* is large and yellowish white. Tubercles are not perceptible.

Plate 5.11. Larva is glistening white to creamy white in colour and has prominent setae and skin folds on the body. Larval body tapers towards the anal end. *Prothoracic shield* is large and yellowish white. Tubercles are not perceptible. The alimentary canal appears as the dark median stripe.

Plate 5.12. Larvae of Root borer on a centimetre scale. A mature larva measures around 25- 30 mm.

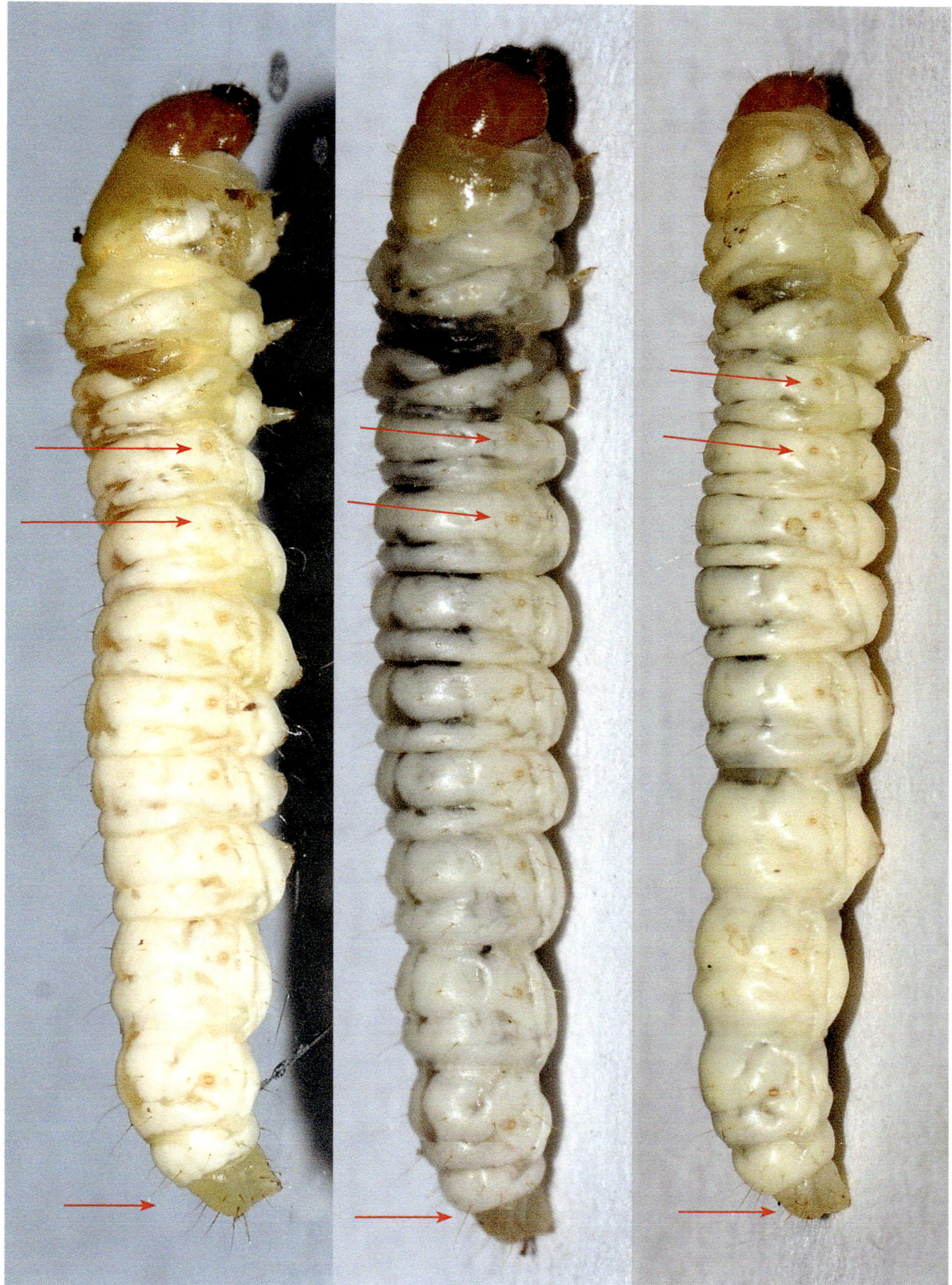

Plate 5.13. Larvae of Root borer (side view). *Spiracles* are visible as orange coloured round area on the white body. Head capsule is orange red in colour. Anal end appears water soaked area (not white like the body).

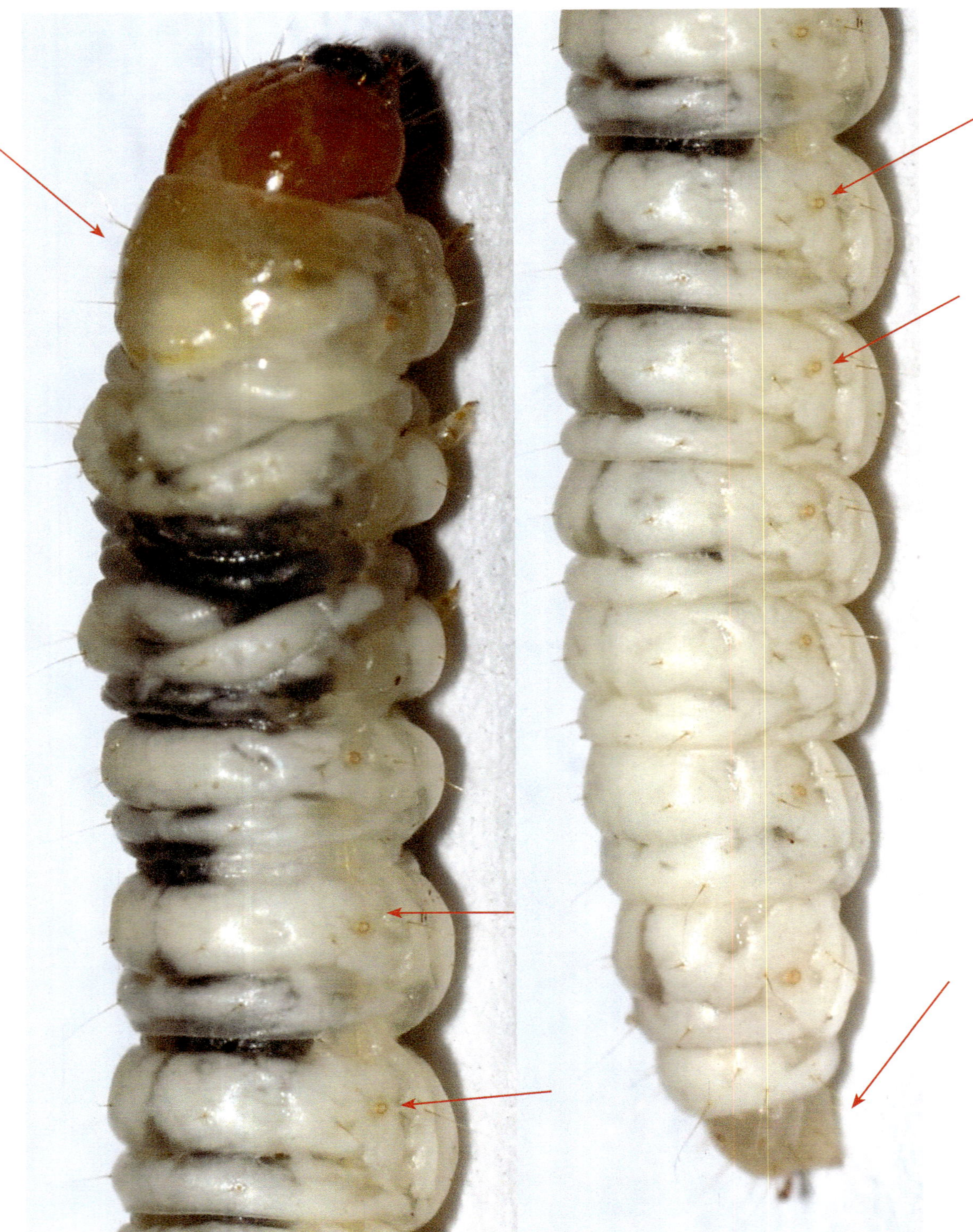

Plate 5.14. Larva of Root borer (side view, close up). *Spiracles* are visible as orange coloured roundish oval area. Head capsule is orange reddish in colour. *Prothoracic shield* is well developed and light yellow in colour. Anal end appears water soaked area (not white like the body).

Plate 5.15. Head and thoracic region of Root borer larva (close up). The *vertex* triangle of head capsule is larger. *Epicranial suture* area is white and extended sideways. Head capsule bears patchy light area. *Prothoracic shield* is well developed and large.

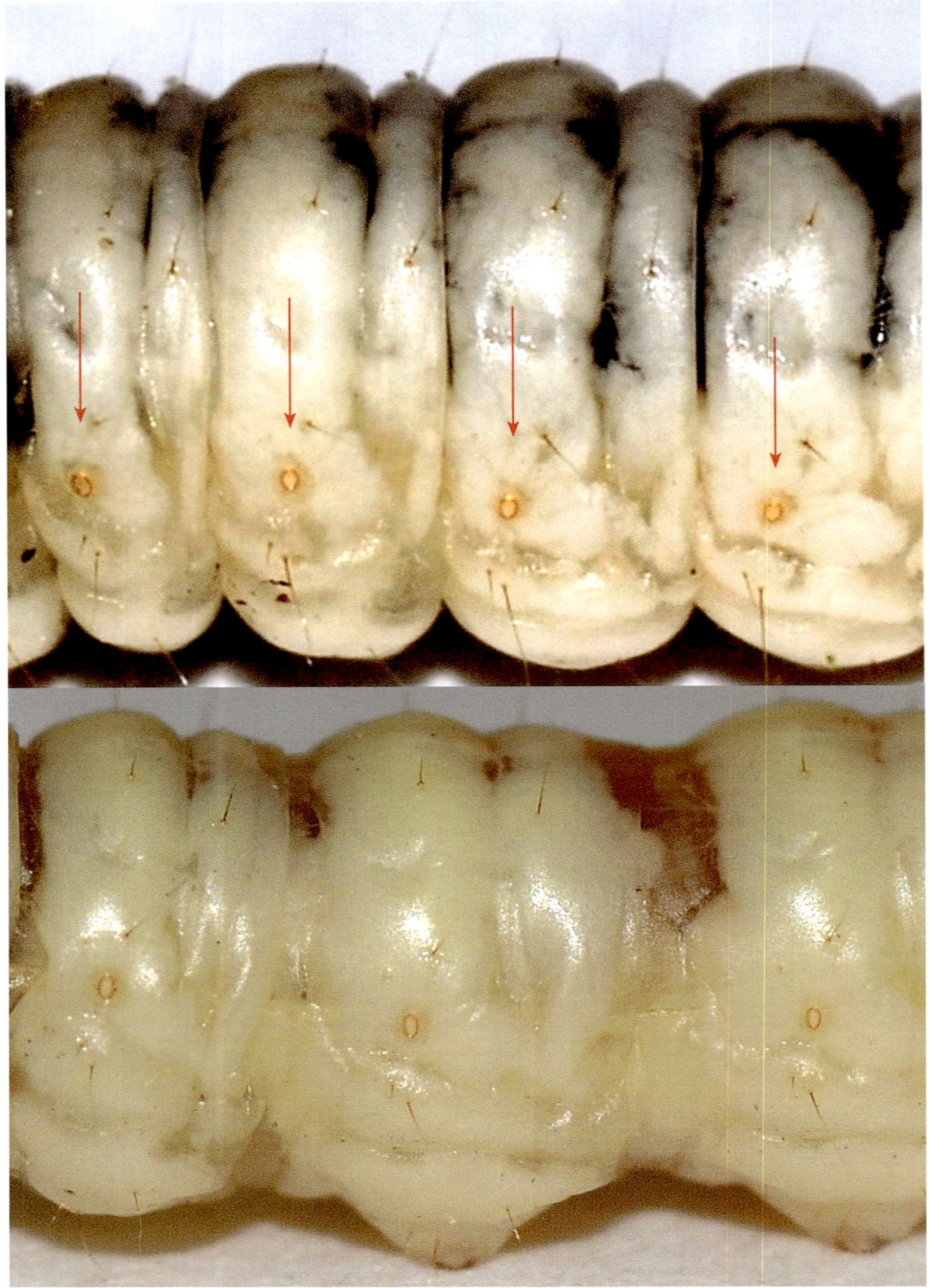

Plate 5.16. Larvae of Root borer (side view, close up). *Spiracles* are visible as orange rimmed round oval area with white centre. Generally *tracheal trunk* is not visible. When faintly visible, it appears as white tube.

Plate 5.17. Larvae of Root borer in the process of *ecdysis*. In a freshly formed larva, head casule is yellowish except the mandibular area, which remains dark brown.

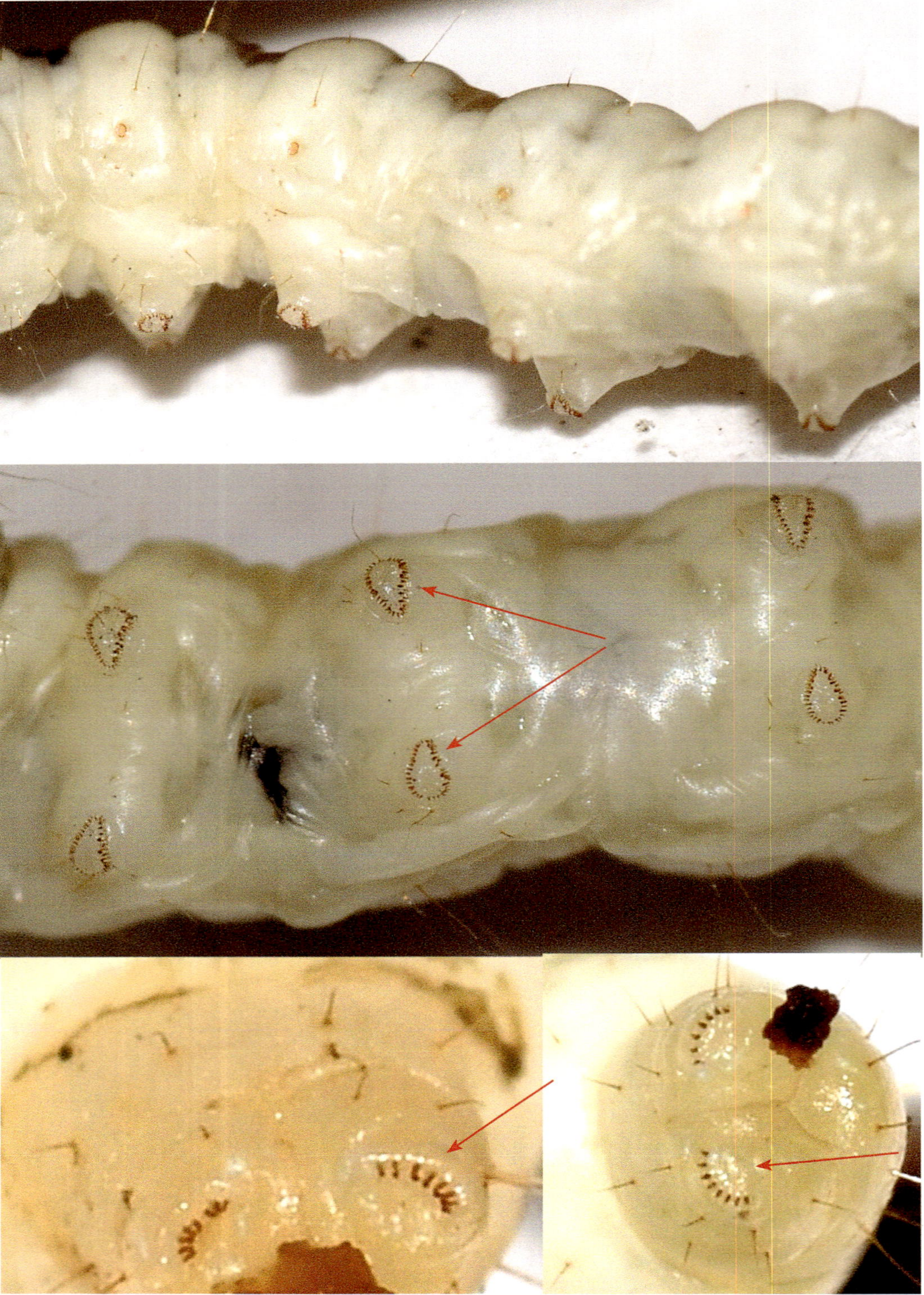

Plate 5.18. Crochets are *uniserial* and *uniordinal*. Crochets on the abdominal prolegs are arranged in an obovate circle (pear shaped) the narrow area is towards *meson*. On the anal prolegs, crochets cover only 25%of the circle and there are only 7-8 in number.

Plate 5.19. A Root borer larva has a habit of taking sharp bend. Larvae taking sharp bend in the feeding tunnel and on extrication from the cane tissue.

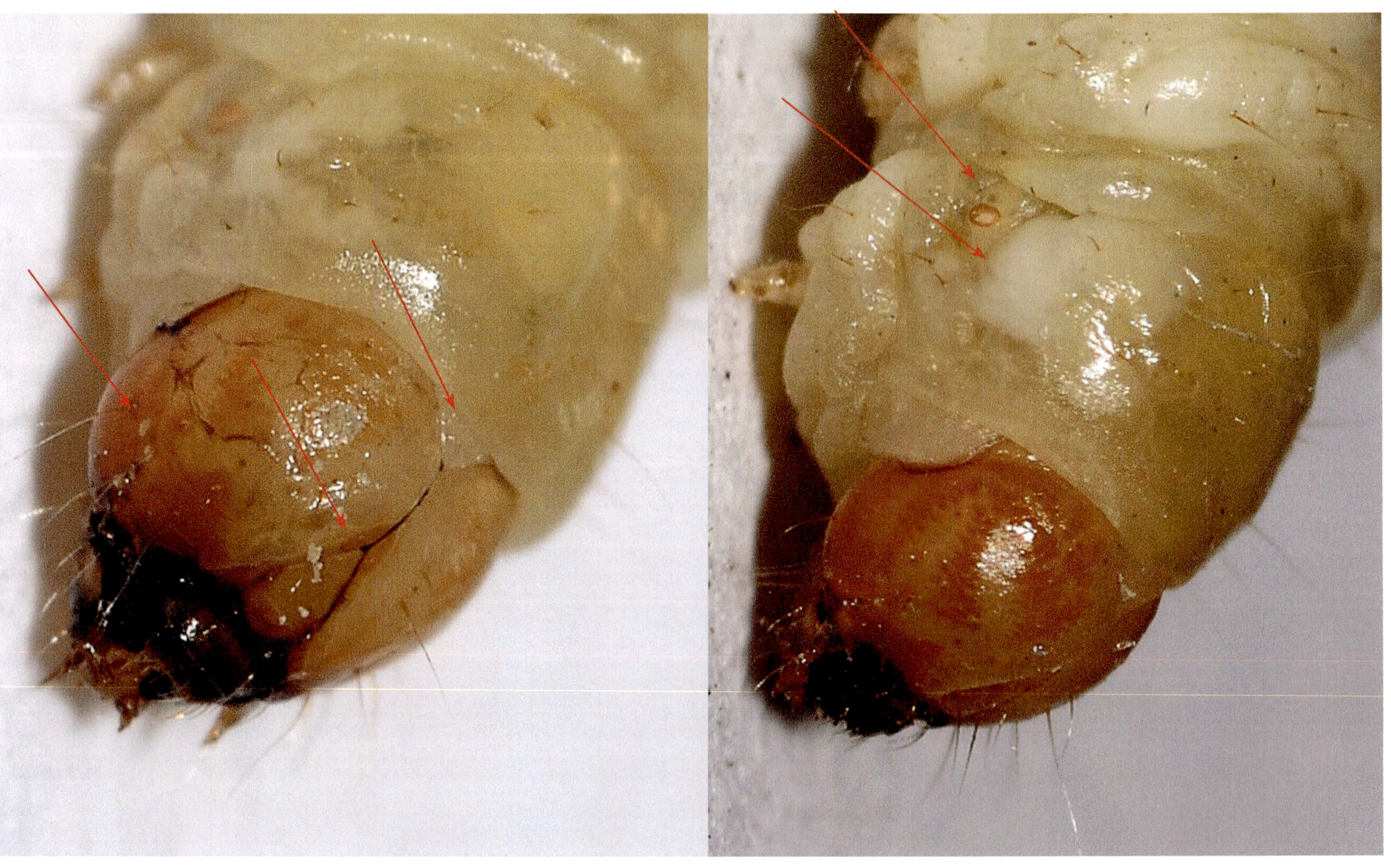

Plate 5.20. Head of the larva. *Epicranial suture* line is dark coloured (not white) and does not continue to the *prothoracic shield*. At the *vertex* of the head capsule, characteristically a large triangular area is present.

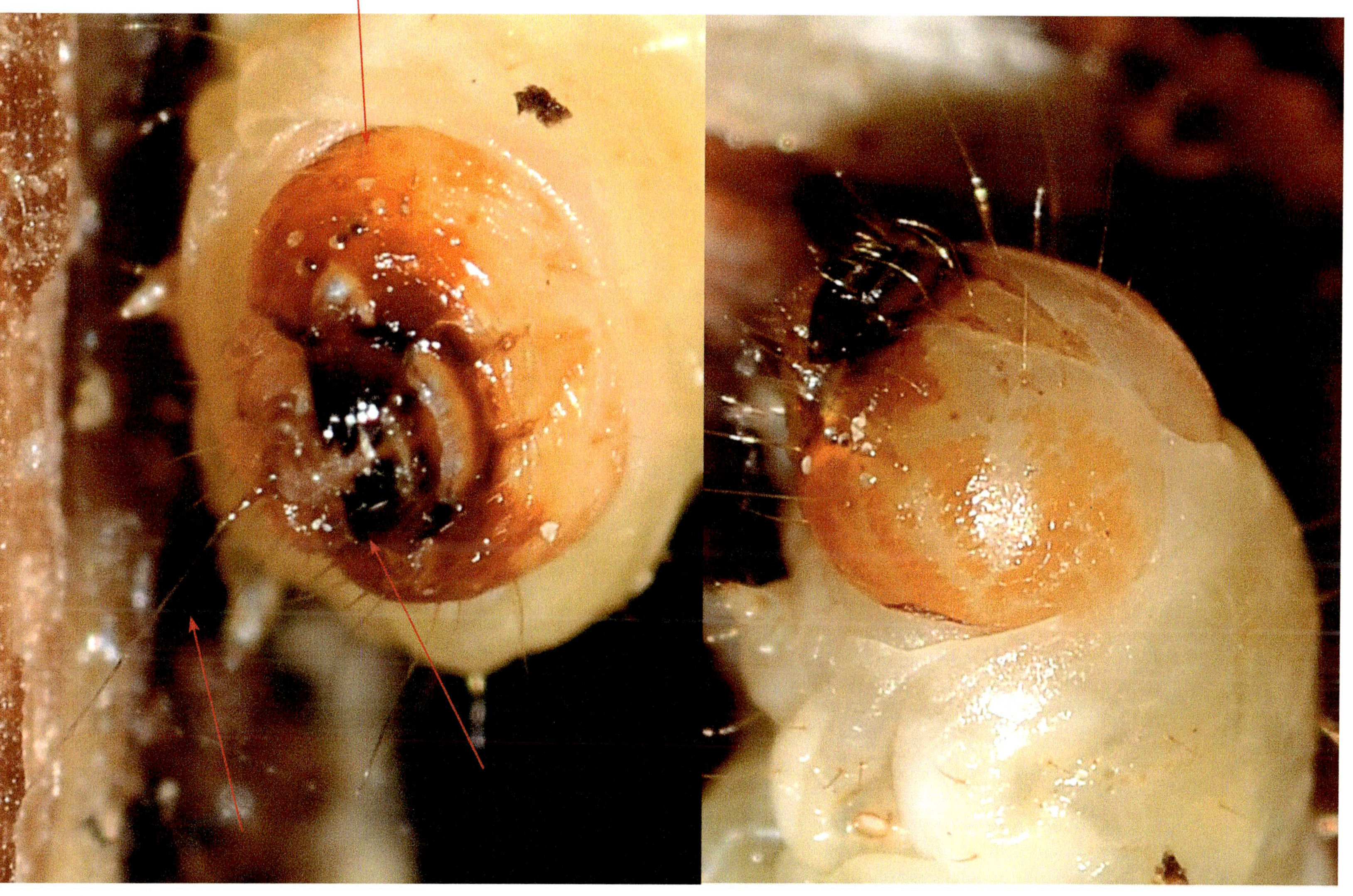

Plate 5.21. Head of the larva. *Epicranial suture* line is dark coloured (not white) and does not continue to the *prothoracic shield*. At the *vertex* of the head capsule, characteristically a large triangular area is present. Six ocelli are arranged in groups of 1+4 +1. Larva is aggressive, tries to attack, if disturbed with four sharp teeth.

Plate 5.22. Formation of pupa in the leaf sheath portion of the cane shoot. Prepupal larva makes a silken tunnel for the safe exit of the moth.

Plate 5.23. Prepupal larva is making an exit tunnel in the leaf sheath area of cane shoot. Pupal transformation in the mature cane after placing the larva inside the cane using IISR inoculator. Prepupal larva makes a silken tunnel for the safe exit of the moth. This silken tunnel is camouflaged with tiny bits of host tissues.

Plate 5.24. Formation of pupa inside the feeding tunnel of a cane. Pupal transformation in the cane after placing the larva inside the cane using IISR inoculator. A larva is preparing the chamber and later on transformed into to a pupa.

Plate 5.25. Formation of pupa in the lower portion of the cane. Sometimes a larva does not prepare a long silken pupation chamber/tunnel, especially when a larva stays above ground portion of the cane.

Plate 5.26. Pupae of Root borer (different views). Thoracic region is broader and the body gradually tapers towards the anal end.

Plate 5.27. A freshly formed pupa is yellowish at the wing pad area and whitish at the abdominal area. However, at the anal end a dark hump like protuberance develops on the dorsal side.

Plate 5.28. Gradual changes take place in colour of the pupa (dorsal and ventral views of a male pupa). A pupa turns yellowish with dark abdominal band on the 5th , 6th and 7th abdominal segments.

Plate 5.29. Gradual changes take place in colour of the pupa (dorsal and ventral views of a female pupa). A pupa finally becomes dark brown in colour.

Plate 5.30. Male and female pupae. Male pupa is a bit shorter in length and narrower in width.

Plate 5.31. Size of the pupae (on a centimetre scale). Male pupa is around 14 mm and female pupa is around 16 mm.

Plate 5.32. Root borer pupa does not possess any specialized ornamentations on 5th , 6th and 7th segments. A pupa bears characteristic depressions throughout the body. On the ventral side, marks of prolegs are prominent . In 5th and 6th segments characteristic depressed mark develops in each segment between the marks of prolegs.

Plate 5.33. Ventral side of the pupal abdomen (close up). Ventral side is smooth except the depressions and marks of prolegs with a characteristic long depressed area connecting the marks of the proleg pairs of the segment. The wing pad is broadly round.

Plate 5.34. Dorsal view of pupa (close up). The median line of *prothoracic shield* continues to the *mesothoraic shield* and even extends to *metathoracic shield*. The *mesothoraic spiracle* is small and two shades darker than the body. *Spiracles* on 2nd and 3rd abdominal segment show prominent arch formation. *Spiracles* are elongated oval with raised dark rim.

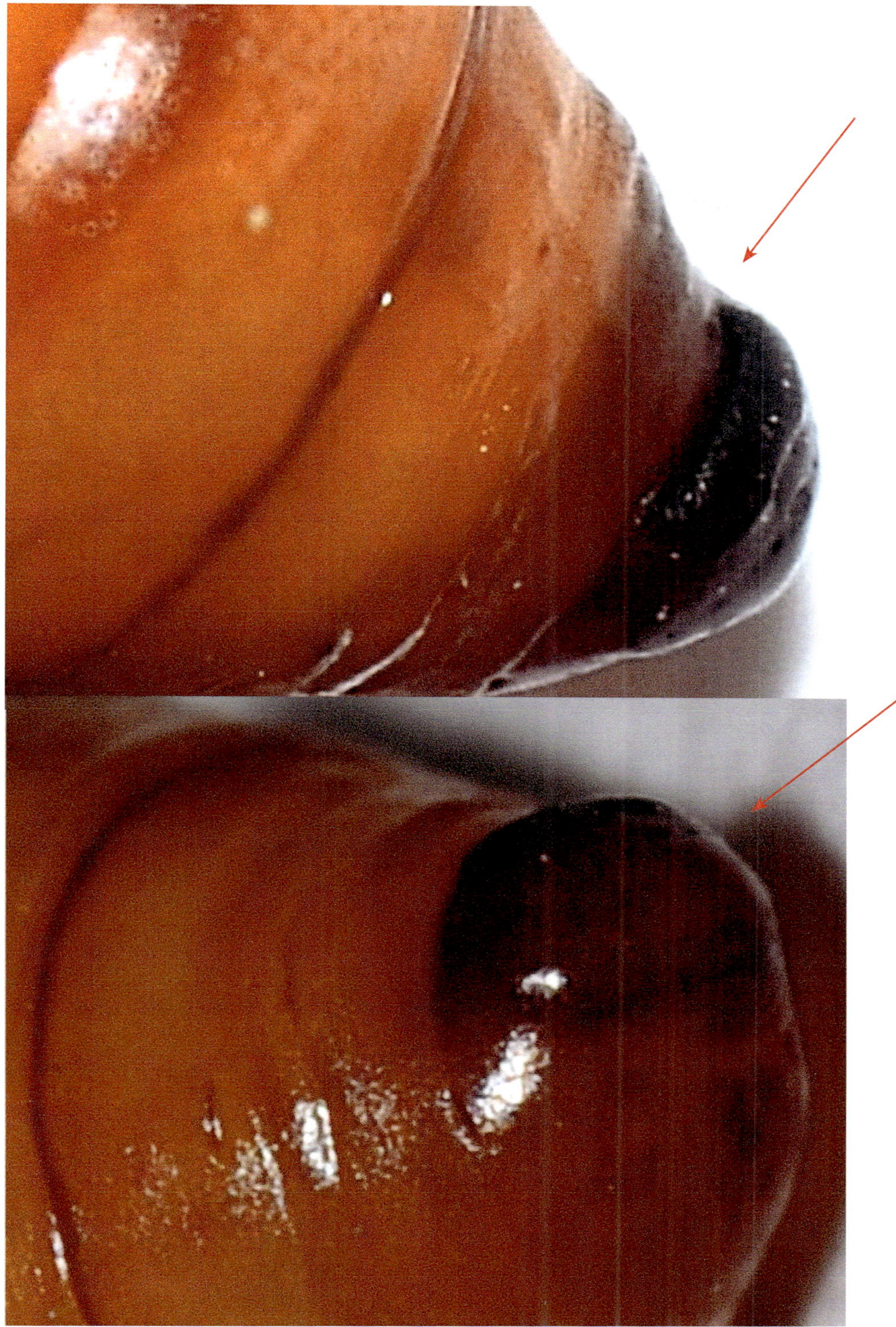

Plate 5.35. Dorsal side of the anal end of a pupa(close up). Presence of a solid hump like protruberance with darker terminal end.

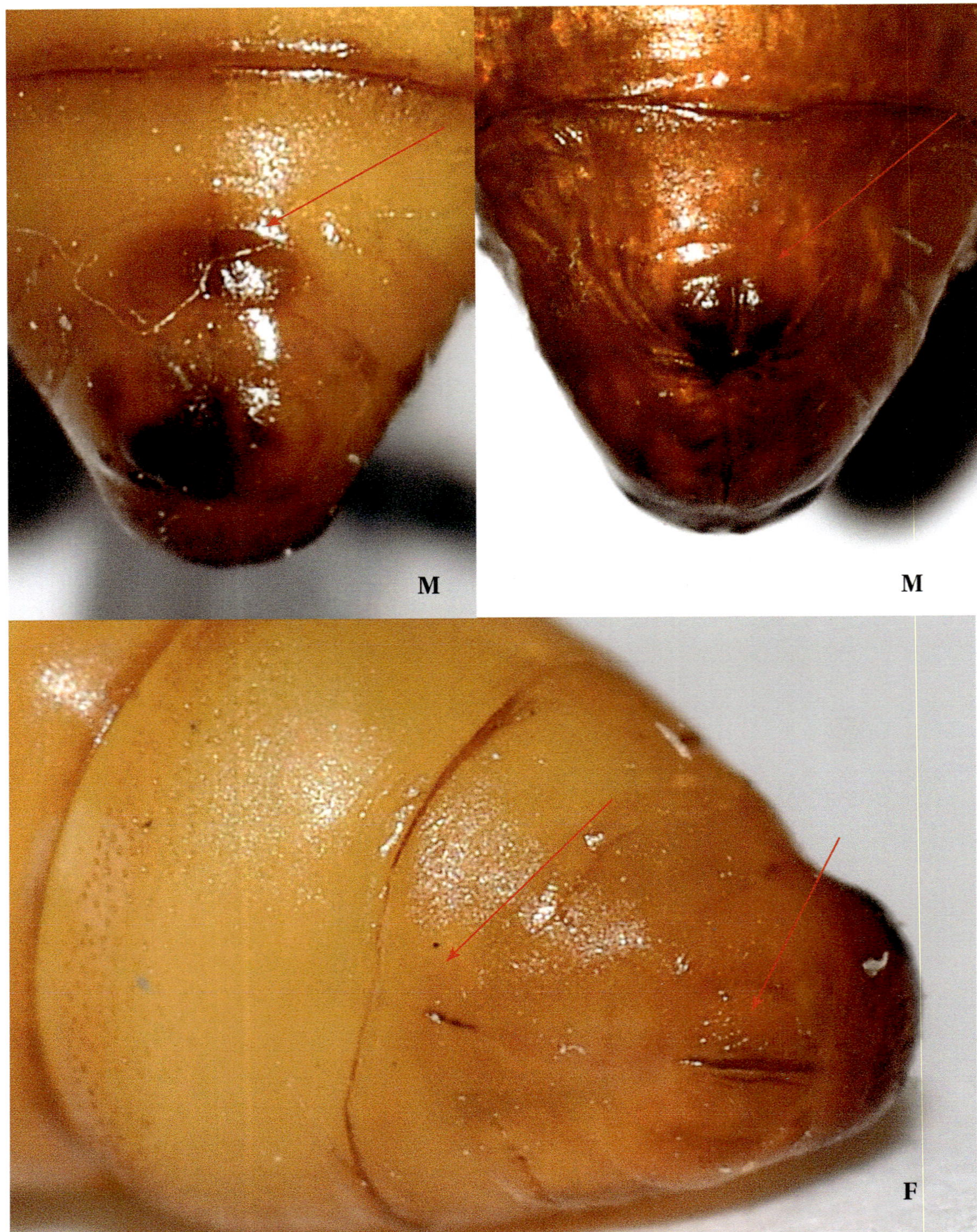

Plate 5.36. Anal end of pupa (close up). In male, genitalia is represented by the formation of two slightly raised flat lobes with central depression on the 9th segment (depression is close to the anal markings). In female, genital area is marked with a small depression at the middle of the 8th segment. Anal marking is represented by a straight deep furrow.

Plate 5.37. Female moth of Root borer. Antennae are long and *filiform*. Moths with pinkish tinge. *Labial palpi* porrect and cone shaped.

Plate 5.38. Male moth of Root borer. Antennae are distinctly *pectinate* (comb shaped) with ladle shaped bottom (*scape*). Moths with pinkish tinge. *Labial palpi* cone shaped and projected upward.

Plate 5.39. Female and male moths of Root borer on a centimetre scale showing relative sizes. Female about 16-17 mm and male about 14 mm. Ventral side of the female moth is pale yellowish pink.

Plate 5.40. Female moth of Root borer. Typical sitting habit. *Labial palpi* porrect, cone shaped and projected forward with slight upward curvature.

Plate 5.41. Female and male moths of Root borer (side view). Typical sitting habit. *Labial palpi* porrect and cone shaped and projected forward in female, whereas in male, *labial palpi* take upward curvature and remain appressed on head.

Plate 5.42. Female moth of Root borer. Typical sitting habit. *Labial palpi* porrect and cone shaped and projected a bit upward. Antennae are *filiform* and long.

Plate 5.43. Female moth of Root borer. Typical sitting habit. *Labial palpi* porrect and cone shaped and projected forward. Antennae are *filiform* and longer. Presence of dark scales extending from the radial region to the apex. Face is smooth.

Plate 5.44. Female moth of Root borer (ventral view). Moths also feign death in danger (left) and after death (right). Note the relative position of the legs.

Plate 5.45. Male moth of Root borer (ventral view). Moths also feign death in danger (left) and live moth (right).

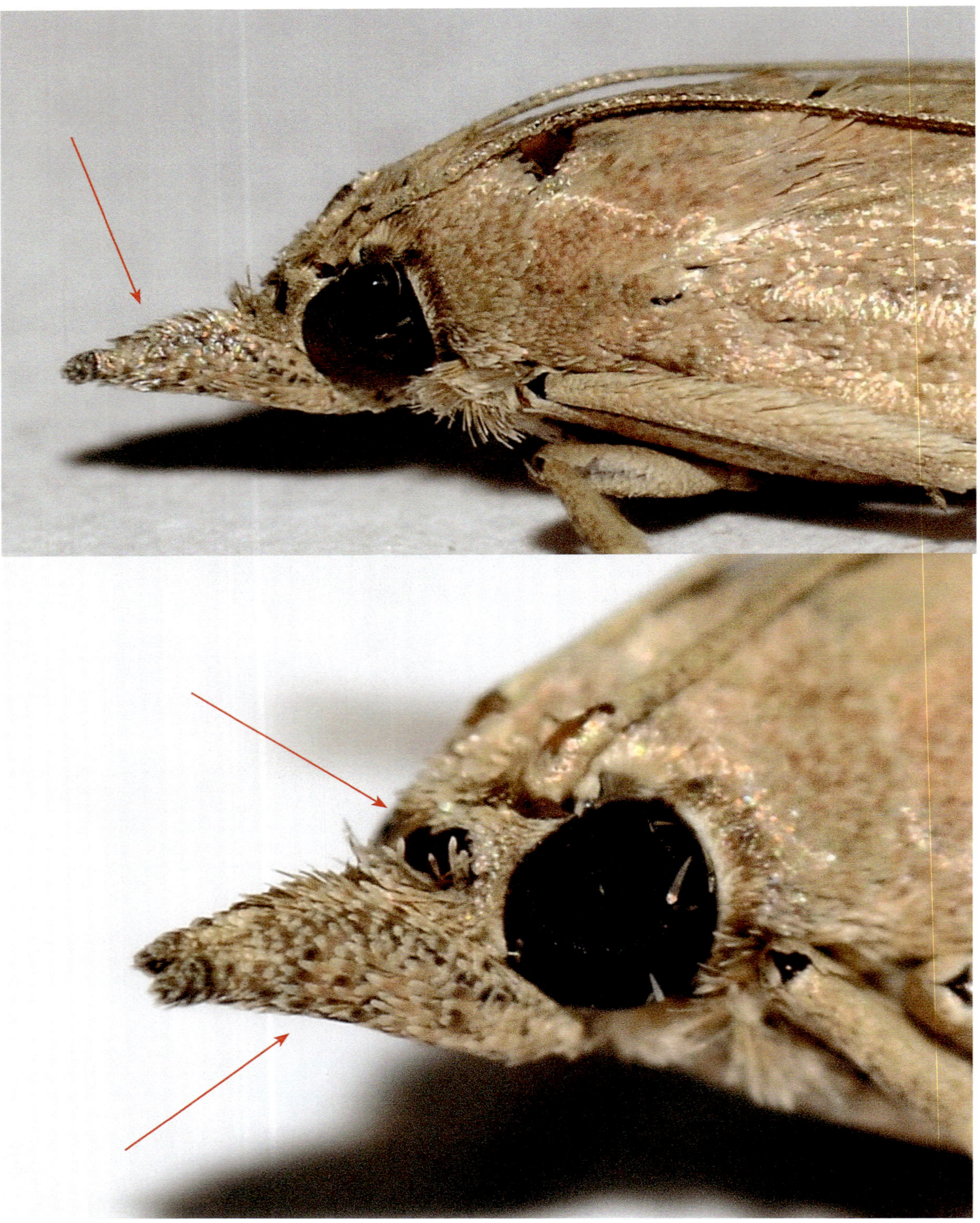

Plate 5.46. Head of a female moth (side view, close up). *Labial palpi* porrect, conical and slightly raised upward. Presence of a large scale less area on the face, just in front of the antennae. Eyes are dark coloured. Antennae are relatively narrow, longer and *filiform*. Presence of pinkish scales on the thoracic region imparts a distinctive appearance.

Plate 5.47. Head of a female moth (ventral view, close up). Porrect, conical and slightly raised *labial palpi*. *Proboscis* is scaled. Eyes are dark coloured.

Plate 5.48. Head and thoracic region of a male moth. *Labial palpi* are porrect, conical upwardly raised and appressed on the face. Antennae are distinctively *pectinate* at the outer side. Dorsally antennae bear scales. Antennae are ladle shaped at the base with small broad and curved area bearing long scales.

Plate 5.49. Head (close up). Antennae are distinctively *pectinate* at the outer side. Dorsally antennae bear scales. *Scape* of the antenna is ladle shaped with broad inwardly curved area bearing the scale (form sinus). At beginning of the pedicel there is a dark tuft of scales.

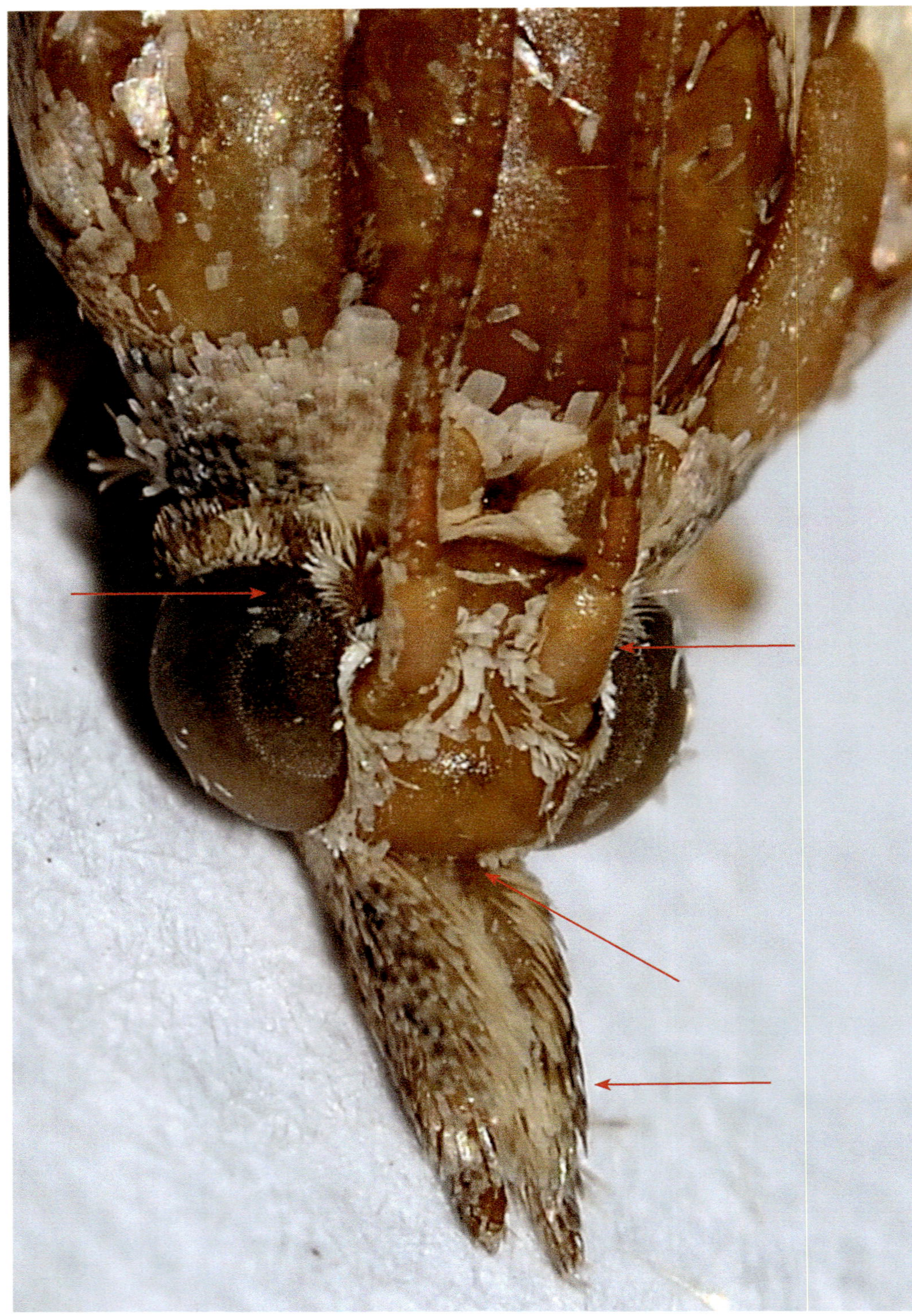

Plate 5.50. Head of a female moth (close up). *Labial palpi* are porrect, conical slightly upward. Face is round and smooth. Antenna is *filiform* with ladle like *scape*.

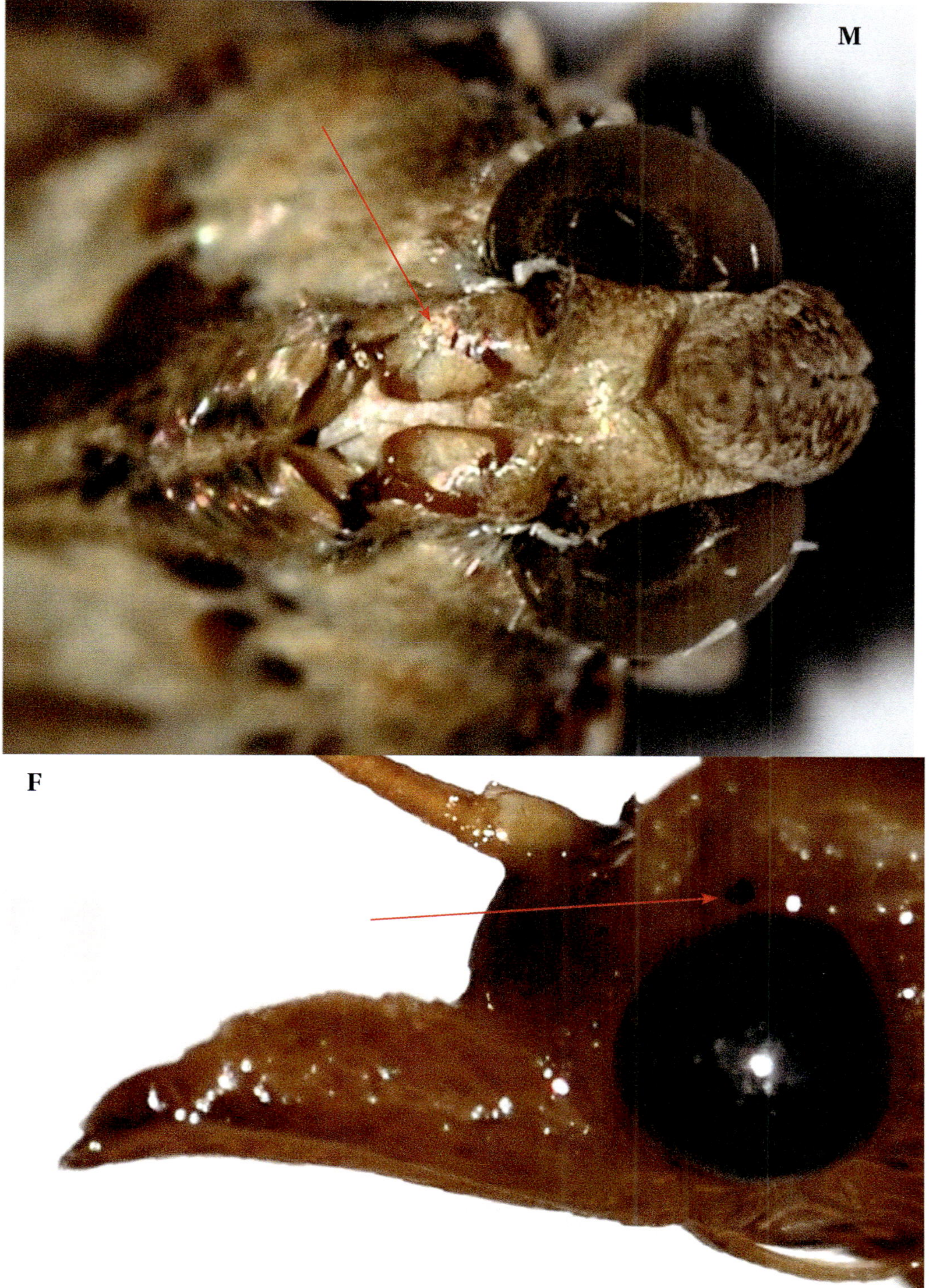

Plate 5.51. Close up of head. In male, antennae are ladle like at the base with rows of scale from both dorsal and ventral side in the sinus area. *Ocellus* is also present just above the eyes.

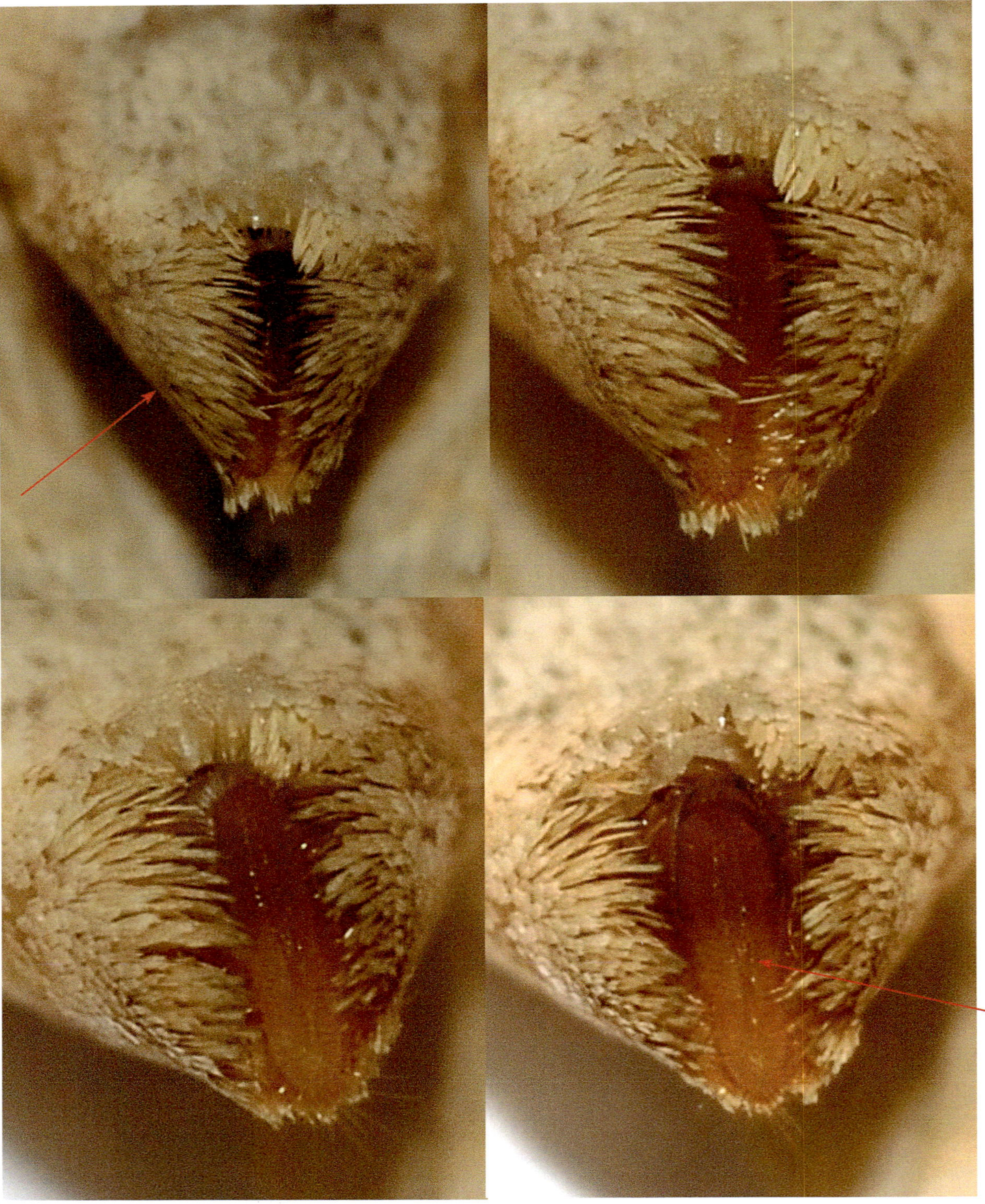

Plate 5.52. *Papillae anales* or ovipositor mostly remains hiding within the anal scales. Ovipositor is setose, relatively narrow and has a long median depression .

Plate 5.53. Forewing venations (stained and unstained). Sc is is thicker. R_1 and R_2 are free, R_3 and R_4 are stalked. R_5 absent. M_1 is free. M_2 and M_3 are on short stalk. Cu1a and Cu1b well developed. A_1 missing, A_2 is thick and well developed, A_3 is thin but poorly developed.

Plate 5.54. Forewing venations (stained and unstained, variations). Sc is is thicker. R_1 and R_2 are free, R_3 and R_4 are stalked. R_5 and M_1 are stalked. M_2 and M_3 are stalked. Cu1a and Cu1b well developed. A_1 missing, A_2 is thick and well developed, A_3 is thin but poorly developed.

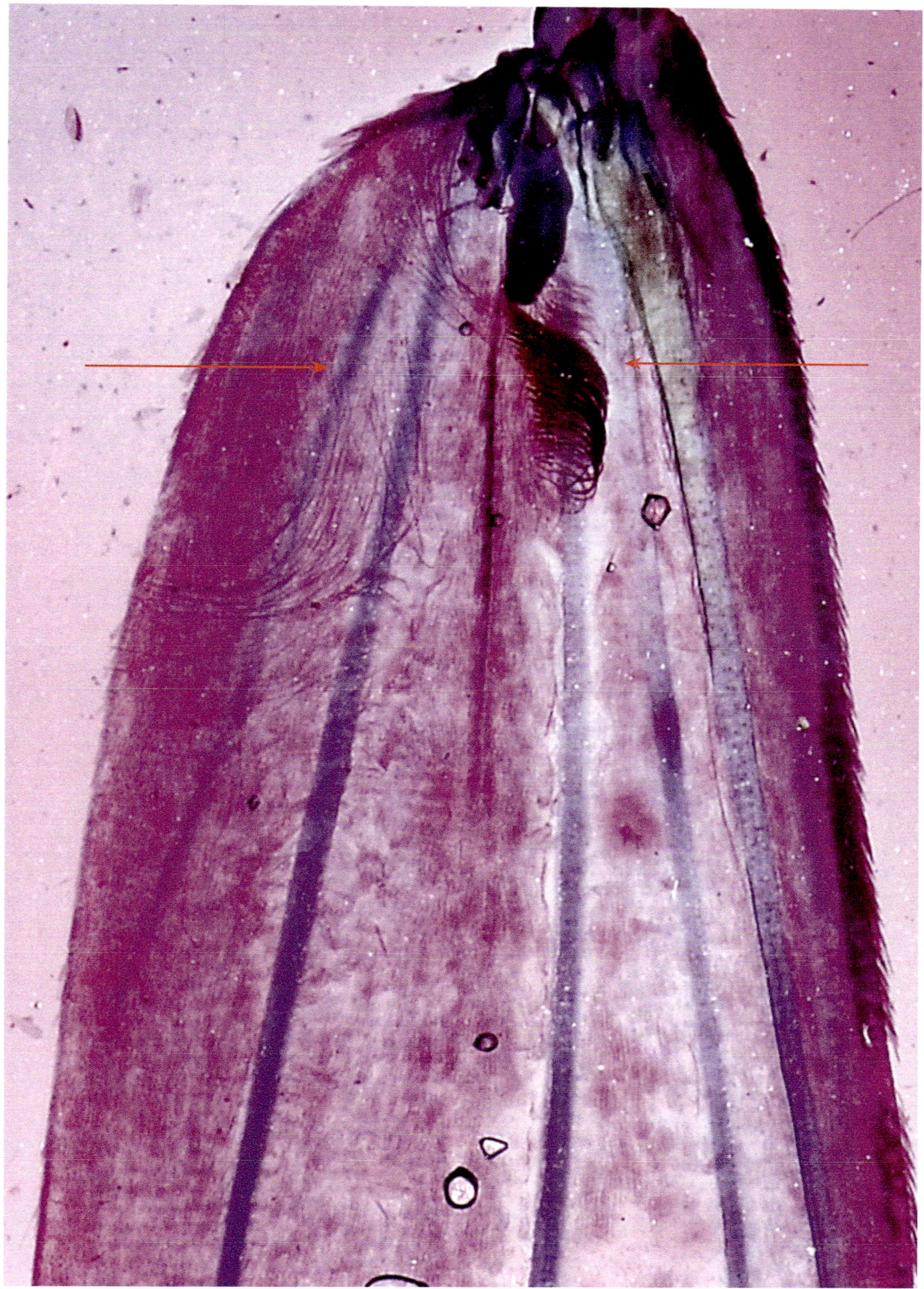

Plate 5.55. Close up forewing (female). Presence of A_3. The scales of the *retinaculum* are nicely curved.

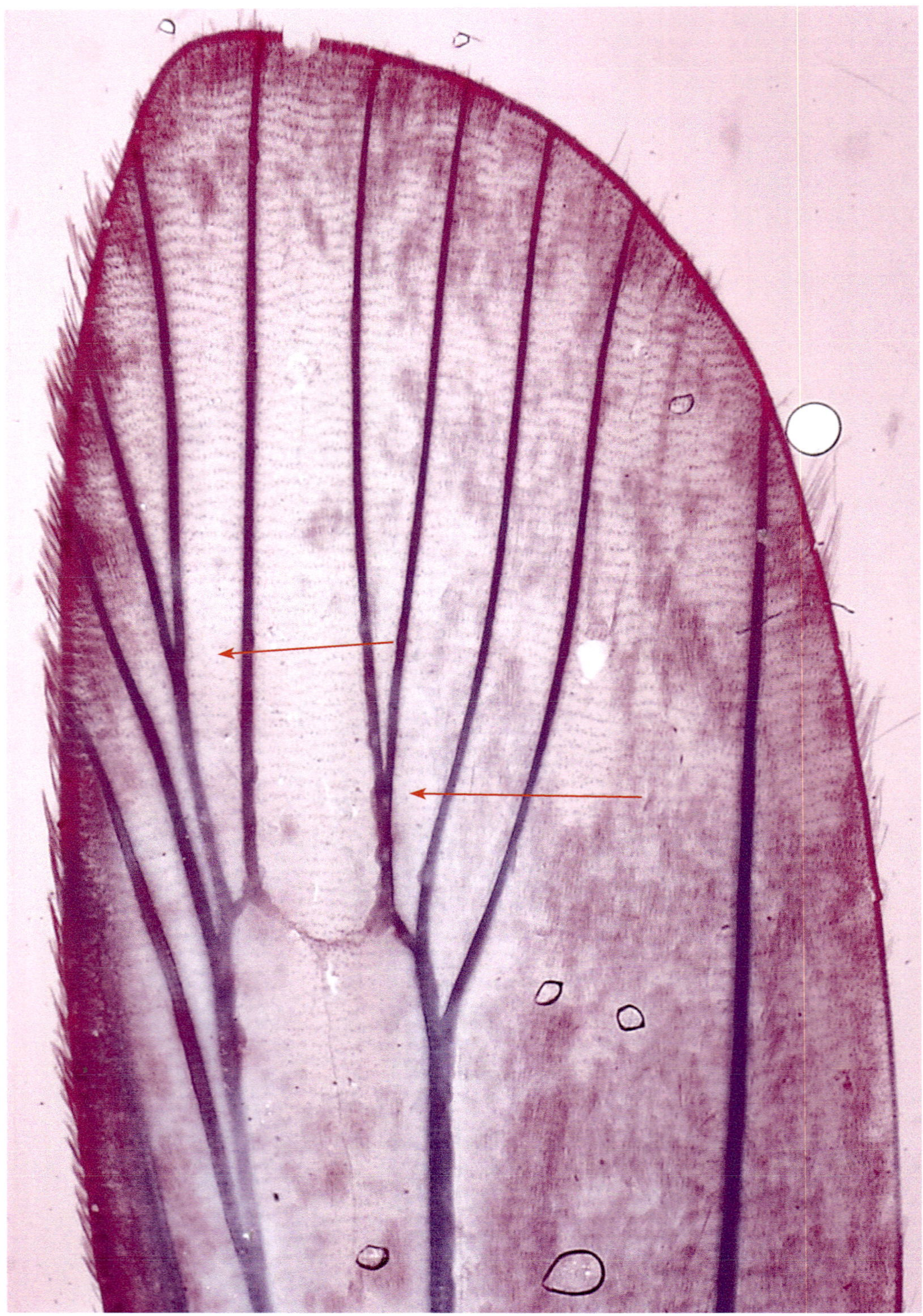

Plate 5.56. Close up of forewing (above discal cell). Typical absence of R_5 (This position is typically the position of R_5 and here it is designated as M_1). However, the vein M_1 receives the *tracheal* support from Rs.

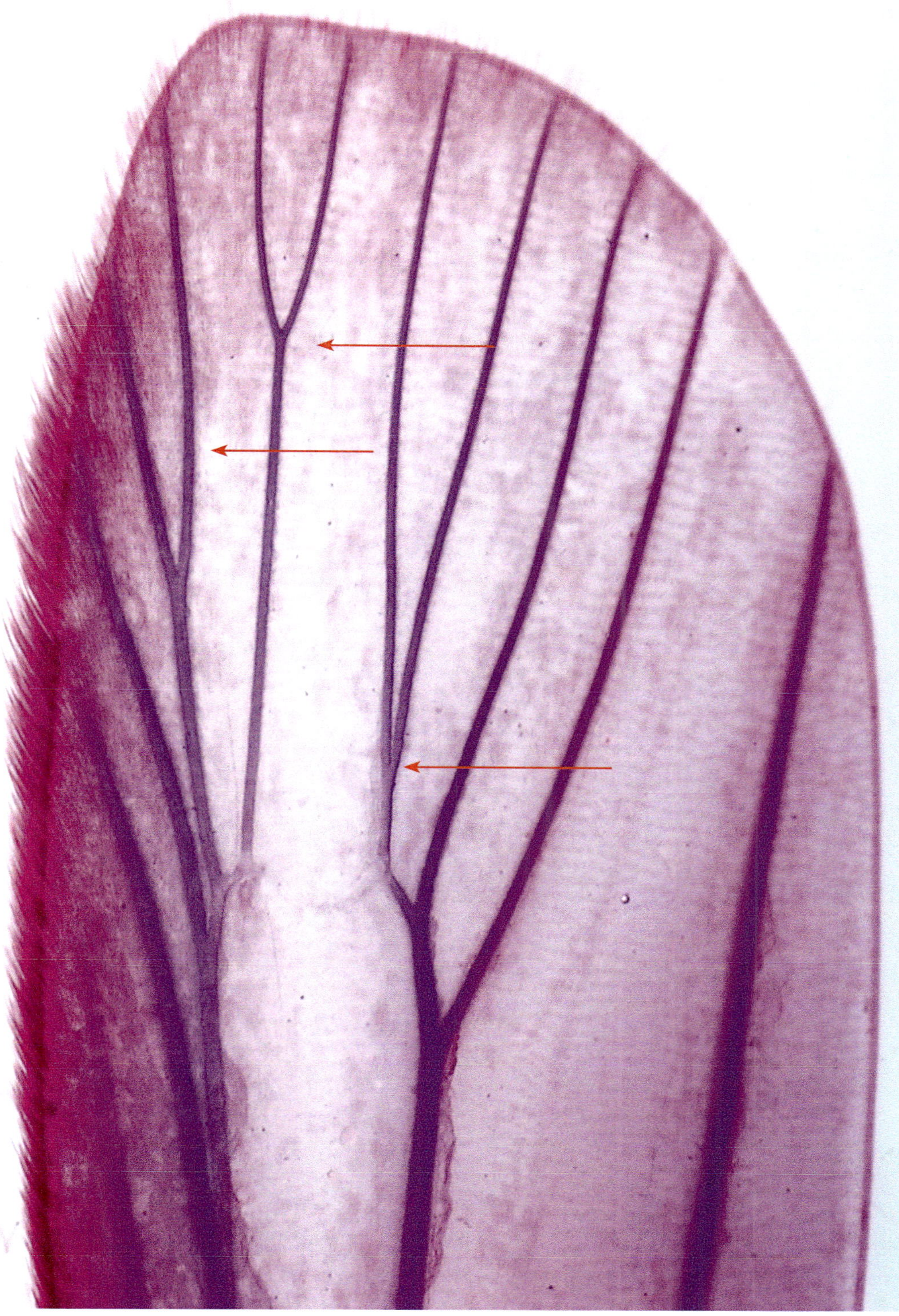

Plate 5.57. Close up forewing (above discal cell, variations). Typical stalk formation of R_5 and M_1.

Plate 5.58. Eggs of Root borer. Eggs are laid singly and naked. Female moths of Root borer also lays egg on any hard surface.

6

Green borer, *Raphimetopus ablutella* (Zeller)

Between the two important coloured moth borers (larva) of sugarcane viz. Green borer and Pink borer (body of the larva is fully coloured and without any stripe), Green borer is specialized to sugarcane, whereas the Pink borer is *polyphagus* (infests many Graminaceous plants). The incidence of Green borer, due to its characteristic habit of hibernating in soil near the clump base, is mainly observed in the neglected fields (with almost negligible interculture operations, soil remains undisturbed) and in fields with limited irrigation facility. It is also observed in the crop planted in the sandy soil around the riverbeds where water hardly submerges the fields in the rainy season and the crop gets very little or no irrigation. Green borer is a true shoot borer of sugarcane as it occurs only during the shoot stage of the crop and feeds only the immature developing leaves of the crown of a young cane shoot.

In the recent years, the incidence of Green borer has gone down considerably. During last five years (2014-2019), but once, the author even after repeated surveys could not catch hold a live specimen of Green borer, *Raphimetopus ablutella* (Zeller) in the sugarcane areas of Uttar Pradesh from where it was reported earlier. As a result, only a little description from the earlier experiences with this borer is provided here.

Lefroy (1906) in his book "*Indian Insect Pests*" on page 136 recorded its presence in Bihar. Please refer the box below for the details.

> Three borers have been found by Mr. M. Mackenzie in Behar, and these are as yet little known. The larva of the Gold-fringed Moth [1] attacks cane much as the moth-borer does; the Green Borer [2] attacks the portion of the cane below the soil, and a third borer [3] is found in the stalk.
>
> [1] 122. *Chilo auricilia.* Ddgn. (Pyralidæ.) | [2] 146. *Anerastia ablutella.* Zell. (Pyralidæ.)
> [3] 149. *Polyocha saccharella.* Ddgn. (Pyralidæ.)

From the description of Lefroy(1906), it is very clear that the name "*Green borer*" to this species infesting sugarcane was assigned much earlier (not at a later date, as has been mentioned in different publications). It is obvious that the name was assigned due to the very distinctive coppery-green or green colour of the larvae.

Fletcher and Ghosh (1919) furnished a preliminary description of this borer on page 385 with a supportive plate (Plate 44) in the *Proceedings of the Third Entomological Meeting*, held at Pusa (please refer ***Appendix IV***). In the Proceedings, Fletcher and Ghosh wrote,

Raphimetopus (Anerastia) ablutella (C. S. 1801).

Plate 44, fig. 1.

Foodplant—Sugarcane.

In the course of the two seasons 1917 and 1918, the caterpillars of this species were observed to occur in very small numbers in young sugarcane in April 1918 and to cause "dead heart" like the other borers.

The full-grown larva is about 16 mm. long and about 2 mm. across the middle of the body which is cylindrical and slightly tapering at the ends. The head is yellow-brown, glossy. The prothorax green; the meso-and metathorax and the abdomen copper-coloured dorsally and greenish ventrally. The spiracles are round, yellow-brown. The hairs are brown, thin and longish. The hooklets on prolegs are in a complete circle.

One pupated on 3rd May and emerged as moth on the 11th May. The pupa is about 8 mm. long and tapering hindwards, the hind end having a few long circinate hairs. The colour is greenish brown.

This insect is stated by Messrs. Mackenzie and Lefroy to have been found in great numbers at Siripur, Hathwa, Saran, in young sugarcane when about three weeks to a month above the ground. (*Agri. Journ. Ind.* III, p. 104). The larva observed by them was completely green.

The moth borer was first described by Zeller as *Anerastia ablutella* Zeller in 1839. Later Hampson (1918) transferred this species to *Raphimetopus* as *Raphimetopus ablutella*.

(2) RAPHIMETOPUS ABLUTELLA.

Anerastia ablutella Zell. Isis, 1839, p. 17; Herr.-Schäff. Eur. Schmett. iv. p. 109, Tin. f. 89; Rag. Rom. Mém. viii. p. 403; Hmpsn. Moths Ind. iv. p. 56; Staud. Cat. Lep. pal. ii. p. 12.

†*Anerastia stigmatella* Rag. Nouv. Gen. p. 49 (1888); id. Rom. Mém. viii. p. 403, pl. 40. f. 4.

Anerastia bimaculata Rag. Nouv. Gen. p. 49 (1888); id. Rom. Mém. viii. p. 404, pl. 40. f. 5; Hmpsn. Moths Ind. iv. p. 56.

Anerastia majorella Roths. Nov. Zool. xx. p. 138 (1913).

SPAIN; SICILY; ALGERIA; N. NIGERIA; CONGO, EGYPT; DAMARALAND; TRANSVAAL; ORANGE R. COLONY; ADEN; ASIA MINOR; PERSIA; W. TURKESTAN; JAPAN; KASHMIR; PUNJAB; MADRAS.

In India, a good deal of information on the biology of this borer infesting sugarcane was furnished by Siddiqi and Singh (1959) and morphology of the larva was described by Avasthy and Chaudhary (1962). Of late, another name *Coenochroa ablutella* (Zeller) has also been proposed for this moth borer.

6.1 Currently accepted name

Raphimetopus ablutella (Zeller)

6.2 Synonyms

= *Anerastia ablutella* Zeller

= *Anerastia bimaculata* Ragonot

6.3 Symptoms and brief biology

In affected sugarcane plant, the symptoms of infestation of Green borer are indistinguishable from that of the symptoms of infestation of Early shoot borer (ESB). Both the borers share almost same space and time in infesting at shoot stage of the sugarcane crop. During the crop season (subtropical sugarcane belt), the Green borer appears a fortnight earlier, and disappears from active life a month earlier with respect to the ESB. At the early part of the crop season, Green borer predominates in infestation of ratoon sprouts and also in the formation of *dead heart* whereas, in later part of the shoot stage, ESB predominates the infestation and formation of *dead heart*, especially in the plant crop.

The Green borer produces *dead heart* identical to that of ESB but their distribution remains predominant in the ratoon sprouts and at the periphery of the fields. The *dead heart* comes out easily when pulled. The lower end of the *dead heart* mostly remains rotten and emits foul smell. Usually, a larva of Green borer restricts itself above the apical dome and eats only the immature developing leaves. In many instances, the apical dome remains undamaged. This type of situation often observed in the field, as the Green borer larva migrates at this stage of feeding to infest another adjoining shoot or if the larva is mature enough, it moves down to the shoot base to pupate in the soil by making a specialized pupal chamber. Thus, in certain situations the affected plant often grows despite the formation of a *dead heart*.

Between the two shoot borers (Early shoot borer and Green borer), the Green borer is the first borer to come out from hibernation. Usually, moths emerge in February and caterpillars infest mostly the young ratoon sprouts. The larva of Green borer feeds on the cane shoots and manifests itself only during the summer months (pre monsoon period). As the crop proceeds to the grand growth phase, its infestation goes down considerably. With the arrival of premonsoon showers, a mature larva prepares itself to proceed for hibernation. For the hibernation, it comes out of the shoot and goes down to the clump base. Reaching inside the soil it prepares an impervious shell with the silk and attaches soil particles at the outer side. Finally, it takes recluse inside the shell. The larva cleverly camouflages the pupal shell by attaching soil particles around it and due to this, the pupal shell is often considered (erroneously) as an earthen shell. Green borer larva hibernates during both the rainy and winter seasons (July onwards) and comes out in February-March, the following year after tiding over the cold winter months in larval stage.

6.4 Egg and neonate larvae

Eggs are a bit oblong with greenish tinge and measure around 0.87 × 0.49 mm. The Green borer moth does not lay eggs on the leaf or *psuedostem* of a cane shoot like other moth borers of sugarcane; rather it lays eggs at the vicinity of clump base in loose masses of soil, in the soil cracks and crevices or in loose clods of soil. A moth lays eggs in clusters, but eggs do not overlap like that of *Chilo* moth borers. Eggs are dull white in colour when freshly laid. The neonate larva is creamy white with dark brown head. Transverse grey bands run on the dorsal side of each abdominal segment.

6.5 Larva

A larva of Green borer is characteristically very active; it is even more active than the larva of Early shoot borer. In fact, an extricated larva of Green borer often jumps (due to sudden lateral movement) to shift its position when disturbed. At rest, a larva appears relatively smaller in length and much thicker in breadth; however, when it moves, it stretches it body considerably and appears like any other moth borer larvae. A mature larva at rest, measures around 18-20 mm in length, whereas in a stretched situation (like when moving fast) it measures around 30 mm. The colour of a mature larva is very characteristic of this species. A mature larva has completely bluish green *prothorax*; *meso* and *metathorax* remain partially green and rest of the dorsal area of the abdominal portion remain dark coppery with greenish tinge in between. The alimentary canal area (inside the body) appears as a centrally located dark band or stripe on the dorsal surface of the body. In a stretched larva, the dorsal inter-segmental area also appears green. The *prothoracic shield* is well developed and completely bluish green in colour and thus, imparts a very distinctive appearance to this thoracic segment of the larva. Moreover, the ventral side of all the three thoracic segments are also bluish green. The ventral side of the abdomen supports different patterns of colours; these colours remain intermixed with pink and green. However, base of the prolegs always remains bluish green. In the ventral abdominal side, middle portion of each abdominal segment including the prolegs appear green or bluish-green and the inter-segmental area remains pink to coppery in colour. A larva is apt to change its body colour, and depending on the situation, it changes its body colour to suit its immediate need. In a threat perception, a larva changes its colour to fiery red, probably to avoid preying. However, if a larva remains partially submerged in water, it turns completely bluish green, leaving all the traces of pink or coppery colour from the body.

Prominent long setae are present on the body. Thoracic appendages appear white. Each thoracic leg also bears a long seta. Head capsule is yellowish to reddish brown and forms a prominent 'V' shaped notch at the *vertex*. *Epicranial suture* is prominent. *Epicranial* area is much wider and takes bluish to greenish white colour. Six *ocelli* are present, and these *ocelli* follow a 1+4+1 arrangement. The first seven abdominal segments are uniformly broad, while the last three taper gradually. *Spiracles* are broadly oval with brownish rim and appear as very distinct bright area on the lateral side of the larval body. The 8th pair of *spiracle* is placed laterally rather than the usual subdorsal placement in the larva of other moth borers of sugarcane. At the lateral side, it bears prominent skin folds, like that of Gurdaspur borer. The greenish colour of the abdominal segment appears below the *spiracle* line or the *tracheal trunk* and the skin folds retain the coppery colour. *Tracheal trunk* is not usually visible except when the larva stretches out; it appears as a faint white band running through the bright *spiracles*. Crochets or hooks on the four pairs of abdominal *prolegs* are stout and arranged in a complete circle. Crochets are *uniserial* and *biordinal*. Crochets or hooks have a peculiar arrangement; a much larger hook immediately follows a smaller hook. As the hooks are placed in one row, the role of smaller hooks is not immediately apparent.

6.6 Pupa

A mature larva (prepupal larva) of Green borer moves to the ground and spins a cocoon attaching loose soil at the periphery. Inside the cocoon, it pupates. Near the small end of the cocoon, the larva leaves a loose cap and the moth emerges by pushing through the cap. A freshly formed pupa is also greenish in colour and with the passage of time, the colour changes to dark brown (greenish brown). Like the Root borer, pupa of Green borer is also much broader at the thoracic area and tapers towards the anal end. Abdominal segments are lighter in colour and have a pitted appearance due to the presence of numerous oval pits. These pits are quite different from the pits that are observed in Root borer or Pink borer. No specialized ornamentations are present on the 5th, 6th and 7th segments. Anal segment is made up of a lightly chitinized dorsal half with two lobes pointing posteriorly and carrying usually six (six to ten) characteristic long brownish *circinate* hairs (the distal end of a hair is rolled over in a circular fashion). A pupa measures around 10-11 mm in length.

6.7 Moth (adult)

Like the moths of Root borer, forewings of the moths of Green borer are also elongated. Forewings are straw coloured, whereas the hindwings are whitish. In addition to the size variation, sexual dimorphism is also prominent in the antennae. Female moths have *filiform* antennae, whereas male moths have *pectinate* antennae (comb shaped).

6.8 Forewing venation

The wing venation is also quite similar to that of Root borer and differs in M_2. In the forewing, the *subcosta* (Sc) reaches the costal area. Both the radial veins i.e. R_1, R_2 are free. R_3 and R_4 are stalked. R_5 is absent. M_1 occupies the position of R_5. Characteristically the vein M_2 is absent in the Green borer. M_3 and Cu1a appear from the *cubitus* cell angle. Cu1b is present. Among the anal veins, A_1 is absent, A_2 is well developed, and A_3 is rudimentary.

List of plates on Green borer, *Raphimetopus ablutella* (Zeller)

Descriptions of plates

Plate 6.1. Larva of Green borer eating away the immature portion of the crown leaves. Dorsal side of the larva appears coppery with greenish tinge and ventral side appears greenish with coppery tinge.

Plate 6.2. Larva of Green borer appears short and stout at rest. Dorsal side of the larva appears coppery with greenish tinge and ventral side appears greenish with coppery tinge. *Prothoracic shield* as well as ventral side of the thoracic region appears bluish green.

Plate 6.3. Larva of Green borer in movement. (Dorsal and lateral view) Length of the active larva appears similar to that of the other moth borers. A stretched larva measures around 30 mm.

Plate 6.4. Variations in the pattern of colours on the body. A larva can change the colour of the body quickly (as per the need of the situation). It may turn completely green, if placed in water (partial submergence).

Plate 6.5. Variations in the pattern of colours on the body. A larva in motion. A stretched larva measures around 30 mm.

Plate 6.6. Dorsal and lateral views. Dorsal side of head and thorax. Head capsule is reddish yellow with prominent 'V' notch. *Epicranial suture* appears greenish. Hairs are very prominent on the body. *Prothoracic* shield appears bluish green, *meso* and *metathoracic* area show variable colour formation. *Spiracles* appear as bright coloured areas on the *tracheal trunk*; 8th pair of *spiracle* is more lateral in placement on the body.

Plate 6.7. In any threat perception, a larva may change its colour quickly to impart a fiery appearance to the body. Probably to scare the predators.

Plate 6.8. Variations in the pattern of colours on the body. A larva in motion showing green area interspersed with coppery area. Larva under partial submergence turns almost bluish green. The body fold are quite similar to that of Root borer.

Plate 6.9. Variations in the pattern of colours on the body. Ventral side of the thoracic area remains bluish green. Colour of the ventral side may turn pinkish. The base of the prolegs always remain bluish green. Crochets on the prolegs are arranged in a complete circle.

Plate 6.10. Colour of the ventral side may turn pinkish. The base of the prolegs always remain bluish green. Crochets on the prolegs are arranged in a complete circle. Crochets are *biordinal* and *uniserial*.

Plate 6.11. Head capsule appears reddish yellow. *Epicranial suture* area is much broader and remains partially greenish white. A larger *wage* shape is formed at the *vertex*. Mandible with three prominent teeth. Presence of frontal projection above mandibles. Six *ocelli* in 1+4+1 arrangement.

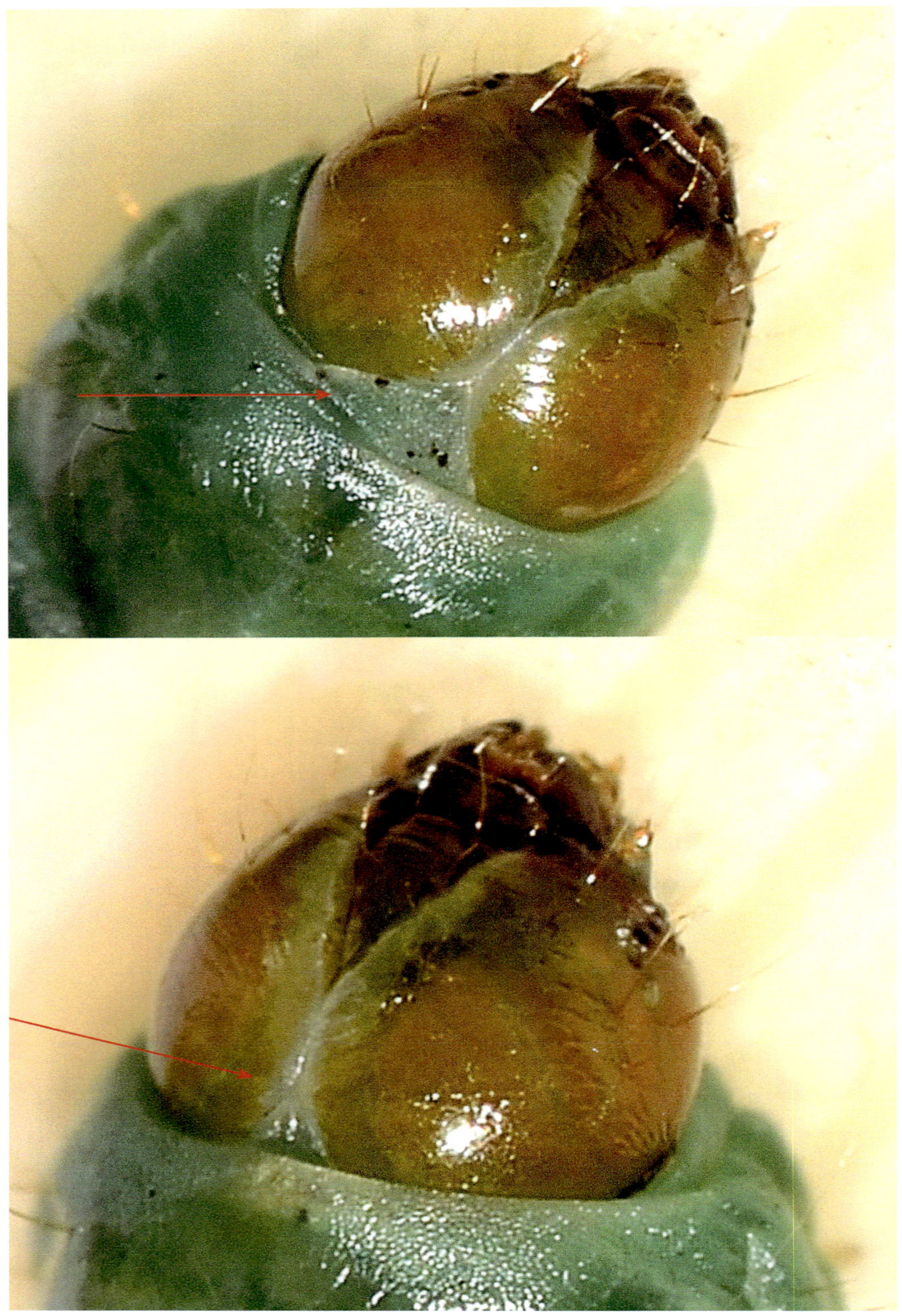

Plate 6.12. Head capsule appears reddish yellow. *Epicranial suture* area remains partially greenish-white. The projection above the mandibles is more prominent. A larger triangle shape is formed at the *vertex* area.

7

Pink borer, *Sesamia inferens* (Walker)

The Pink borer, *Sesamia inferens* (Walker) is a *polyphagus* pest and attacks many Graminaceous crops viz., rice, wheat, maize, sorghum, sugarcane, finger millet, etc. Due to the availability of one or the other host crops round the year, the Pink borer is found throughout the year. However, during the peak winter its activity is greatly reduced. In a situation when wheat is intercropped with sugarcane, it gets ready access to the adjoining sugarcane shoots during summer months and causes substantial damage to the ratoon and autumn-planted crop. Thus, it becomes more damaging in sugarcane-wheat intercropping system. Like other common cane borers, it mostly damages the crop at shoot stage and results in the formation of *dead heart*. It also acts as a cane borer in later stage of the crop growth but never assumes serious proportion in the monsoon or post monsoon period.

Due to its characteristic '*light pink body colour*' or "*raw flesh colour*" of the larva, it is commonly called as "*Pink borer*". The Pink borer larva does not bear any stripes on the body. Like the other moth borers of sugarcane, it is not a *Crambid/ Pyralid* one but a *Noctuid* one (*Owlet moth*).

The genus *Sesamia* was erected by Guenee in 1852. Walker (1856) described this species as *Leucania inferens* Walker. Hampson (1894) described this species as *Nonagria inferens* Walker (please refer the box furnished below for details). Later, this species was finally transferred to *Sesamia*, retaining the specific epithet as given by Walker(1856).

Sect. I. Hind wing with veins 3 and 4 from the cell.

A. *Antennæ of male ciliated, of female simple.*

1943. **Nonagria inferens,** *Wlk. Cat.* ix, p. 105; *Moore, Lep. Ceyl.* iii, pl. 145, fig. 3; *C. & S.* no. 1654.
Leucania proscripta, *Wlk. Cat.* ix, p. 106; *C. & S.* no. 1655.
Sesamia fraterna, *Moore, Lep. Atk.* p. 103; *C. & S.* no. 1653.

Fig. 153.
Nonagria inferens, ♂. 1/1.

Ochreous. Fore wing with a red-brown suffusion along median nervure and veins 2 to 5; a marginal dark line; cilia paler. Hind wing white.

Hab. Sind; Bombay; Mhow; Ceylon Burma; Flores.
Exp. 28 millim.

7.1 Currently accepted name

Sesamia inferens (Walker)

7.2 Synonyms

= *Leucania inferens* Walker 1856

= *Nonagria gracilis* Butler, 1880

= *Sesamia tranquilaris* Butler, 1880

= *Sesamia albicillata* Snellen, 1880

= *Nonagria inferens* (Walker) Hampson, 1894

7.3 Symptoms of infestation and brief biology

The Pink borer is not a well-adapted borer pest of sugarcane like other specialized borers. It is a *polyphagus* pest and it can also sustain on sugarcane. This is reflected in its feeding habit in sugarcane crop. Feeding of Pink borer is rough and obvious. In causing the damage to the shoot, it produces relatively larger entry hole (face/ external area) and the holes always impart a dirty or dark look (rim of the hole and adjoining area appear black). Shoot infestation of Pink borer is marked by the formation of large *dead heart* i.e., it affects many leaves (3-4 leaves) of the crown. The "*dead heart*" is also easily pullable like that of Early shoot borer (ESB). When a *dead heart* is pulled out from the young affected shoot, foul smell also emits from rotten internal portion of *dead heart* but with lesser intensity. In the stalk infestation, it produces larger holes; sometimes the larva pushes out frass from the feeding tunnel like the Internode borer but the feeding tunnel never remains clean, considerable frass also remains inside the tunnel. Like Stalk borer it can also eat away the nodal tissues. Thus, the stalk damage symptoms of Pink borer falls in between the damage symptoms of both Internode borer and Stalk borer.

7.4 Egg and neonate larva

Pink borer is a *Noctuid* and its eggs are quite different from that of the *Chilo* moth borers. In comparison to the scaly (nearly flat) eggs of *Chilo* moth borers (*Pyralid/Crambid*), eggs of Pink borer are round, hemispherical i.e., flattened at both the sides (appears just like a small round *peda* or '*poppins*' candy). Eggs also have beautiful corrugated and vertically arranged ornamentations at the periphery. An egg measures about 0.6-0.7 mm in diameter. Eggs are laid in batches of 30-100, in rows (single to multiple, regular to irregular). A female moth sits at the junction of lamina and leaf sheath, uses the gap in ligule area for laying eggs. After laying an egg, it carefully slides the egg down on the glossy inside surface of the leaf sheath (rice, wheat). Thus, a sheet of eggs attached linearly is formed within the leaf sheath. Due to the concealed nature of the egg mass, it is difficult to trace the eggs in sugarcane shoots. However, in rice, due to thin and light colour of leaf sheaths, egg mass becomes apparent at later stages of development, especially at the black head stage of the egg mass. Freshly laid eggs are water soaked creamy to light yellowish-white in colour.

A neonate larva is about 1.2-1.5 mm in length, pale yellow to diffused pink in colour and has light to dark reddish brown head with prominent *prothoracic* shield. Neonate larvae are *gregarious* in habit and initially feed by scrapping the leaf sheath. After certain growth, larvae devour the internal tissues of the leaf sheath. On attaining requisite growth, a larva ventures out (usually a third instar larva) for new areas and enters into the base of the central portion of the crown and feeds on the developing immature leaves and immature stalk portion. The damage due to eating the base of the spindle portion of the crown results in the formation of a large *dead heart*. Like the Early shoot borer, larvae of Pink borer are also migratory in habit and do not restrict themselves to one shoot. After feeding for a certain period in one shoot, it migrates to nearby healthy shoot for feeding. Thus, a larva usually damages a number of cane shoots before it reaches maturity and proceeds for pupation.

7.5 Larva

After the first moult, a second instar larva assumes the characteristic pinkish colour, which deepens in older larvae. A mature larva becomes pinkish (flesh coloured) dorsally and creamy yellowish-white ventrally. The larva is very agile and on extrication from cane shoot, quickly moves away to take shelter under any available cover, like paper etc. A mature larva attains a length of about 30 mm. Head of a mature larva is reddish brown (dull red) in colour and its body is tapered towards the anal end. The *tracheal trunk* is not easily discernible but the black rimmed *spiracles* are very prominent on the pink body. *Spiracles* are elongated-oval with prominent black rim. The four pairs of abdominal *prolegs* are stout, sufficiently long and bear stout crochets. In comparison to the *Chilo* moth borers, the crochets or hooks on the *prolegs* are also quite different in size, number, and in arrangement. Number of hooks are few (only 16 to 18), present only in the inner side (*meson*) of the *proleg*, and hooks cover only one third of the imaginary circle of the *proleg* (*proleg* appears like a foot of a duck). Crochets are *uniserial* and *uniordinal*.

7.6 Pupa

In sugarcane, a Pink borer larva usually pupates in the space between a mature leaf sheath and stalk. However, pupation within the feeding tunnel made in shoot or in the cane is not uncommon. One may also find formation of pupa at the crown leaves especially at the shoot stage, when the larva utilizes the dead heart and the adjoining green leaf for pupation (due to lack of available space in small shoot). In case of pupation within the tunnel in the stalk, a prepupal larva makes a large exit hole and seals it loosely with silk and host materials. The pupa of Pink borer appears stout, light to dark brown in colour and bears several irregular rows of characteristic depressions (pits) on each abdominal segments of the body. The 5th, 6th and 7th segments are smooth (except the pits and marks of *prolegs*) and do not possess any other ornamentations. The anal projections (*cremaster*) are dark brown to black and relatively smaller. Characteristically four triangular spiny (spear shaped) projections are present at the dorsal side of the anal end. Eye area of the pupa is bulged and the head area is marked with a bluish tinge. Occasionally, upper portion of the pupal body is covered with ashy white substances. Sometimes, the white powdery substance may extend up to the genital area. Abdominal *spiracles* are slightly raised, prominent and with slit like opening. The *spiracle* of the 8th segment is rudimentary and appears as light slit area (not raised and prominent like the rest abdominal *spiracles*). The terminal portion of the wing pad area is bit characteristic; it appears like the collar of a shirt i.e., a bit projected like a triangle and the terminal portion of *metathoracic* legs appear like a small knot of a necktie between the collar ends.

Male pupa is smaller and the genital area is marked with two prominent raised lobes with a central depression on the 9th segment. The anal orifice in both sexes is broad and well marked as a large depressed area at the ventral side. Female pupa is larger and develops several concentric light triangular markings on the 8th and 9th segments. A prominent depressed area at the tip of the triangular area on the 8th segment marks the female genital area.

7.7 Moth (Adult)

The moths are also stout and reddish straw coloured. A central longitudinal broad dark brown triangular area runs through the forewings. As a member of the *Noctuidae*, Pink borer moths do not have *porrect* snouts (like other important sugarcane moth borers those belong to the Pyralidae/Crambidae). Like the Top borer moths, it is also apt in feigning death in any threatening situation.

Sexual dimorphism is evident in the antennae, and in size of the moths (male moths are slightly smaller than the female moths). The antennae of male moths are finely *pectinate* (comb shaped, teeth are small), whereas antennae of female moths are *filiform*. During rest, both the male and female moths cleverly hide their antennae within the body hairs and thus, male and female moths mostly remain indistinguishable at rest. However, with slight disturbance, these moths move the antennae out of hiding and thus, their gender becomes immediately evident. The presence of tuft of scales at the head and thoracic region provides its distinctive appearance. *Papillae anales* or ovipositor is relatively smaller and remain in hiding under the long scales. The *Papillae anales* are further remain protected under the two conical *sclerotized* covers. These covers also bear hairs. Probably these relatively hard conical covers facilitate to make room for the ovipositor for depositing eggs on the glossy surface of the closely attached leaf sheath (a moth has to place the eggs up on its dorsal side so that it can attach them on the glossy surface of leaf sheath).

7.8 Forewing venation

The wing venation is typical *Noctuid* type i.e., *subcosta* (Sc) and R_1 is free and both run parallel for a quite long distance and finally both end at different locations of costal area. R_2 is forked. The outer branch is considered as R_2 and it ends in *costal area*. The inner branch of R_2 forms a part of the stalk of R_3 and R_4. The stalk of R_3 and R_4 receives a short support branch from R_5, the other part of the stalk. The combined vein becomes the stalk of R_3 and R_4. In this way, an elongated closed area develops among the radial veins. The formation of this characteristic *elliptic "O"* among the R veins helps its quick identification. Both R_3 and R_4 end at the *costal area*. R_5 ends at the *termen* but on its way sends a short support branch to the stalk of R_3 and R_4 . The three median veins i.e., M_1, M_2 and M_3 are present. M_1 is very close to the vein R_5 and originates at a point very close to the *radial peak*. M_3 originates from the *cubital peak*. The two *cubital veins*, Cu1a and Cu1b are present. Among the anal veins, A_1 is absent, A_2 is present and very robust and A_3 is rudimentary.

List of plates on Pink borer, *Sesamia inferens* (Walker)

Description of plates

Plate 7.1. An infestation of Pink borer at shoot stage results in the formation of a large *dead heart* (involve many leaves usually 3 to 5).

Plate 7.2. Pink borer infestation results in the formation of a large *dead heart* (involve many leaves usually 3 to 5). Hole made by the larva is usually rough, messed up with dark frass. Hole area becomes dark and imparts a dirty look.

Plate 7.3. Pink borer larva voraciously eats away the immature leaf tissues in the central whrol and everything remains messed up with wet dark- frass.

Plate 7.4. Pink borer larvae do not relish sugarcane stalk tissues. Like stalk borer, it also crosses the nodal barrier easily and like internode borer pushes out a lot of frass outside. Exit hole prepared by a Pink borer larva is irregular and large.

Plate 7.5. Pink borer larva appears flesh coloured with prominent black *spiracles* on the body (dorsal and side view). The ventral side of the larva is yellowish-white.

Plate 7.6. Pink borer larva (variations). Body gradually tapers towards the anal end. Prominent presence of dorsal folds. Hairs on the body are insignificant except the anal end.

Plate 7.7. Tapering of larval body. Hairs are not prominent on the body except the few anal segments. A mature larva attains a size of 25-30 mm.

Plate 7.8. Thoracic region of larva (side view). Only prothoracic shield is well developed. *Ocelli* are prominent. Thoracic *spiracle* is dark rimmed and elongated. *Tracheal tube* and finer air tubes are also visible.

Plate 7.9. Abdominal segment (side view). *Tracheal trunk* is not usually visible. *Spiracle* is elongated oval with very dark to black rim. Tubercles are present on the body but not prominent. Hairs are also present on the body but not prominent.

Plate 7.10. Anal side of the abdominal segment (side view). *Tracheal trunk* is not usually visible. *Spiracle* is elongated oval with very dark or black rim . Tubercles on the 8th and 9th segments are dark and prominent. On the 9th segment, all the four dorsal and subdorsal setae remain separate. Hairs are also prominent in these segments. Anal plate is prominent.

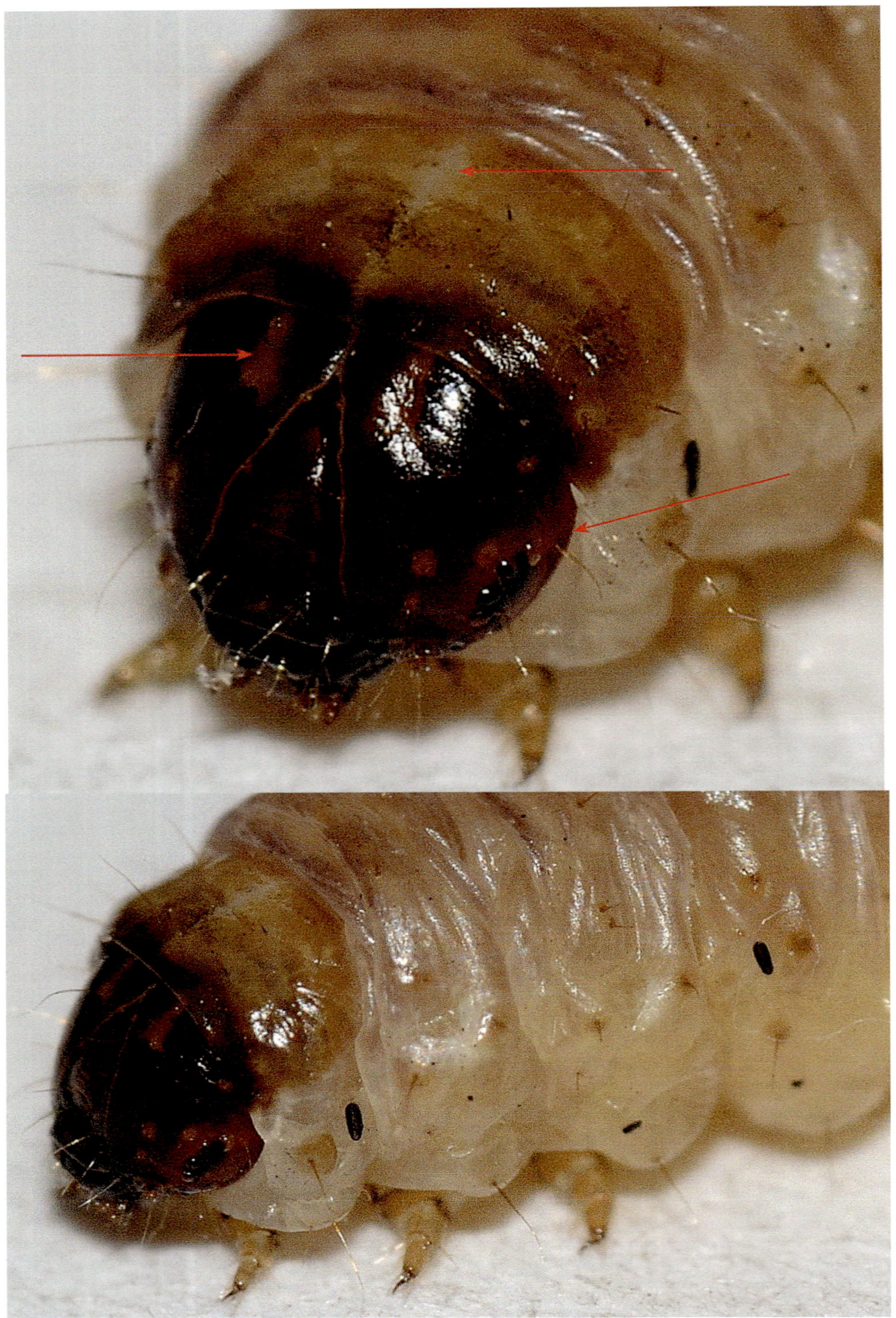

Plate 7.11. Head capsule is dark with light patches. The base of the setae are light coloured and thus, imparts a distinctive appearance to the dark head. *Epicranial suture* with sharp white line.

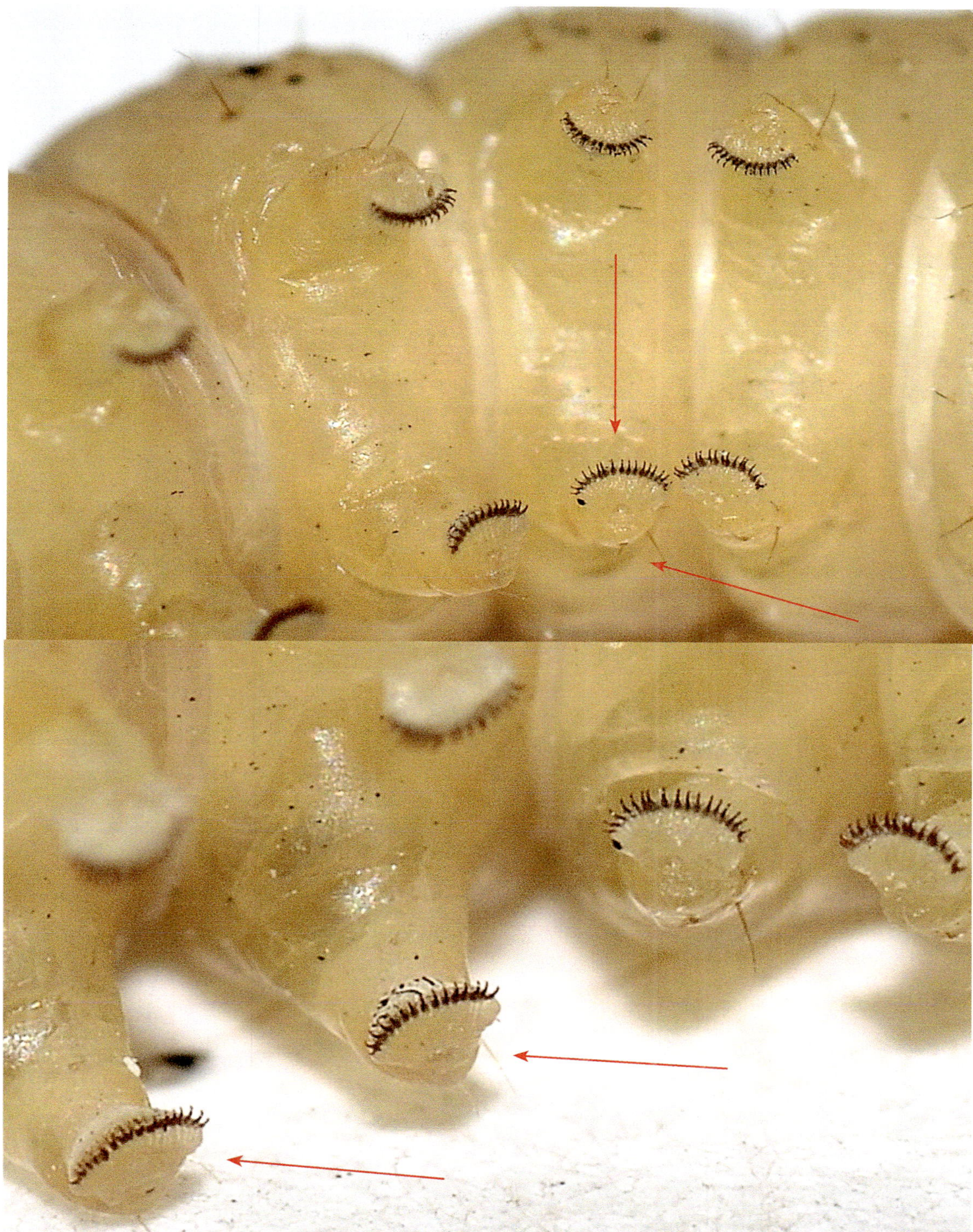

Plate 7.12. Abdominal *prolegs* of a Pink borer larva are duck foot like. Crochets are sharp with stout base and are arranged in an incomplete circle (only covers $1/4^{th}$ of a circle).

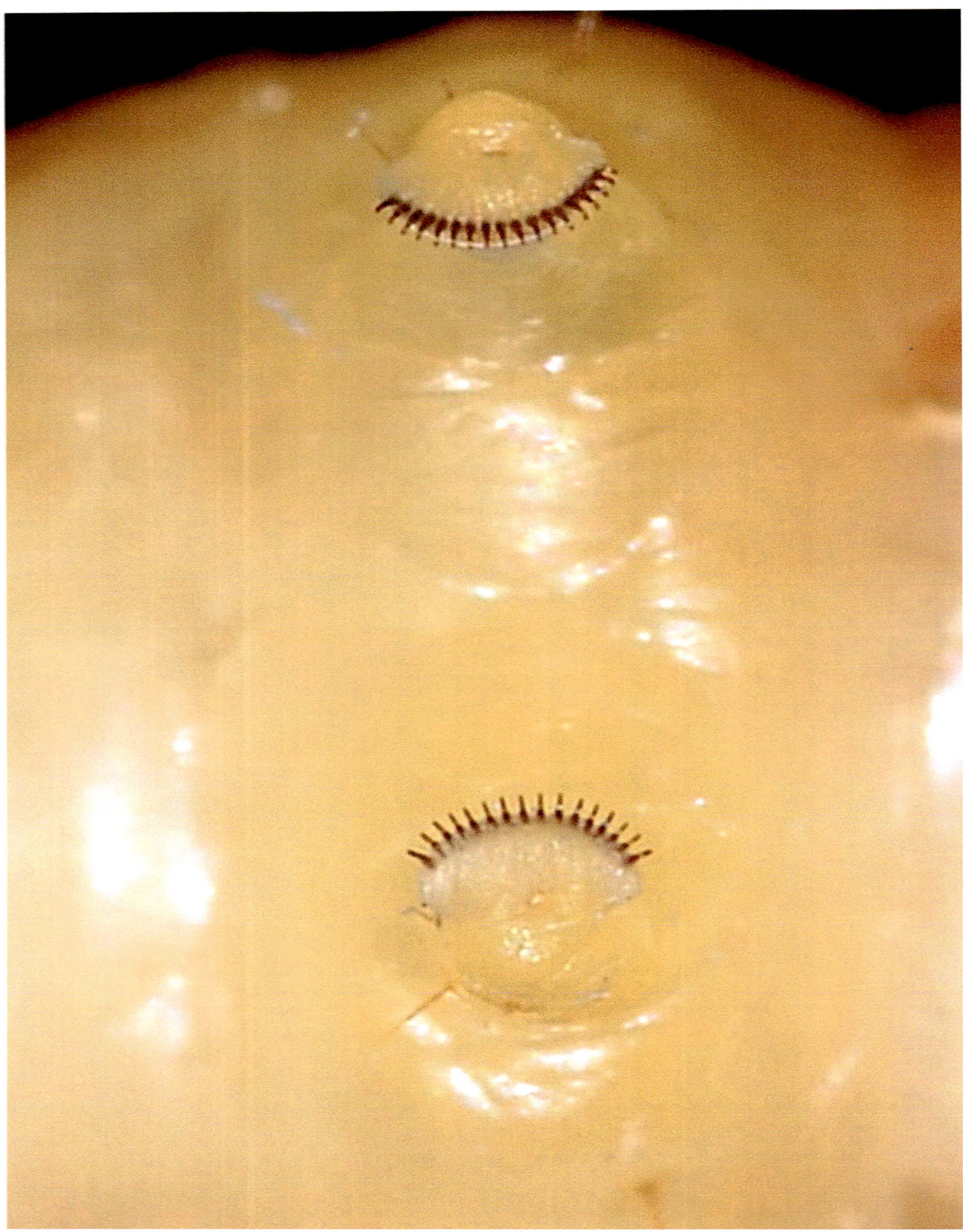

Plate 7.13. Close up of *prolegs* showing *uniserial* and *uniordinal* crochets (15-18) which covers only 1/4th of the imaginary circle. Crochets are present towards the *meson*.

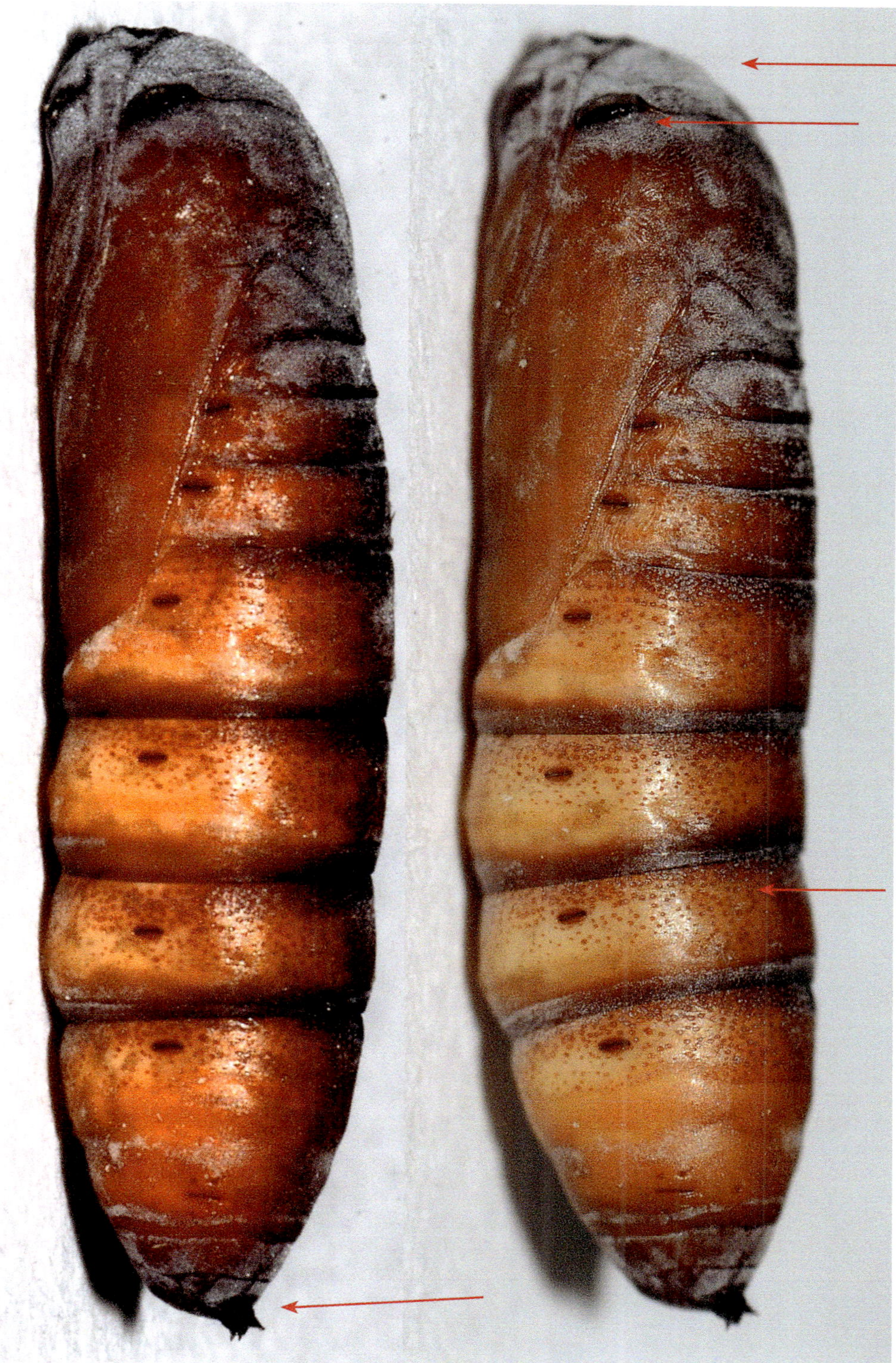

Plate 7.14. Pupa of Pink borer (side view). A mature pupa is light to dark coloured with a light cover of ashy white substances concentrating towards the head region. Body is smooth except the presence of characteristic pits on the body. *Cremaster* is dark coloured and has four triangular dark brown to black spine like projections.

Plate 7.15. Pupa of Pink borer (dorsal and ventral view). A mature pupa is light to dark coloured with ashy white substances. Body is smooth except the presence of characteristic pits on the body. *Cremaster* is dark coloured and has four triangular dark brown to black spine like projections.

Plate 7.16. Male and female pupa of Pink borer. Variations in size and thickness

Plate 7.17. Size of a immature female pupa (around 18 mm).

Plate 7.18. Pupae are stout, light to dark coloured with numerous pits on the body. White powdery coating appears on the body and mostly concentrate at the head, thoracic and anal region. Development of bluish tinge on the face often takes place.

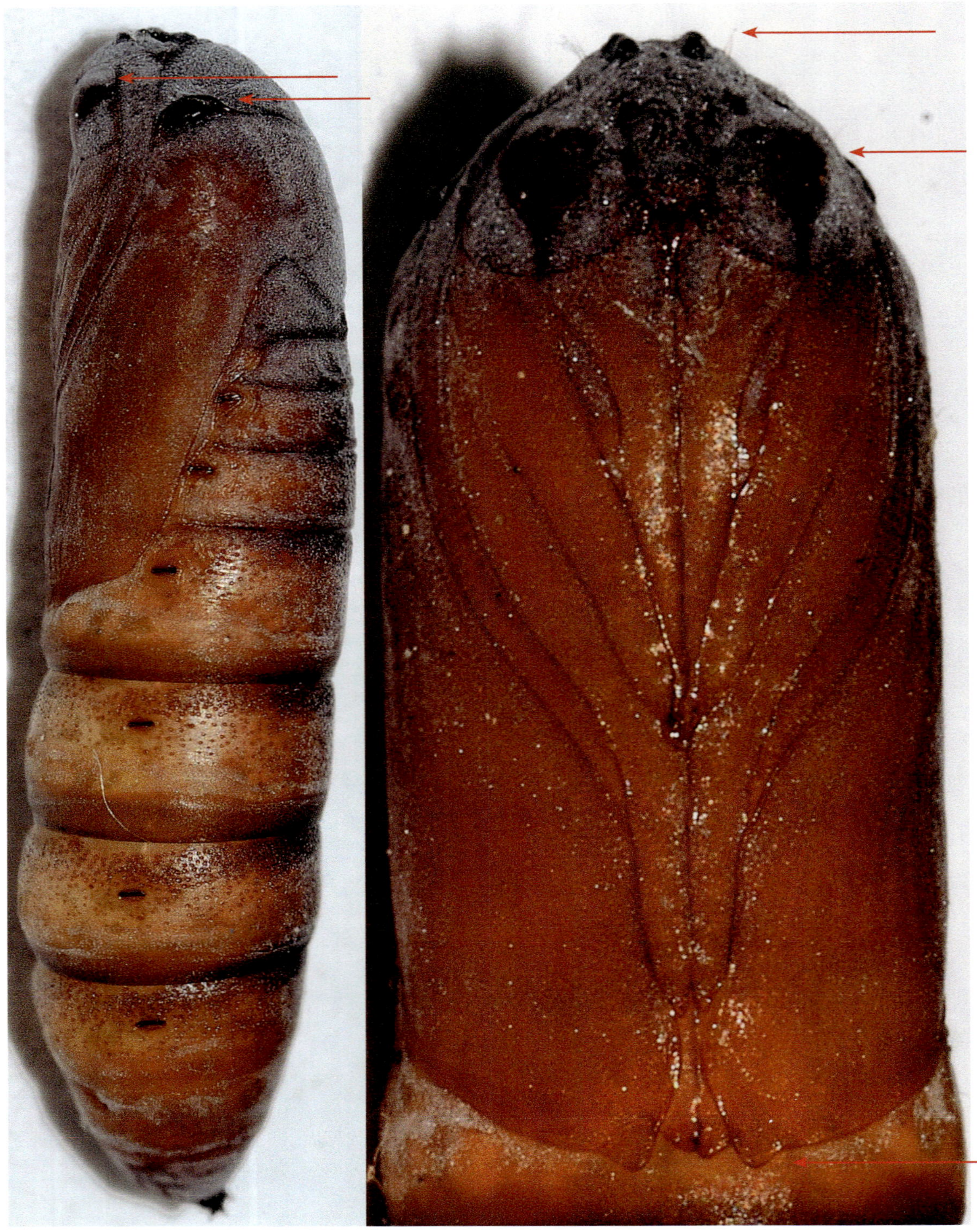

Plate 7.19. The ventral side of the thoracic region (close up). The terminal portion of the wing pad appears cone type. These cones of the wing pad and hind pair of legs impart an appearance of the corner of a collar of a shirt with short neck tie-knot.

Plate 7.20. Head of the pupa (close up). There are two dark projections on the head (like eye). On the face two prominent setae are present. *Mesothoracic spiracles* are very prominent. Median ridge is prominent on *prothoracic* shield and continues further to the *mesothoracic* shield The white substances concentrate towards the face.

Plate 7.21. Anal end of the pupa (close up). *Cremaster* with four triangular dark spines. *Spiracle* on the 8th segment is rudimentary but marking is visible . Density of pits is more towards the beginning of a segment. White substance also present at the anal side.

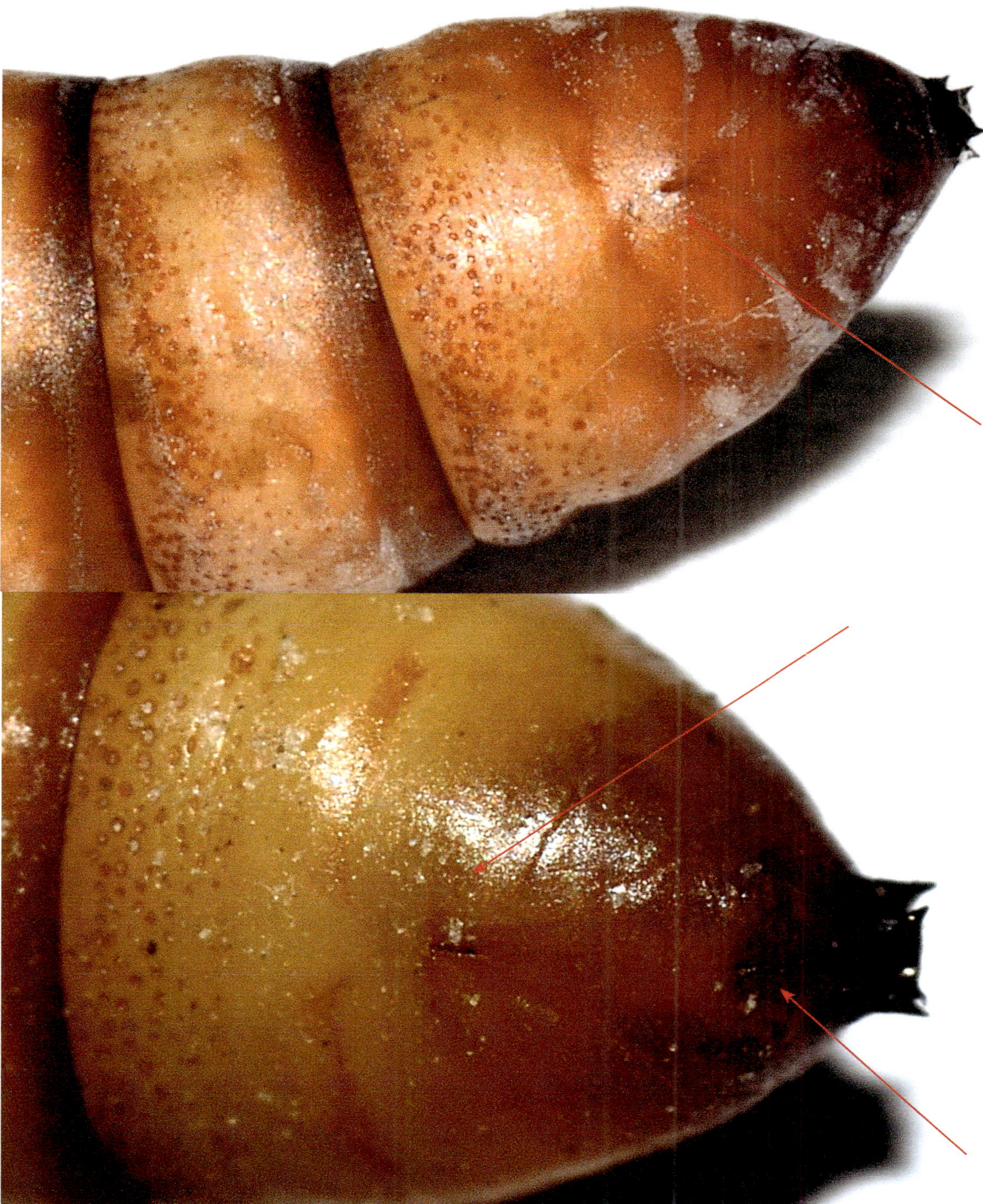

Plate 7.22. Genital area in a female pupa is marked by the presence of a small depression at the end of triangles on the middle of the 8th segment.

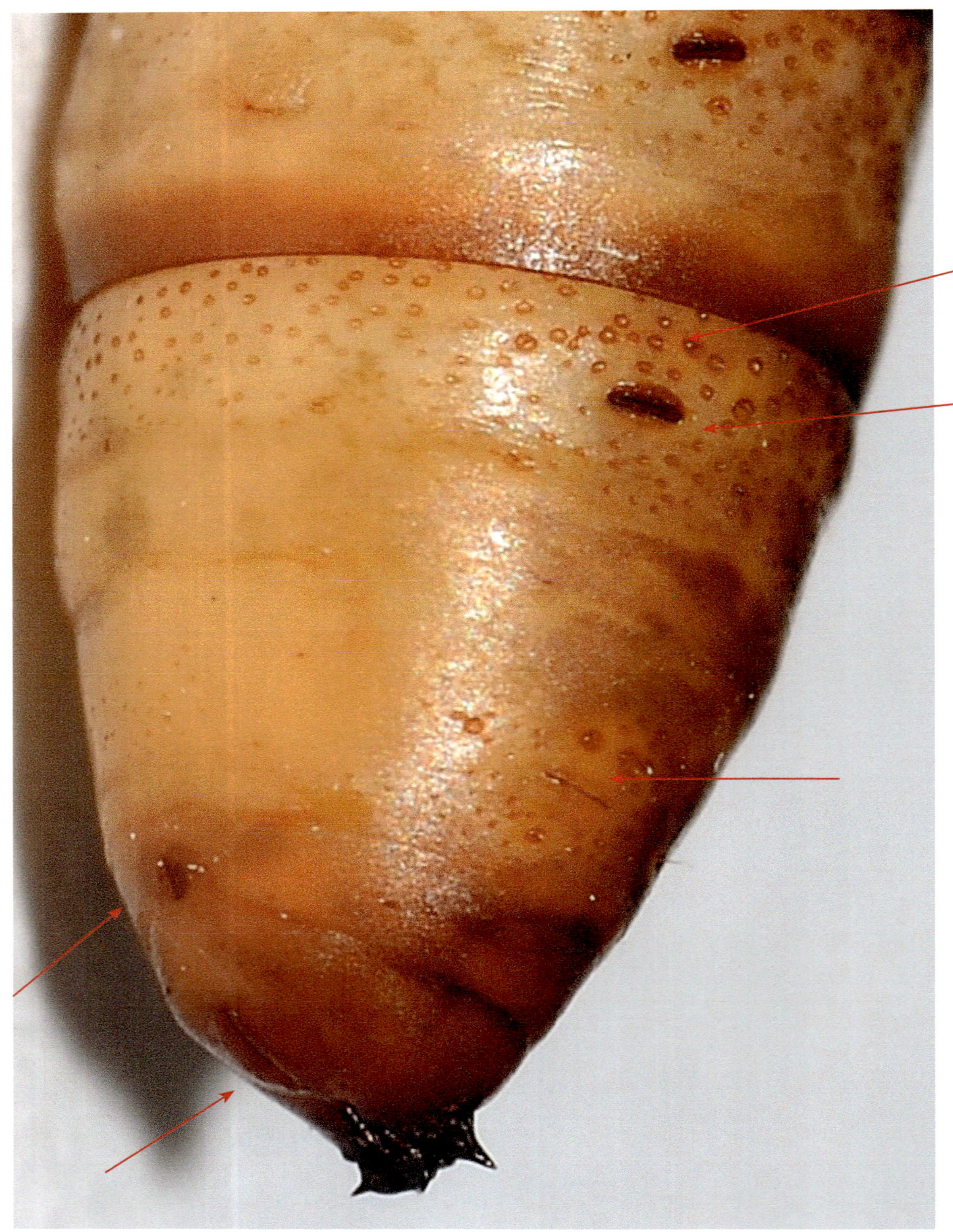

Plate 7.23. In male, genital area is marked with small raised lobes with a prominent depression on the 9th segment. Anal area is straight and very prominently depressed.

Plate 7.24. Female moth of Pink borer (dorsal and side views). An area of dark scales run through the middle of the forewings covering area of anal veins . Another dark area may present on the costal side, in line with the discal boundary. Head bears very prominent tuft of hairs.

Plate 7.25. Male moth of Pink borer (dorsal view). Similar to female moths. Head and thoracic region bears very prominent tuft of hairs. Antennae are *pectinate* (comb like).

Plate 7.26. Ventral view of a female moth feigning death. Face appears owl like with dark black eyes. Labial palpi are raised upward and proboscis is present. Occasional black scales present on the body of the abdomen. Ventral view of a male moth sitting on a plastic plate.

Plate 7.27. Male and female moth (front view). Head with tuft of prominent hairs. Antennae mostly remain hiding in the hairs of head and thoracic region. Antennae are *filiform* in female and *pectinate* in male.

Plate 7.28. Male and female moth (front view, close up). Head with tuft of prominent hairs. Antennae mostly remain hiding in the hairs of head and thoracic region. Antennae are *filiform* in female and *pectinate* in male.

Plate 7.29. Antennae of male and female moths (close up). Antennae are *filiform* in female and *pectinate* in male. Each tooth of the *pectinate* segments bears a tuft of sensitive hairs.

Plate 7.30. Dorsal and ventral view of head and thoracic region of moths (close up). Head and thoracic region bear characteristic tufts of long scales. Antennae remain hiding in these scales. Ventrally *labial palpi* are upwardly curved. *Proboscis* is visible. Large dark eyes imparts an owl like appearance to the head.

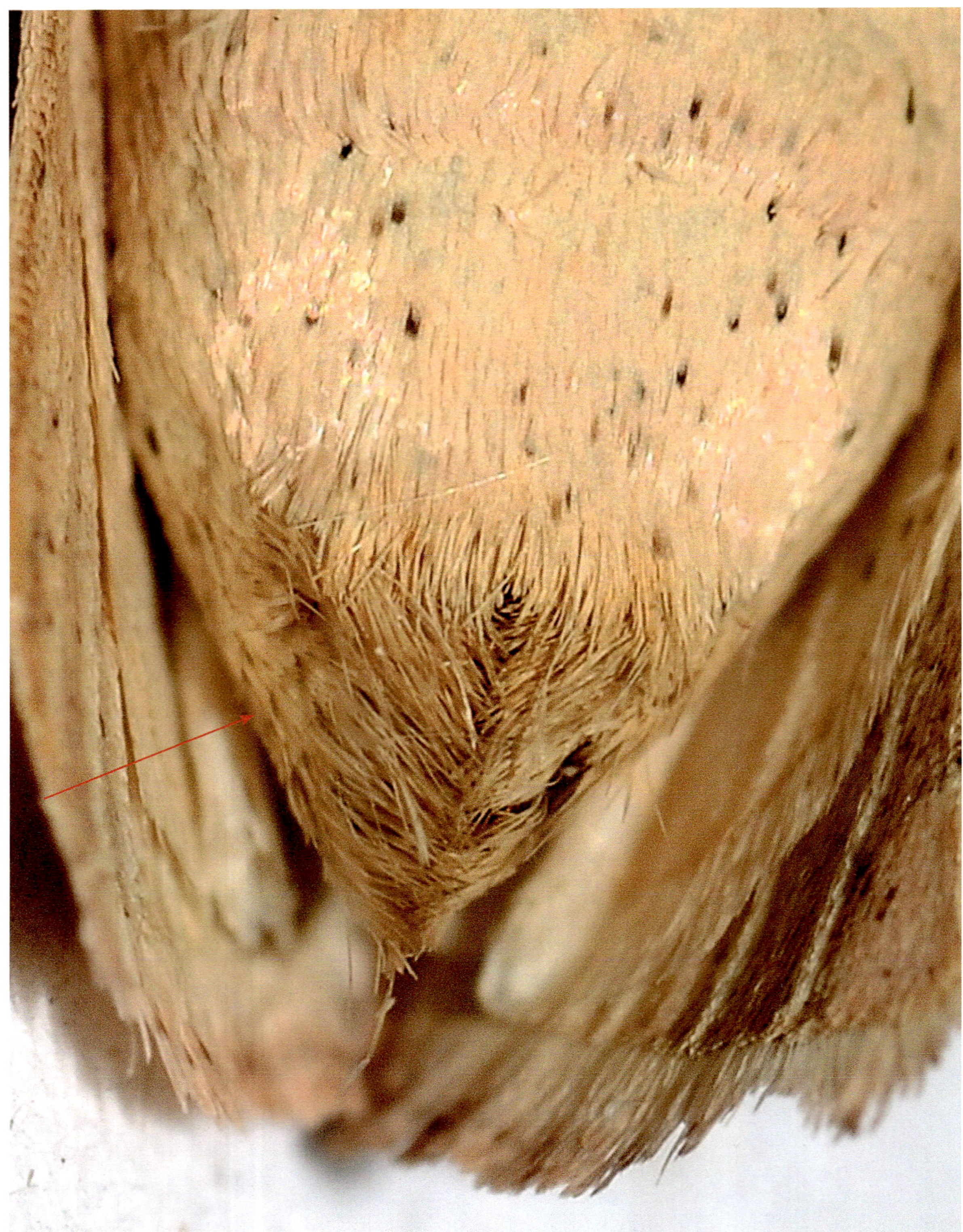

Plate 7.31. Anal end of a female moth. Female genitalia and the *papillae anales* (ovipositor) remain covered with tough scales.

Plate 7.32. Anal end of the female moth. *Papillae anales* are cone shaped and remain covered with two conical *sclerotized* plates

Plate 7.33. Forewing venation. It has typical elongated *'O' formation* in the radial veins with the support from discal cell. Sc is long and strong. R_1 runs parallel to Sc for a long distance, both end separately. R_2 and R_3 are stalked. Stalk is relatively smaller. R_5 is also stalked. A small branch of R_5 joins the branch of R_2 and forms the stalk for R_3 and R_4. M_1 arises near the radial angle. A_2 is very prominent.

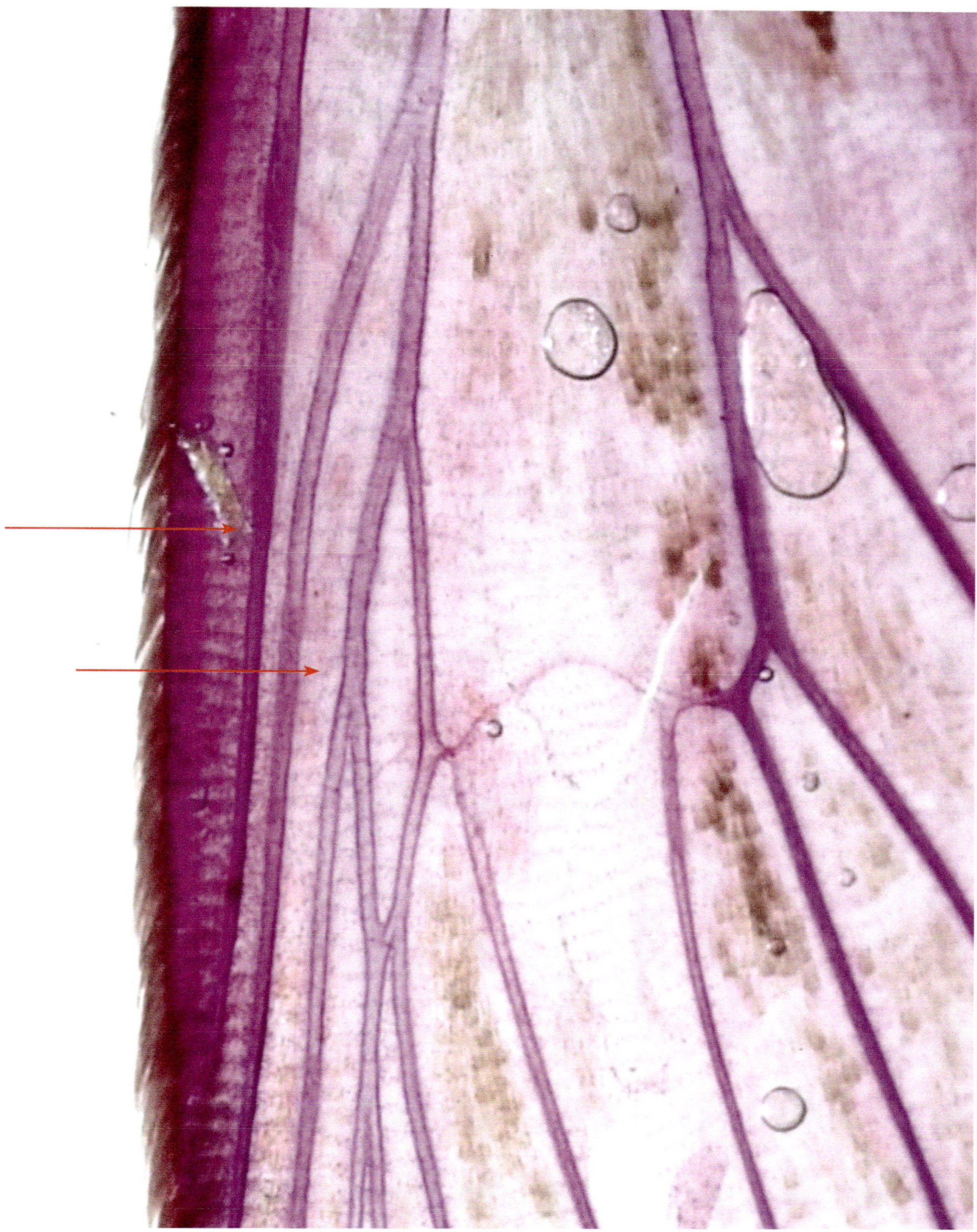

Plate 7.34. Discal cell and adjoining area (close up). R_1 runs parallel to Sc for a long distance. R_2 and R_3 are stalked. Stalk is relatively smaller. R_5 is also stalked. A small branch of R_5 joins the R_3 and forms the stalk for R_3 and R_4. Thus, a characteristic 'O' formation takes place. M_1 arises near the radial angle. Cu1b and R_1 share almost the same plane.

Plate 7.35. Egg masses of Pink borer on leaf sheath of rice. Female moth pushes the eggs from the ligular area into the space inside of the leaf sheath. Eggs are yellowish white in colour and remain attached to the glossy surface of the leaf sheath. Eggs are hemispherical (*peda* or *poppins* like) with fine striations vertically run through the periphery.

8

Early shoot borer, *Chilo infuscatellus* Snellen

The actual identity of the Early shoot borer remained a controversial issue among the entomologists working in India. Since its recognition as *Diatraea saccharalis* F. by Cotes (1889) the confusion continued for over three decades. In the *"Indian Insect Pests"* written by Lefroy (1906), the Early shoot borer or *shoot borer* was recognised as *Diatraea saccharalis* Fabricius. Obviously, that was a misidentification. The publication of Fletcher and Ghosh (1919) provided excellent drawings (Plate 48; page 387; Please refer ***Appendix IV*** for the details) of the larva and pupa of this species, though they also misidentified the Early shoot borer as *Diatraea auricilia*. Hampson (1919) gave another name, *Argyria sticticraspis* to this species. This anomaly was later corrected by Fletcher (1928) himself after making a comparison with the museum specimens of different moth species kept at the British Museum (of Natural History), London. Hereafter, the name *Argyria sticticraspis* Hampson, for the Early shoot borer (ESB) became popular in the Indian literature.

Snellen (1890) first identified this moth borer from Java (now Indonesia) as *Chilo infuscatellus* Snellen. One can find a good account of this borer along with excellent drawings in the publication of van Deventer (1906) from Java. Kapur (1950) erected the genus *Chilotraea* and placed this moth in this genus as *Chilotraea infuscatellus* (Snellen) Kapur. He considered appearance of the face of moth and venation of forewings as the two important distinguishing characters for the genus *Chilotraea* (in the fold of *Chilo*). In *Chilotraea*, subcostal vein (Sc) anastomoses with *radius* (R) and give rise to the combined vein $Sc+R_1$ or these two veins run very close together. However, while revising the world species of *Chilo*, Bleszynski (1970) did not agree to this further division of genus *Chilo*, into *Chilo* and *Chilotraea* and he justifiably retained all the species in the genus *Chilo*. Arora (2000) however was in agreement with Kapur(1950), but many authors subsequently felt that it would be better to retain this species in the genus *Chilo*.

As mentioned in different publications, that the symptom of borer infestation in sugarcane was first reported in India as *"dhosah"* from Bengal by Baboo Joykissen Mookerjee (1857) is not appropriate (Original communication to the Agri-Horticulture Society, Calcutta is appended here as the ***Appendix I***). It was a report indicating total destruction of sugarcane crop of var. Bombay in Rangpur, Hoogly and Burdwan districts that appeared in the publication of *The Agri-Horticulture Society*, Calcutta with the title *"Remarks on a disease affecting the Bombay or Red sugarcane in certain districts of Bengal"*. In that communication, Mr. Mookerjee reported on page 356 that

> "About four years ago the disease (the natives call it *dhosah*) which proved fatal, and totally destroyed the culture of one of the much esteemed and principal harvests of the district, began to make its appearance. A few cane bushes out of a field on one *don* (thirteen cottahs) appeared withered in the first year. The next year more than two-thirds of each of the cane fields shared the same fate. In the first year very little attention was paid to this subject, but the cultivators were alarmed in the next, and commenced enquiring into the matter. On dissection, the pits of the canes were found of a reddish colour, and eaten up by worms of reddish white colour with a black mark or spot on their mouth. In the third or last year, whole fields were destroyed. From the stinking smell of the canes thus destroyed, it became hardly possible for men to pass by these fields, much less their being fit for consumption by human beings, or even by cattle. After some experiments, such as putting less or more manure, and shifting the cultivation from one to another field, the disappointed cultivators gave up the culture of the red cane in despair, and re-introduced that of the country canes (Kajlee and Pooree), which seems to be promising."

However, such occurrence of foul smelling fields is mostly observed when a bacterial disease, known as *Top rot* kills a large number of shoots in the field. From the above narratives, it appears that in all probability, the sugarcane disease now known as 'Red rot' had engulfed the crop at that period. At that time, 'Red rot was totally unknown to the scientific world. The disease came to light only in 1893, when F.A.F.C. Went described this sugarcane disease from Java (now Indonesia). The crop damage was accentuated by the incidence of Top rot that is known for producing a foul smelling field. In addition, the incidence of certain cane borers (Stalk borer or Plassey borer or Pink borer) further aggravated the situation. It is to be noted here that a worm known as "*kansua*" (Early shoot borer) was known to occur in the Karnal district of Haryana during the summer months damaging the young sugarcane crop (Watt, 1893). Around 30 years later, Cotes(1889) wrongly assumed that the total destruction of the sugarcane crop, as reported by Mookerjee (1857), was due the incidence of *Diatraea saccharalis* F. However going through the description of the damage, the actual identity of the borer species involved cannot be ascertained. Therefore, as a conjecture, it would not be appropriate to attribute that the damage was due to the infestation of *Chilo simplex* Butler (as has been done by subsequent authors in many publications).

However, one gets a very vivid account of infestation of Early shoot borer in the pages of Indian Museum Notes and I quote a passage as provided by Mr. E. C. Cotes on page 24 vol 1 (published in 1889) and subsequently by Mr. E. P. Stebbing on page 67 of vol. 5 (published in 1903).

In 1888, the Personal Assistant to the Director of Land Records and Agriculture, North-Western Provinces, wrote that the pest, which appeared in dry seasons, had destroyed as much as a fourth of the sugarcane crop in the neighbourhood of the Cawnpore Experimental farm. A similar report relating to the country adjacent to Burdwan and Seebpore was sent by the Director of Land Records and Agriculture, Bengal. The letter forwarded was from the agricultural officer of the above mentioned places. He wrote—

> " The sugarcane planting season extends from the beginning of February to the end of May. If there be no rains in April or May, and if the cane fields are not frequently irrigated, which from the scarcity of water at this time, is hardly possible, the pest makes its appearance. The pest first shows itself by the drying of the middle stalk of the plant, and is hence called by the ryots the *majera* (a Bengali term meaning relating to the middle); on pulling, the stalk now easily comes out, and its lower end is found to have become a rotten mass. Very soon the whole plant dies away, and from the root stock a number of smaller plants make their appearance, to be in their turn attacked by the worm. If the rains

hold off a long time, or if the fields are not thoroughly irrigated, three or four generations of plants are in this way attacked and destroyed. At last, when the rains set in the fields become free from the insect, and a number of sickly-looking cane plants shoot out, but these make very little progress and never attain the proper size of the cane plant. If only one generation of plants is lost, and if this happens at an early stage of the growth of the plant, the damage done is not much."

8.1 Currently accepted name

Chilo infuscatellus Snellen

8.2 Synonyms

= *Chilo simplex* Butler

= *Argyria sticticraspis* Hampson

= *Diatraea auricilia* (Dudgeon) Fletcher and Ghosh

= *Proceras infuscatellus* (Snellen) Kalshoven

= *Chilotraea infuscatellus* (Snellen) Kapur

8.3 Symptoms of infestation and brief biology

Among the shoot borers that affect sugarcane during the shoot stage (in sugarcane, up to four months age of the crop growth is called shoot stage), *Chilo infuscatellus* Snellen, known as Early shoot borer (ESB) or only Shoot borer is the most important. It has an all India footprint but its infestation is more severe in tropical sugarcane belt. In tropical sugarcane area, due to favourable ambient temperature it damages the crop round the year. In addition, it continues as tissue borer and causes considerable damage to the crop. In subtropical sugarcane belt, ESB causes maximum damage at shoot stage only and that too during the summer months (April, May and June). The activity of this pest abruptly diminishes with the onset of monsoon rains. During the monsoon period, sugarcane crop switches over to the grand growth phase (a phase marked by vertical growth of the sugarcane plant as a result of quick elongation of stalk) and other borers like Stalk borer, Plassey borer, Gurdaspur borer and Internode borer predominate. The overall impact of Early shoot borer gradually declines in the monsoon months and the damage becomes limited or negligible in the mature canes.

The importance of this borer can be judged from the losses or setback it causes during the initial stage of the crop growth i.e., (i) by killing the mother shoots it results in a gappy stand of the crop, (ii) and by attacking the subsequently formed tillers, it reduces the chance of getting good millable canes (NMC). (Tillers are formed late and thus, hardly contribute to the yield as they remain weak and develop into lanky canes with less weight and sugar).

Due to the infestation of Early shoot borer, a typical *dead heart* symptom is produced at shoot stage of the crop. The *dead heart* can be pulled out easily from the infested crown and the terminal end (distal internal portion) of the *dead heart* emits offensive foul smell due to the bacterial rotting of tissues. The easily pullable *dead heart* along with offensive foul smell is very characteristic of the infestation of Early shoot borer. The characteristic *dead heart* is formed due to the feeding of ESB larva in the crown area. It devours the immature portion of the developing leaves and thus, snaps the water flow to the upper portion of the immature crown leaves. The *dead hearts* produced by Early shoot borer, Green borer and Pink borer are quite similar and all these can be pulled out easily however, the foul smell is less offensive in the *dead hearts* produced by the Green borer and Pink borer. Among these borers, presence of characteristic striped larva in the affected shoot immediately confirms the ESB infestation.

The larva of ESB is notorious in affecting a number of shoots during its growth and development. A larva is migratory in habit and after feeding for a certain period in a shoot, it moves on to a new destination i.e., an uninfested shoot in the same or adjacent clump. Thus, a single larva may destroy three, four, or more cane shoots in the same or adjacent clumps, before attaining maturity. Due to this habit of larva, sudden flaring up of the incidence of ESB is often observed in the dry summer months. Because of this migratory habit of a larva, one has to search for a fresh *dead heart* (drying central leaves remain greenish) to catch a live larva inside the damaged shoot. Usually two or more bore-holes are found at the bottom white/ *achlorophyllous* area of the *pseudostem* portion (upright leaf sheath of a shoot) of the immature cane plant. To catch a larva live, a shoot bearing freshly formed *dead heart* has to be split open carefully. The larva is very active, and on extrication from the shoot quickly tries to hide itself under a cover. The larvae are relatively short and stout.

8.4 Egg and neonate larva

A female moth of this species lays eggs in a cluster (egg mass) mostly on the underside of lower leaves. Freshly laid eggs are translucent white, which later on turn creamy white. The eggs are scale like, dorsoventrally flattened like an upturned saucer and remain firmly attached on the surface of the leaf. Fertilized eggs are usually laid in an imbricate row, parallel to the mid-rib. A female moth in captivity can also lay unfertilized eggs. In the laying of unfertilized eggs, the moth does not follow a regular pattern.

The neonate larva is very agile and it is about 1.5 mm in length. It has a prominent dark brown to black head along with the dark *prothoraic shield*. A neonate larva wanders for a few hours on the leaves and leaf sheaths and finally reaches the base of the *pseudostem* by crawling or hanging by a silken thread. A larva generally enters in the space between two immature leaf sheaths and feeds on the soft internal tissues of the leaf sheath like a leaf scrapper. At this place, the neonate larva stays for few days and moults once or twice before becoming aggressive to venture into the inner portion of the shoot.

8.5 Larva

After attaining required growth living as a leaf sheath scrapper, a larva ventures out into the central immature portion of the crown by making a horizontal feeding hole (usually a larva at the 3rd instar stage makes this move and reaches the central core of the immature crown leaves). The larva, staying above the apical dome voraciously feeds the central immature portion of the developing leaves. Due to this feeding habit, a characteristic and easily pullable '*dead heart*' is formed. Larva of Early shoot borer is very agile and appears stout (in comparison to the larva of Stalk borer). A mature larva measures around 18-25 mm. Body colour of the larva depends on the instar a larva is passing through. The larva is light yellowish-brown, creamy to dirty white in colour. However, the colour of the ESB larva is always appears a shade darker than the corresponding larva of Stalk borer.

Tubercles on the larval body are prominent and bear dark brown setae. The large dorsal tubercles (D_1) are a bit rectangular and the smaller tubercle D_2 is bit elongated. In a segment, these large dorsal tubercle pairs D_1 are spaced widely and thus, leaves a large gap in between (D_1 and D_1) at the alimentary canal region. The tubercles present at the lateral side of the *meso* and *meta* thoracic segment are smaller in comparison to that of Stalk borer. The two large dorsal tubercles (D_1 and D_1) on the 8th segment join together to form a bigger tubercle. On the 9th segment, the two dorsal tubercles are also joined together like the 8th segment but the two subdorsal tubercles (SD_1) mostly retain their independence. (In Plassey borer and Maize borer, the four tubercles, dorsal pair and the adjoining subdorsal pair is joined on the 9th segment and together all these tubercles form a large *sclerotized* area.)

To be exact, there are seven stripes present on the larval body. Among these, five stripes are prominent and two stripes are ephemeral. The five prominent stripes are present above the *tracheal trunk* whereas, the two ephemeral stripes are present well below the *tracheal trunk*. These five prominent stripes are arranged as one median, a subdorsal pair and a lateral pair. The ephemeral pair is relatively lighter in colour, and is restricted to the tubercular area. In 2nd and 3rd instar larvae, the three prominent wavy pinkish-violet stripes (median and subdorsal pair) run from *mesothoracic* segment to the 9th abdominal segment.

Characteristically, in the larva of Early shoot borer, the median stripe does not touch the large dorsal tubercles (D_1 and D_1) (as the space between the two large dorsal tubercles is relatively large) in comparison to that of Stalk borer larva wherein the median stripe touches the large dorsal tubercles (D_1) (due to their close proximity). In the immature larva, the dorsal stripe goes beyond the 8th segment, crosses the tubercles of the 9th segment and occasionally touches the last segment. In a mature larva, in the 8th segment, the two subdorsal stripes nearly converge with the median stripe at the two conjoined tubercles and thus, results in the development of '*psi*' (ψ) like appearance, as the median stripe continues to grow beyond the two conjoined tubercles of 8th segment. The two subdorsal stripes are very prominent, wavy and wide, and run through both the larger (D_1) and smaller (D_2) dorsal tubercles. These stripes partially cover the larger tubercles (D_1) at the subdorsal area and run through the smaller tubercles (D_2). The lateral stripes fully cover the lateral tubercles and pass through the area just above the *tracheal trunk*. These lateral stripes delicately circumvent the larger *spiracles* of the 8th segment and go further down to the 9th segment.

The ephemeral stripe pair is present below the *tracheal trunk*. The colour of ephemeral stripe pair is much lighter than the other prominent stripes. This stripe partly remains visible only in the limited surrounding areas around the tubercles. The presence of this pair becomes apparent when the larva is viewed from the anal end. The pink colour of all the stripes of larva usually fades away during hibernation and the colour becomes lighter before pupation.

Head of a mature larva is light to dark brown in colour. There are six prominent *ocelli* at the lateral side of the head and the *ocelli* follow a 1+4+1 arrangement. The *epicranial suture* appear as a diffused white inverted 'Y' in the front. Head capsule bears different intricate colour patterns and these become *prominent* when colour of the head capsule remains lighter. *Prothoracic shield* is well-developed and light yellowish-brown in colour with light coloured median depression. Terminal area of the prothoracic shield is dark brown due to the presence of numerous black patchy areas and spots. *Mesothoracic* plate is also prominent with terminal dark area. A smaller *metathoracic* plate is also present. In many cases, at the abdominal segments of a mature larva, smaller and insignificant tubercle like glistening areas, occupying dorsal central position like the *metathoracic* plate also appear.

The *Tracheal trunk* is visible as a white tube running at the lateral side (*Tracheal trunk* of the Stalk borer larva is very prominent). Occasionally, an extension of a wavy lateral stripe may go down and cross the *tracheal trunk*. *Spiracles* are prominent, black rimmed, elongated elliptical or oval with a white centre and appear open. The *spiracles* present on the 8th abdominal segment are larger and occupy a near subdorsal position on the body.

The crochets in the abdominal *prolegs* are arranged in an *incomplete circle* (characteristically the outer side/lateral side is devoid of hooks). This helps in its quick identification in the field. Around 1/4th of the terminal hook bearing area of the *proleg* always remain without any crochet or hook. The hooks are *triordinal*, reddish brown in colour and follow a *uniserial* arrangement. At the inner side of the prolegs (towards *meson*) hooks are larger and their size gradually diminishes towards the outer side (lateral side). In the field, the relatively dark colour and stout appearance of the larvae, incomplete arrangements of crochets on the *prolegs*, well spaced median stripe and relatively smaller lateral thoracic tubercles help its quick differentiation from that of the larva of Stalk borer.

8.6 Pupa

Early shoot borer usually pupates within the feeding tunnel in a young shoot (*pseudostem*) or in the tunnel made in the cane portion. At shoot stage, a prepupal larva moves up in the young shoot and cuts an exit hole, sufficiently above the soil surface, and covers the exit hole with silk and small fragments of host tissue. At later stage, when it pupates in the cane tissue, it makes an (rather wider) or irregular exit hole (the exit hole is not round like that of Top borer) and covers it with silk and host tissues.

The pupa of Early shoot borer is slender, distinctively yellowish to light brown in colour. The pupa also retains the five prominent wavy stripes as found on the larval body. These stripes are also present on the raised wavy abdominal ornamentations. The three dorsal stripes run through the abdomen and reach the 9th abdominal segment. Occasionally, presence of the third pair of stripes (the ephemeral stripes) or the subspiracular stripes may be observed.

The head of a pupa appears smooth, though it possesses two rudimentary horn like structures. Face area is slightly convex and bears two prominent setae bent inwards (curved). The *mesothoracic spiracle* is arch type.

Each abdominal segment of a pupa bears two tubercles on the dorsal side. Tubercles have one seta each. The 5th, 6th and 7th abdominal segments bear the characteristic wavy ridges (ornamentations), which are extension of body cuticle and develop at about 2/3rd distance away from the lower end of the segment. These ornamentations are in the form of discontinuous, sharp and wavy irregular ridges with terminal dark brown areas. The wavy ridge on the 7th segment of the pupa fully encircles the segment, whereas in the 5th and 6th segments the wavy ridges reach up to the *spiracle* and just cross the *spiracle* (in male two small wavy ridges run beyond the *spiracle* on 6th segment; one wavy ridge runs beyond the *spiracle* on the 5th segment). On the ventral side, a pupa is smooth except for the markings of *prolegs* on 5th and 6th segments and the wavy ridge on the 7th segment. In this species, marking of *prolegs* on 4th abdominal segment is often visible. On the 4th segment, the markings of *prolegs* occur just below the wingpad area, and in conjunction with the unequal hind leg pair impart an appearance of a human face (like two eyes of human face with nose in between; hind legs appear as nose).

The *cremaster* is also provided with two dark brown ridges on the ventral side. In the dorsal side, there are four ridges; two sharp ridges followed by two terminal blunt ridges. Each ridge supports one stout spine. Thus, there are two spines in each dorsal half. The base of the spine on the terminal ridge is broad. The genital opening of the female pupa is present on the 8th abdominal segment as a small elongated dark depression. The development of *papillae anales* in the *pharate adult* may be observed in the 9th abdominal segment. In male pupa, genital opening is present on the 9th abdominal segment as two broadly raised lobes with a central depression. The anal area is well marked with a dark 'Y' shaped depression.

An immature pupa of Early shoot borer remains very sensitive to any external stimulus and quickly responds to any such impulse. On disturbance, it quickly rolls over, rather spins its body rapidly with the help of discontinuous ridges present on the abdominal segments. Thus, an extricated pupa with its fast rolling movement can cover a good distance at one go. When it stays inside the tunnel, it can change its position by rolling movement. Probably this movement helps it to ward off the possible attacks of pupal parasites.

8.7 Moth (Adult)

The female moths of Early shoot borer are typically bright straw coloured (light yellowish to dark yellowish brown) with black spots in the furrow of corrugated terminal area of the forewings. A few white scales precede the terminal black spots (at the inner side of the black spot towards the head). Occasionally scattered dark scales are present on the forewings and impart a metallic appearance. However, these metallic scales are not raised enough and cover only a limited area on the wing. On the forewings, there is no development of collective and raised metallic spot at the *discal cell* area (as found in the moths of Stalk borer). Female moths are slightly bigger than the males. Forewings of the females are a shade lighter and brighter than the males. In the forewings of male, darker shade develops due to the presence of large number of black metallic scales. In the seven terminal black spots of the forewings, the white area is more prominent in male. These white areas/scales appear very distinctive in front of the black spot to the naked eye. (In contrast, in the forewings of Plassey borer, the white scales start from the central area of the black spot, continue up to the terminal area, and thus imparts an appearance of *white cut mark* on the forewings.)

Labial palpus is porrect and around 2.5 times of the diameter of the compound eye. Antennae are *filiform* and ribbed in female and remain clothed with scales and the thickness appears uniform in most part of the *flagellum,* though it tapers towards the apex. Antennae of male are stout, thicker and broader at the side. Dorsally the antennae have scales and appear ribbed (minutely serrated) and laterally they are devoid of scales and appear smooth and light brown. The *flagellum* of male antenna has around 51 segments. These segments are broader at the base and gradually taper towards apex. *Ocellus* is present near the base of the antenna. The black *ocellus* is very distinct and fully developed. Face of the moth is broadly rounded with a smaller corneous point.

Papillae anales or the ovipositor is densely clothed with sensory setae. Ovipositor appears like a pair of bathroom-slipper with flat toe area (ventral side). The two halves of the *Papillae anales* are with prominent space in between with a central depression or concavity.

8.8 Forewing venation

The forewings appear stout in comparison to other moth borers. The relationship between Sc and R_1 is variable. In the forewing, R_1 may anastomose partially with Sc (confluent and runs very close together) or tangentially touches the Sc. Whenever Sc touches R_1, a *tracheal* connection or bridge develops from Sc to R_1, especially from the point of touch. Usually both the veins terminate separately at the *costal* area with separate pairs of *mechanoreceptors* (circular structures with sensory roots). R_2 is free and very close to the Rs. R_2 ends at the *costal area.* R_3 and R_4 are stalked and both the veins end at *costal area.* R_5 is free and ends at the *termen.* The three *media* veins i.e., M_1, M_2 and M_3 are present and more or less spaced equally. The *cubitus* veins Cu1a and Cu1b are present. Two anal veins are present. A_1 is absent, A_2 is very prominent, A_3 is feeble but runs up to the middle of the anal area. The first angle of *discal cell* (radial angle) is almost at a right angle with Rs. The setae present at the *termen* are lighter in colour.

List of plates on Early shoot borer, *Chilo infuscatellus* Snellen

Descriptions of plates

Plate 8.1. Incidence of Early shoot borer. Mother shoot showing formation of *dead heart* symptom in a plant crop.

Plate 8.2. Incidence of Early shoot borer. Ratoon sprouts showing formation of *dead heart* symptom.

Plate 8.3. Incidence of Early shoot borer. Mother shoot showing formation of *dead heart* symptom. Formation of more than one holes near the base of the affected shoot. Holes always remain above the apical dome.

Plate 8.4. A shoot showing formation of *dead heart* symptom. A larva actively feeding inside the central whorl of the leaf (*pseudostem*).

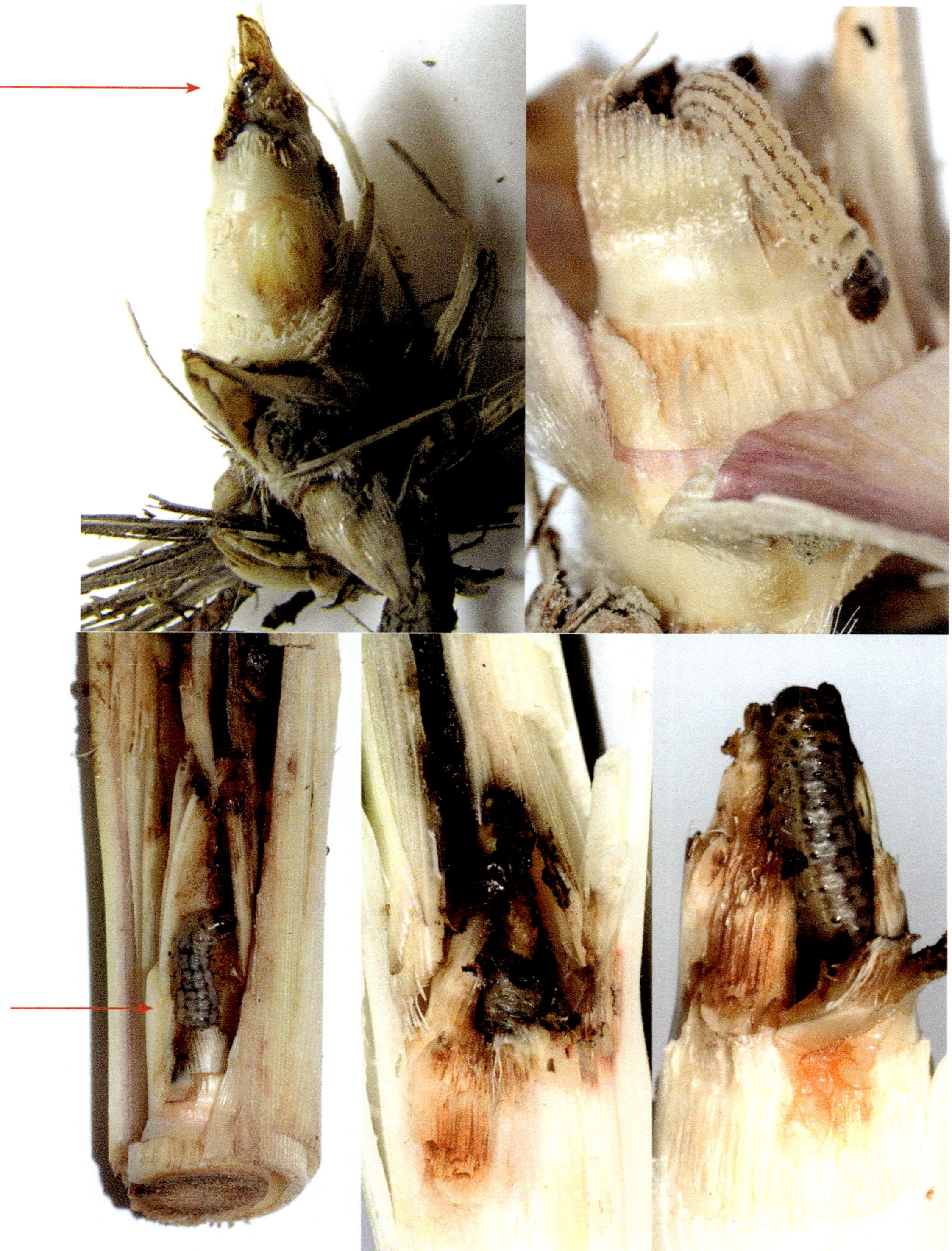

Plate 8.5. Damage of the immature stalk portion. Larva eats away the apical dome of the shoot and feeds inside the immature cane.

Plate 8.6. Immature larva showing prominent prothoracic shield and five dark stripes on the body. Three dorsal stripes (median and two subdorsal) cross the 8th abdominal segment.

Plate 8.7. Size of an immature (3rd instar) larva and a mature larva (5th instar) of Early shoot borer

Plate 8.8. Larvae of Early shoot borer at different instars. (**A**) 3rd instar, (**B**) 3rd instar in the process of *ecdysis*, (**C**) 4th instar , (**D**) 5th instar and (**E**) Prepupal.

Plate 8.9. Position of stripes in relation to dorsal tubercles. Size and shape of tubercles. In a mature larva terminal portion of *prothoracic shield* is dark. Formation of lighter central shields on the abdominal segments. Immediately after moulting, head and thoracic shields remain white.

Plate 8.10. Side view of mature larvae of Early shoot borer showing black rimmed *spiracles*, slight visibility of *tracheal tube* and the placement of prominent lateral stripe above the *tracheal trunk* and the running of subdorsal stripe.

Plate 8.11. Side view of larva (close up) showing black rimmed, elliptical *spiracles* with white centre. Visibility of *tracheal tube* and placement of lateral stripe above it. The ephemeral pair of faint lateral stripes run through the tubercles below the *tracheal trunk*.

Plate 8.12. Side view (from anal side) of the larva (close up) showing large *spiracle* of 8th abdominal segment and its placement. Visibility of *tracheal tube*. Placement of median, subdorsal, lateral and ephemeral stripes.

Plate 8.13. Stripes from the anal end (close up). The second pair or ephemeral pair of lateral stripes are faint but visible below the *tracheal trunk*.

Plate 8.14. The ephemeral pair of lateral stripes are lighter in colour (faint) and run through the tubercles and occupy the space below the *tracheal trunk*. Also note the visibility of *tracheal trunk* and wavy pattern of stripes and running of both median and subdorsal stripes.

Plate 8.15. Close up of head and thoracic region of Early shoot borer larva showing the *epicranial suture, prothoracic shield* with terminal dark colouration. *Mesothoracic* shield with notch formation and prominent *metathoracic* shield

Plate 8.16. A. Side view of head and thoracic regions. The lateral tubercles at the thoracic region are smaller in comparison to that of Stalk borer. **B**. Front view of the head.

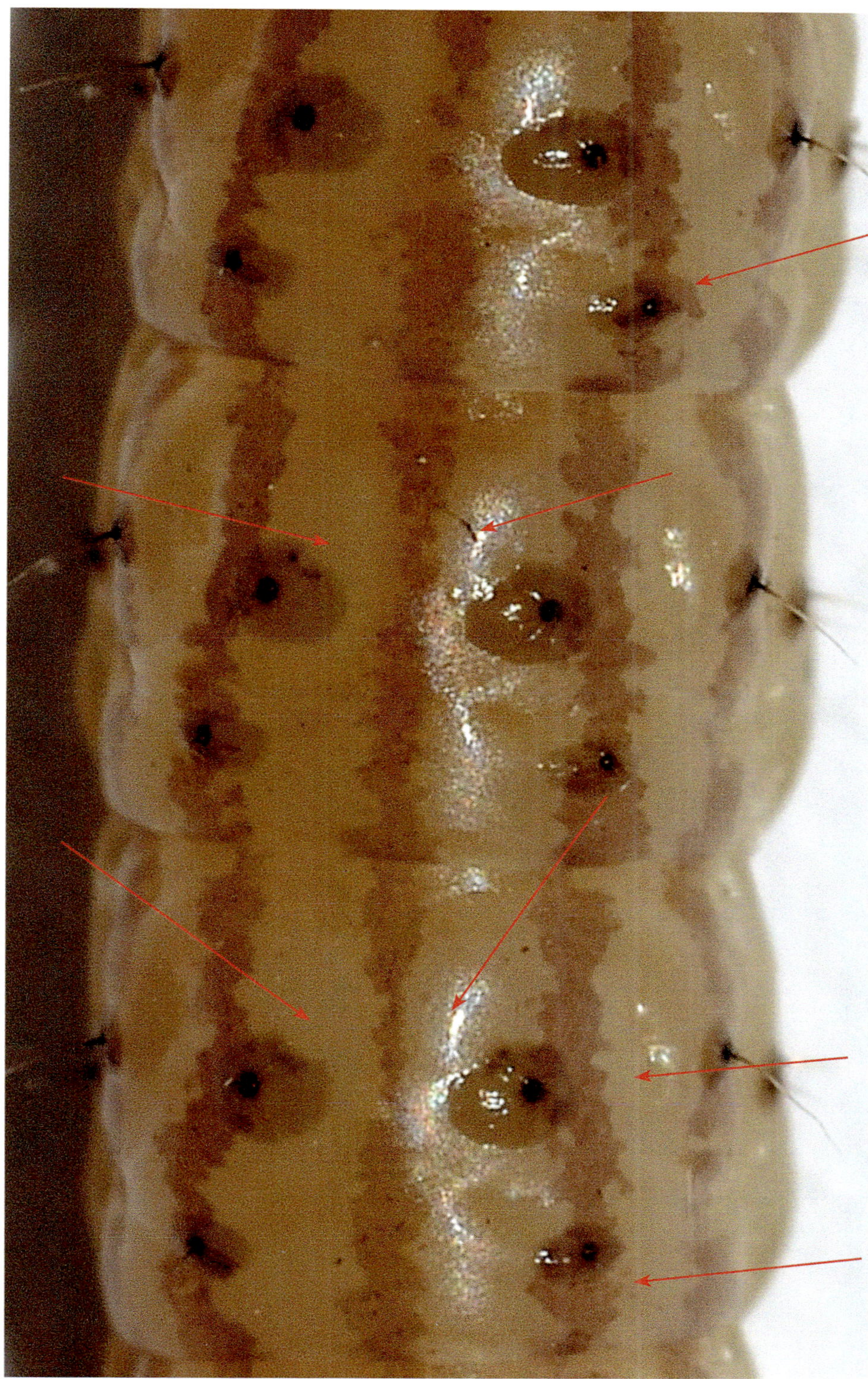

Plate 8.17. Median and dorsal stripes are pinkish and wavy. Median stripe does not touch the dorsal tubercles (D_1). Subdorsal stripes run through the smaller tubercles (D_2) and almost fully cover them or cover beyond the setae. It also partially touches the D_1.

Plate 8.18. Median stripe does not touch the large dorsal tubercles (D_1). Subdorsal stripes run through the smaller tubercles (D_2) and almost fully cover them or cover beyond the setae. Lateral stripe runs through the tubercle above the *spiracle*.

Plate 8.19. Placement of stripes on the body viewed from anal end (close up). Median stripe cross the 8th abdominal segment. On the 9th abdominal segment, the tubercles retain their identity. Often light spotting develops on the *sclerotized* anal plate.

Plate 8.20. Anal end of the larva. Lateral stripes, tubercles and *spiracles* on the 8th segment. Median stripe cross the 8th abdominal segment. On the 9th abdominal segment the tubercles retain their identity. Often light spotting develops on the *sclerotized* anal plate.

Plate 8.21. Arrangement of crochets on the four pairs of abdominal prolegs. Side view of prolegs showing incomplete arrangement of hooks and formation of rim.

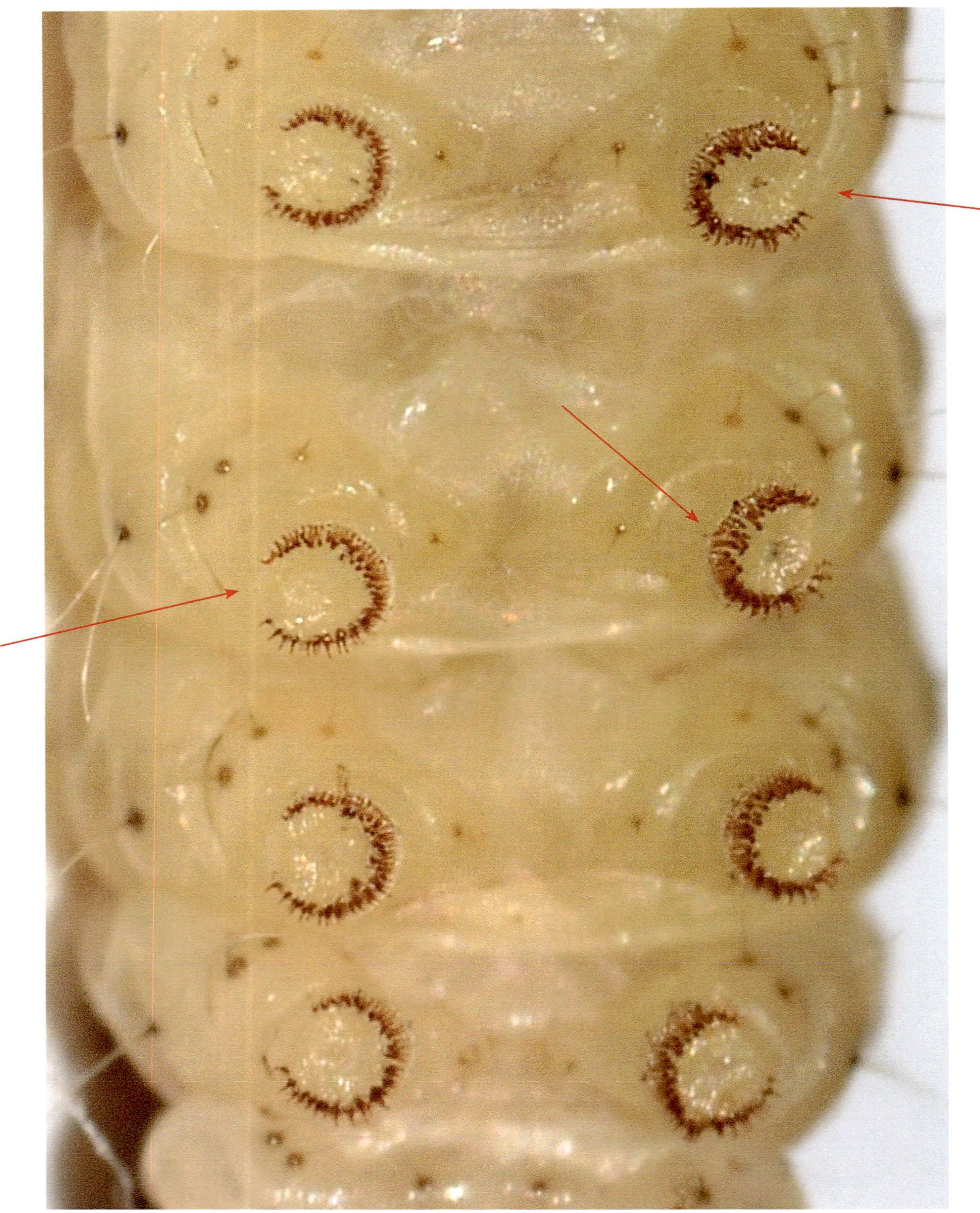

Plate 8.22. Arrangement of crochets on the four pairs of abdominal prolegs. Arrangement is an incomplete circle (20-25% area near the lateral side remain free), hooks are *triordinal.*

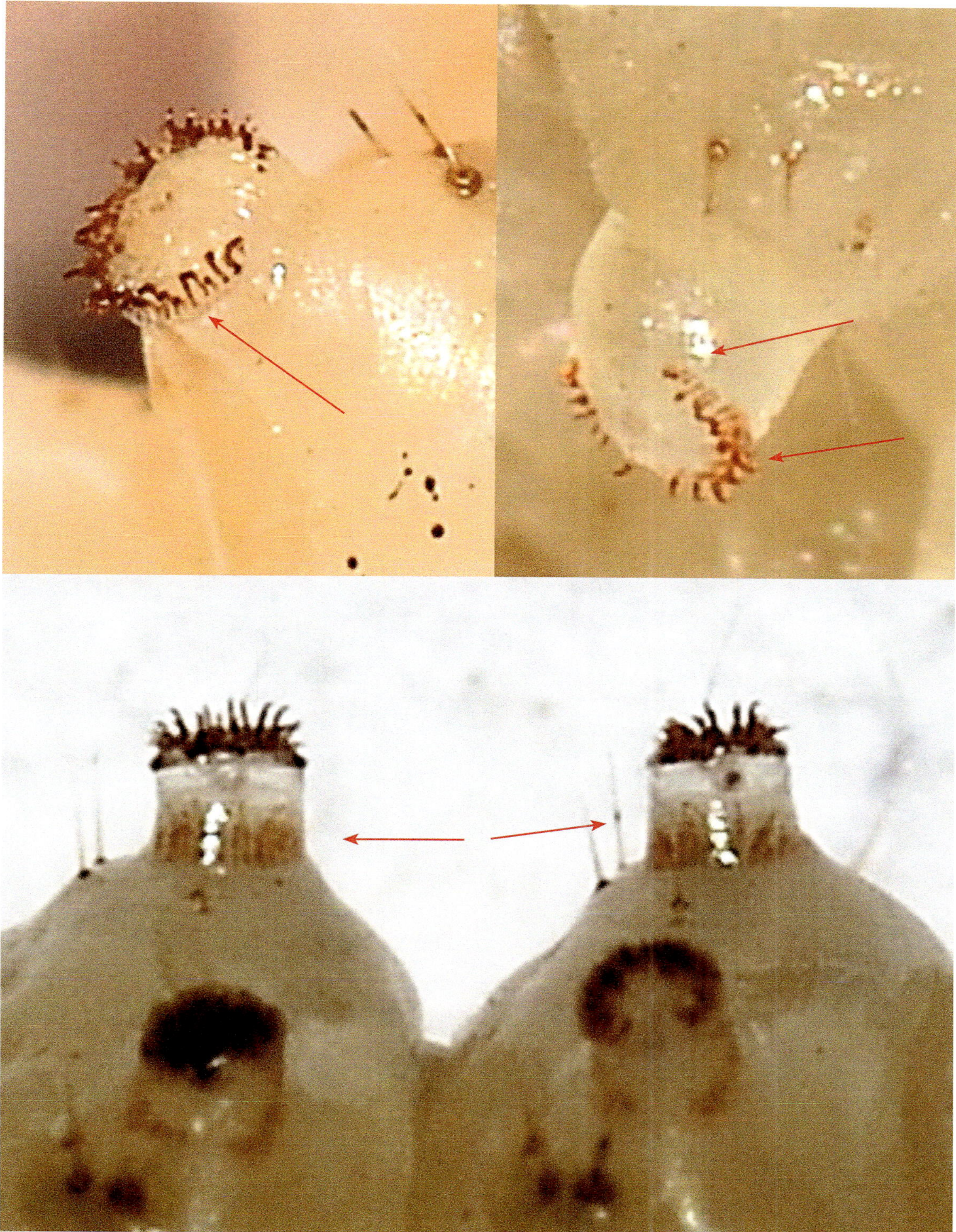

Plate 8.23. Close up of prolegs showing incomplete and *triordinal* arrangement of hooks. Development of hooks at the time *ecdysis*.

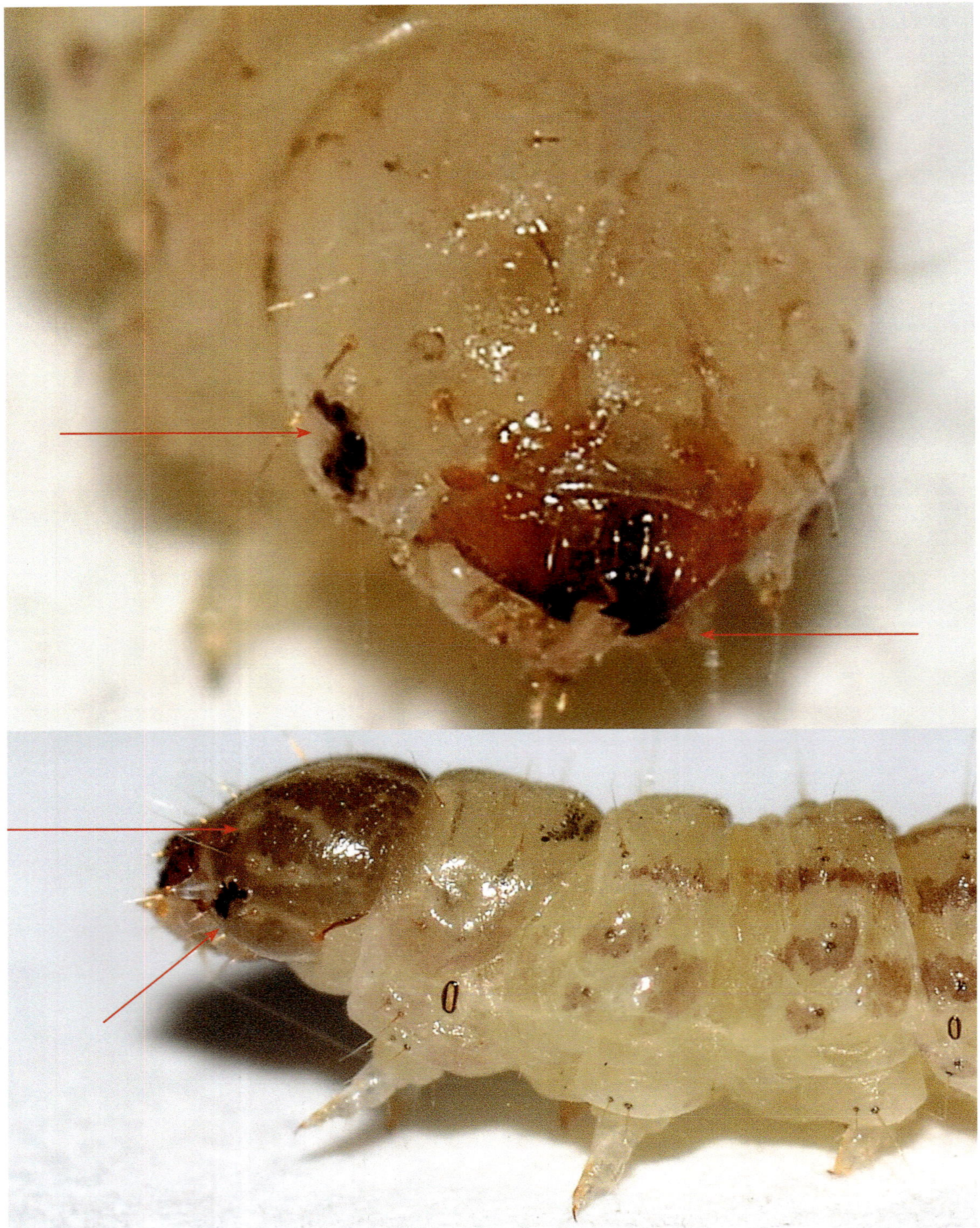

Plate 8.24. Head of the larvae. A freshly moulted head capsule is white except the mandibular area. Six *ocelli* present in 1+4+1 arrangement. *Epicranial suture* is marked with prominent to diffused white line (three levels of cone formation on the face). Four teeth are prominent and sharp. Gradually colour of the head capsule changes with different light coloured patterns.

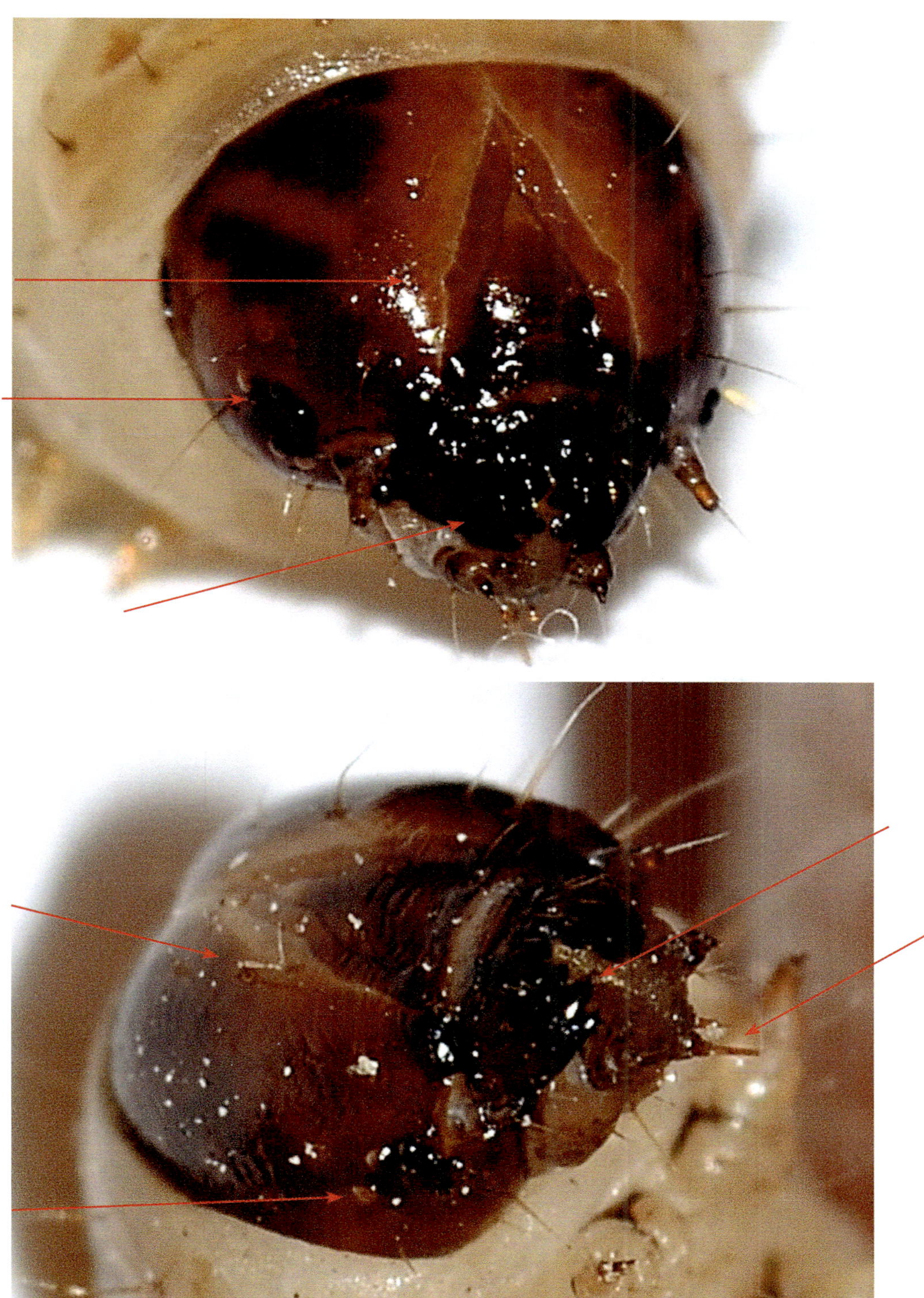

Plate 8.25. Head of the larvae. In a freshly moulted larva, the head capsule is white except the mandibular area. Six *ocelli* are present in 1+4+1 arrangement. Epicranial suture is marked with prominent to diffused white line (three levels of cone formation on the face). Four teeth are prominent and sharp.

Plate 8.26. Pupation takes place within the feeding tunnel or cavity. Exit hole is broad and covered with host tissue and silk. No specialized silken chamber is formed

Plate 8.27. Pupae of Early shoot borer. **A**. Side view **B**. Dorsal view & **C**. Ventral view.

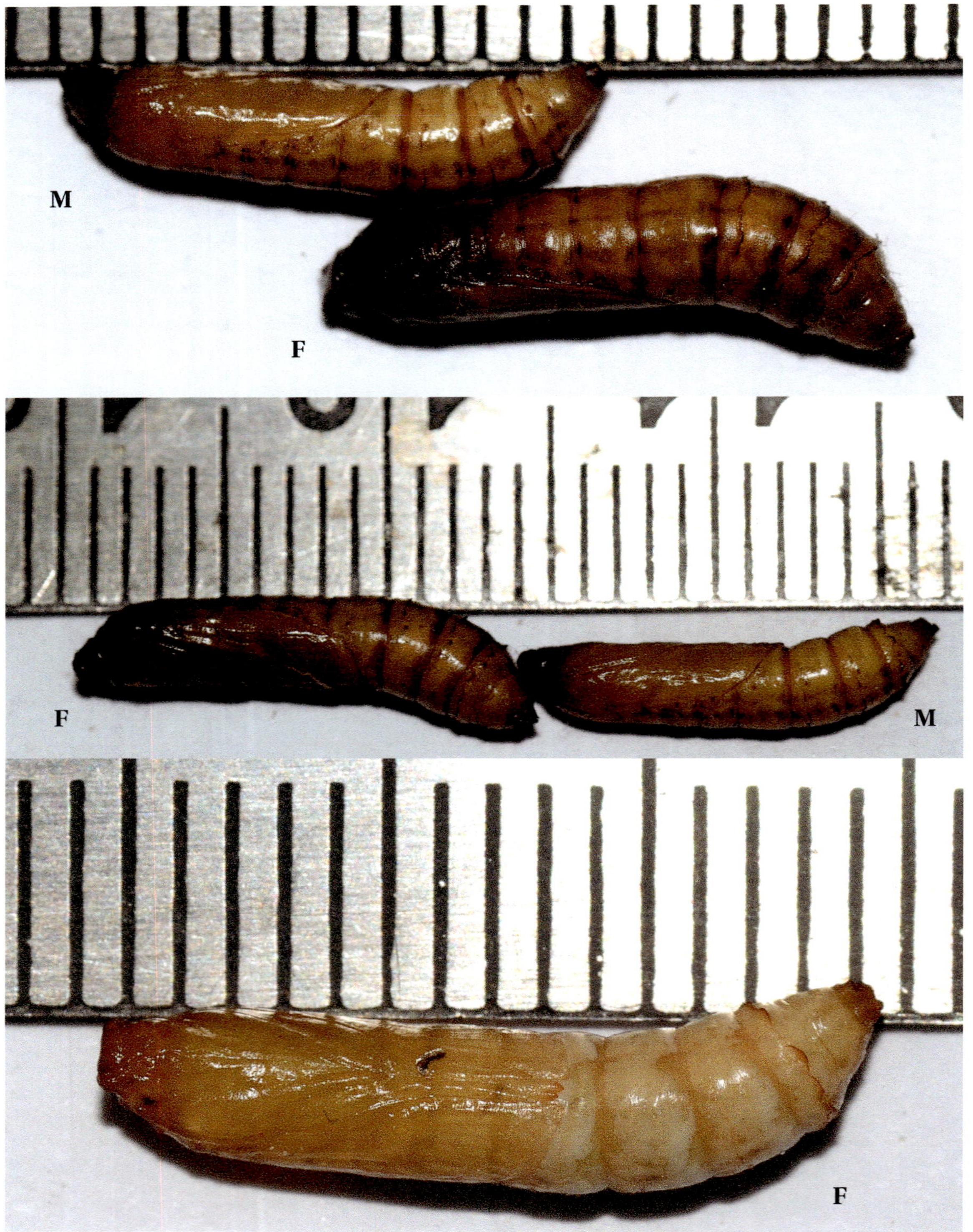

Plate 8.28. Pupae of Early shoot borer. Size of the male and female pupae. Fresh female pupa is around 18 mm and male pupa is around 13 mm in length.

Plate 8.29. Dorsal view of the pupa. Median ridge prominent on *pro* and *mesothorax*. A pair of raised area at the subdorsal portion of the *mesothorax*. Presence of a pair of spines on each abdominal segments. Prominent dorsal stripes.

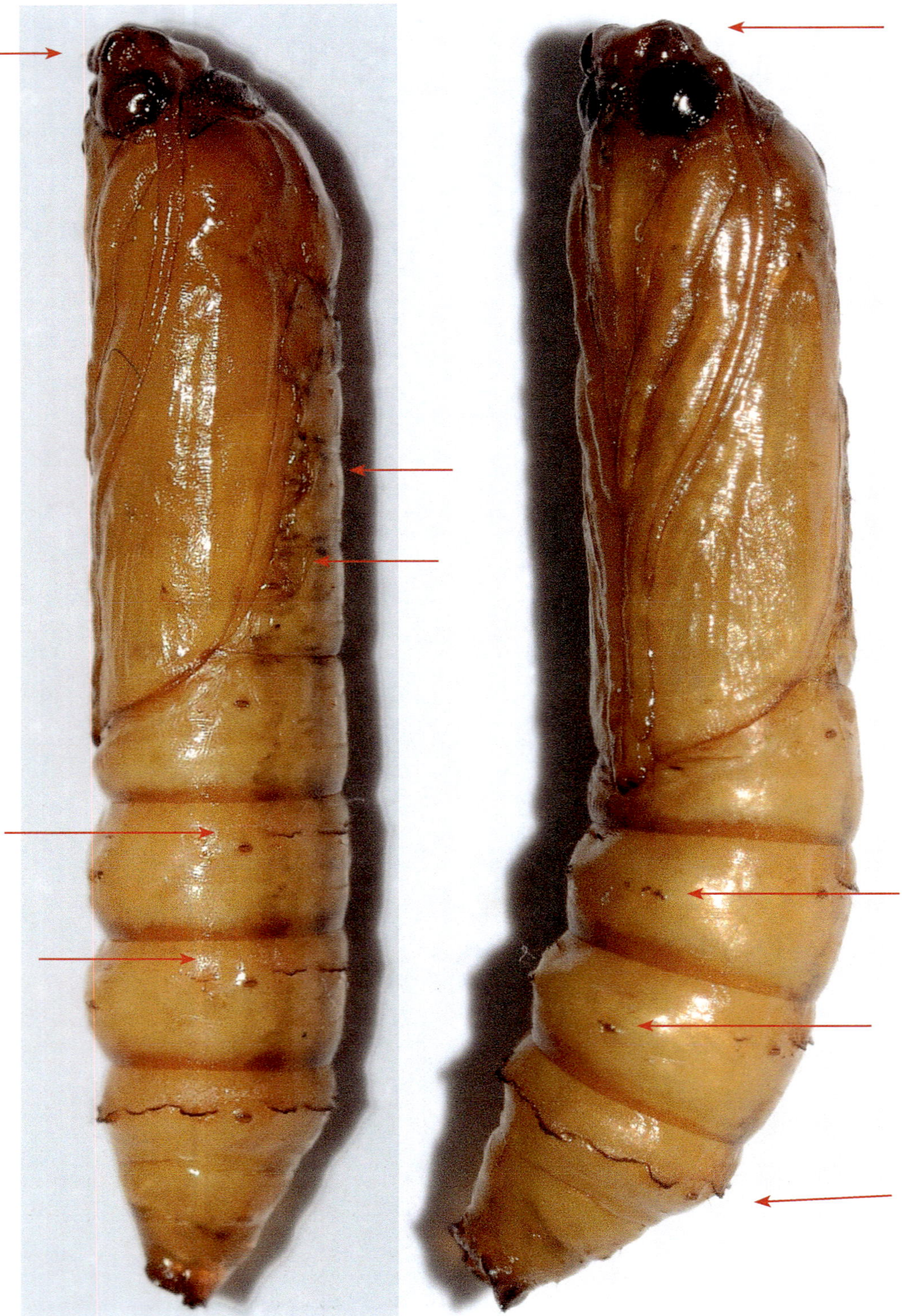

Plate 8.30. Side view of the pupa. Lateral side is smooth except on the 7th abdominal segment, which bears wavy ridges. Insignificant development of wavy ridges below the *spiracles* on 5th and 6th segments. Two prominent bulge on the head is visible from side view.

Plate 8.31. Ventral view of head and thoracic area. Head is rather smooth except the buldges. Unequal development of the third pair or hind of legs. Formation of a *happy face* at terminal portion of the wing pad area..

Plate 8.32. Ventral view of abdominal portion. Ventral side is smooth except for the marks of prolegs and presence of ornamentations (wavy ridges with dark terminal area) on the 7th abdominal segment. Ventral side bears prominent marks of prolegs. Unequal hind legs.

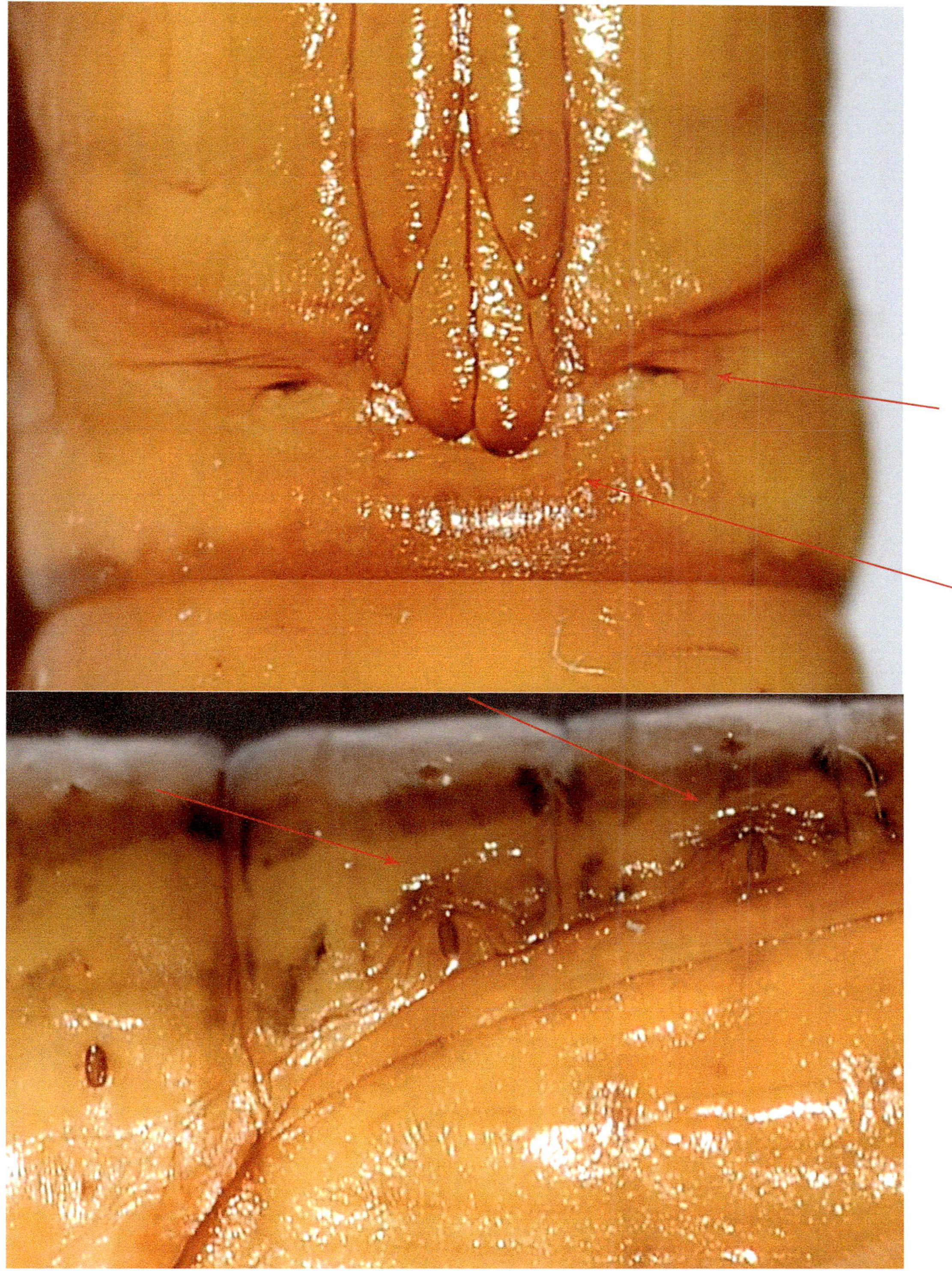

Plate 8.33. Part of ventral view (close up) showing *human face* formation due to visibility of proleg marks on 4th abdominal segment. Side view showing arch formation above the *spiracles* of 2nd and 3rd abdominal segments. Unequal hind legs.

Plate 8.34. Anal end of the pupa. Presence of all the seven stripes (including the ephemeral pair). Stripes run through the wavy ridges. Presence of prominent cut mark at the anal end, dorsal half is longer than the ventral half.

Plate 8.35. On the 5th abdominal segment one wavy ridge is present after the *spiracle* and two wavy ridges present on 6th abdominal segment. These segments become more visible with development of *pharate* adult.

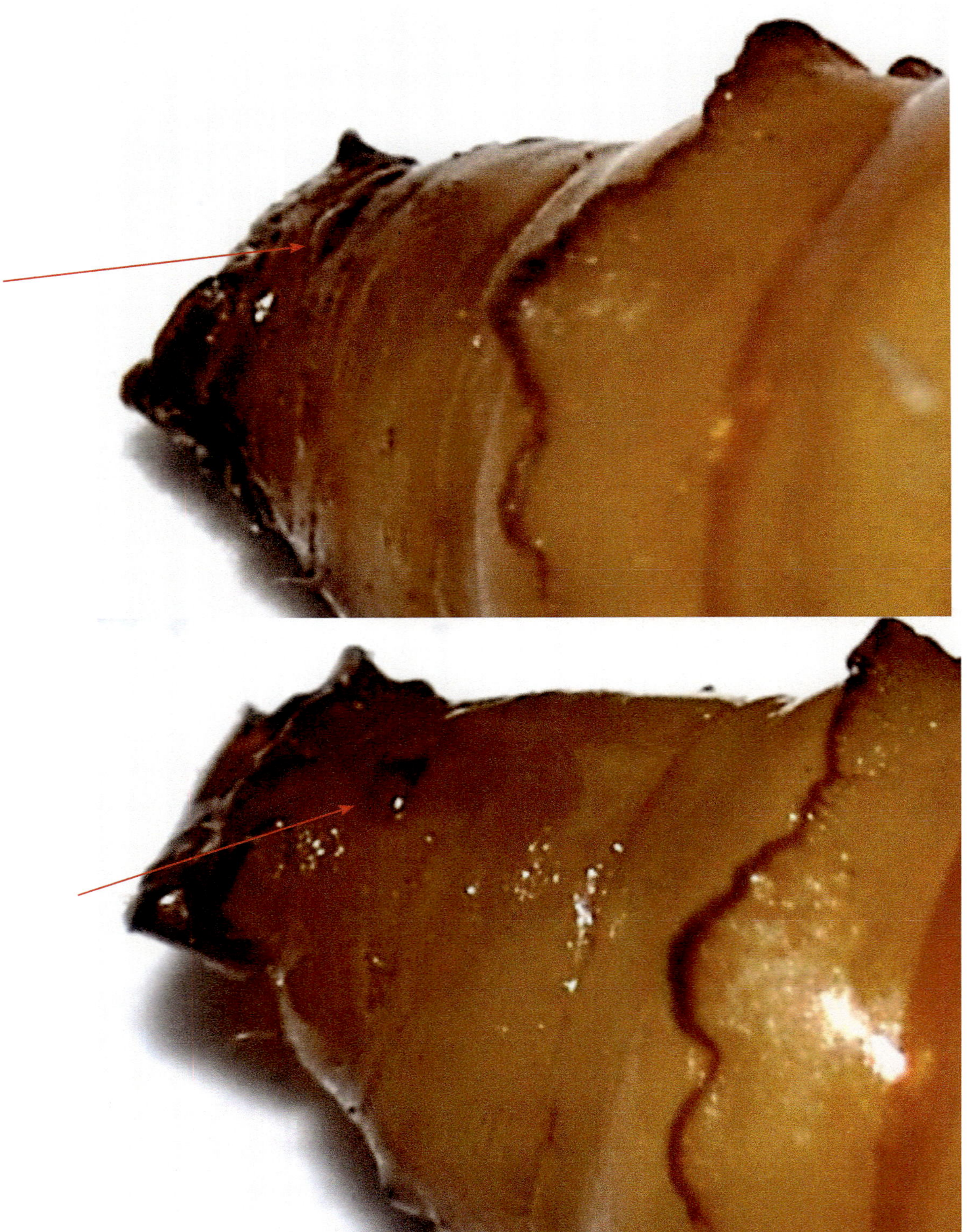

Plate 8.36. Anal end of male pupae showing genital area as two broad raised lobes with median depression on the 9th abdominal segment.

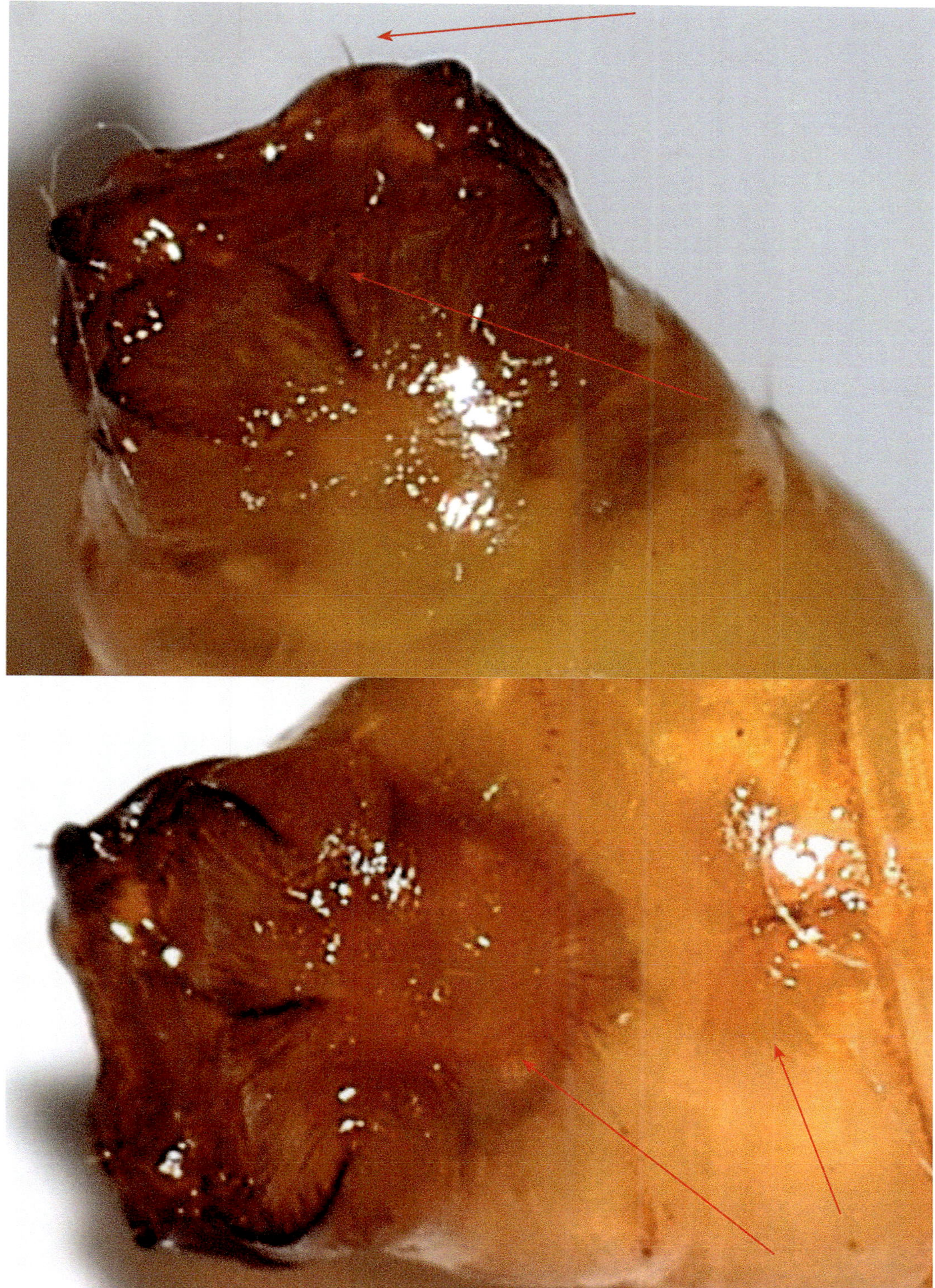

Plate 8.37. Anal end of female pupae showing the genital area as a short depression on the 8th abdominal segment. Development of *papillae anales* on the 9th abdominal segment. Anal marking is 'Y' shaped.

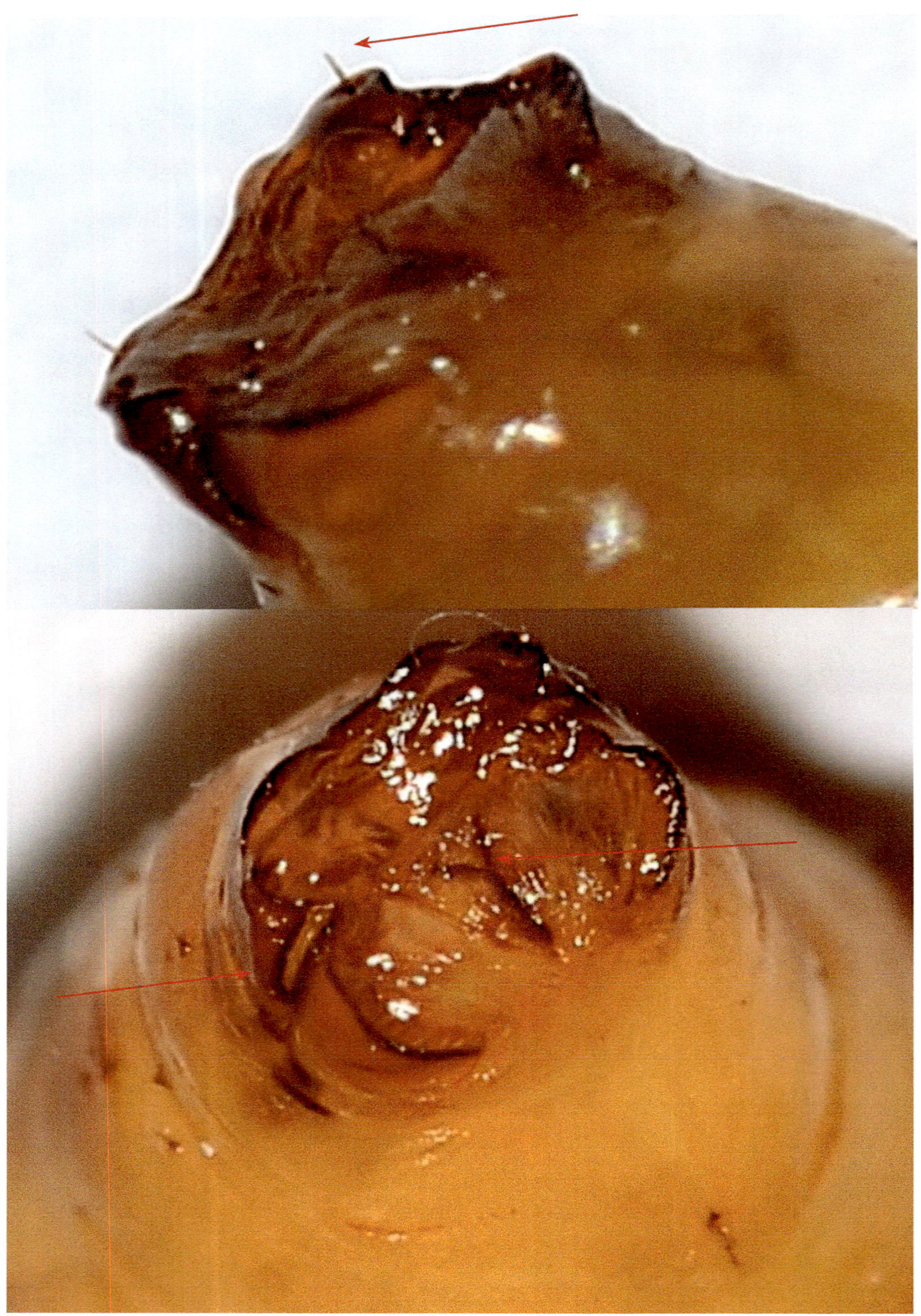

Plate 8.38. Anal end of female pupae showing a depression on the 8th segment, finer cut mark separating dorsal and ventral area. Raised development of dorsal projections with two spines on each half. Ventral half has no spine. Anal marking is 'Y' shaped.

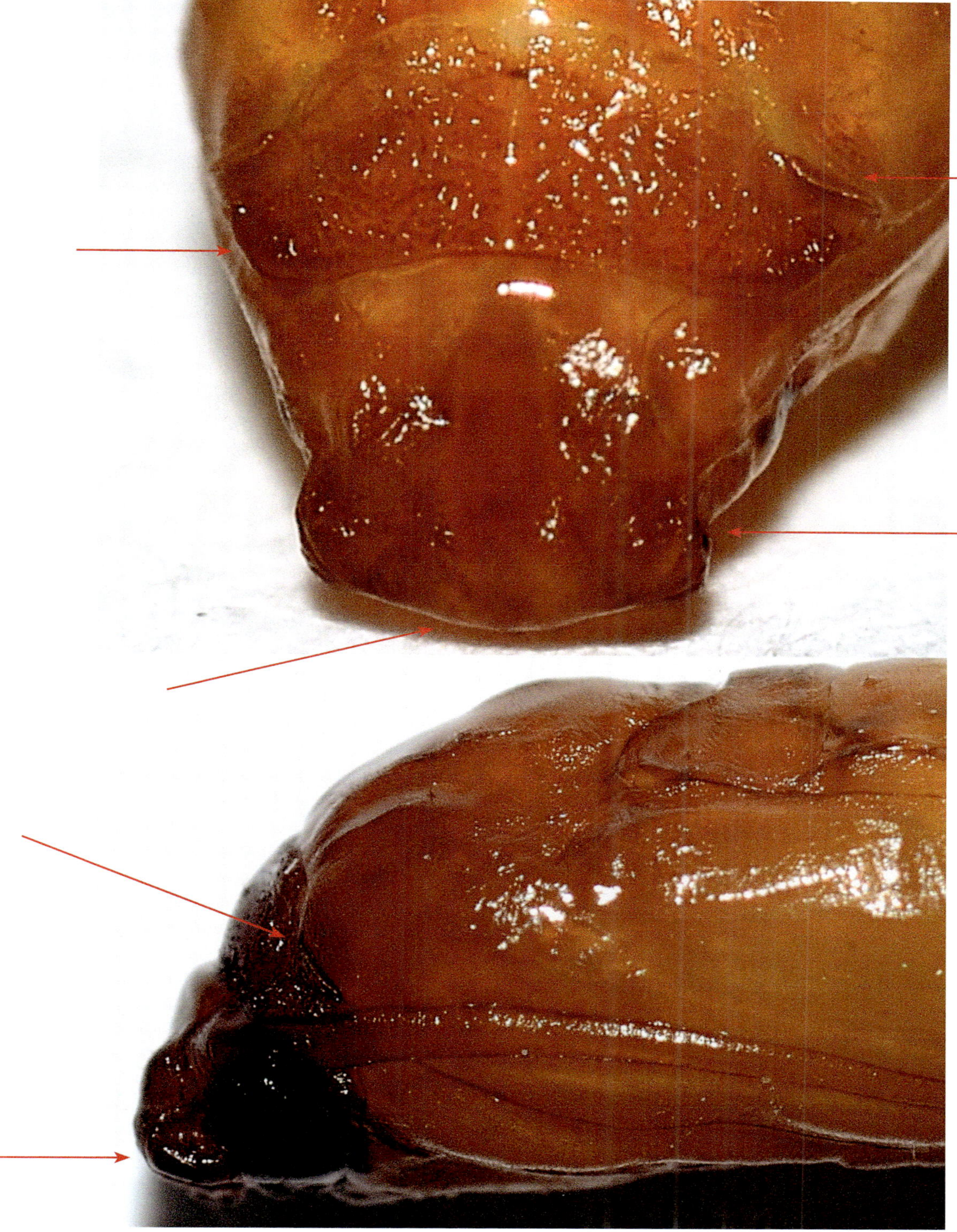

Plate 8.39. Head of the pupa. (Front and side views). Opening of *mesothoracic spiracle* directed towards anal end. Note the raised area on *mesothoracic plate*. As well as at the side of head.

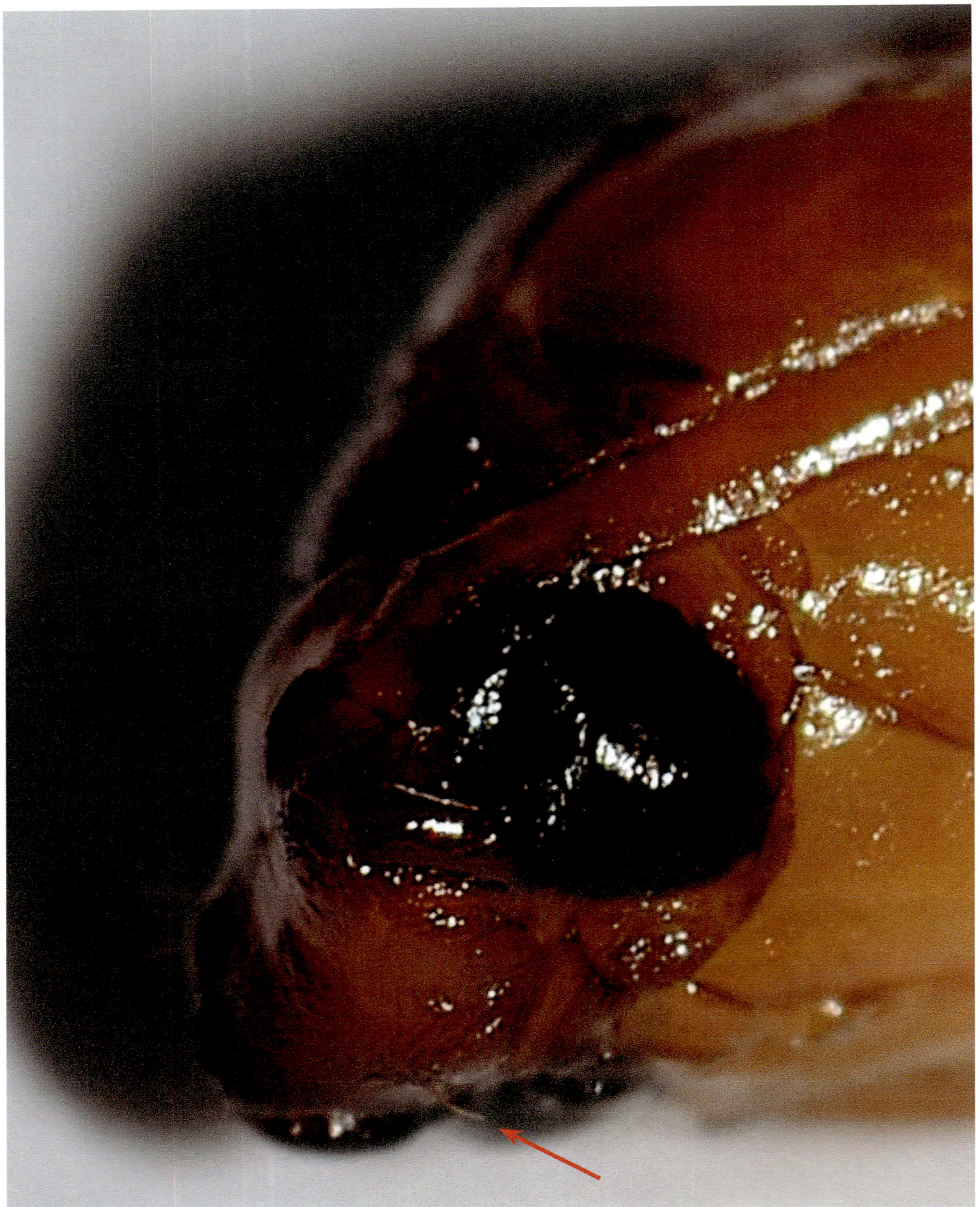

Plate 8.40. Face of the pupa is smooth with slight convex development bearing two curved spines, one at each half.

Plate 8.41. Dorsal view of female moth. Sitting habit and placement of antennae. Moths are distinctly yellowish-straw coloured. Prominent black terminal spots on the forewings with a few white scales preceding the spots. Discal spots are present. Numerous individual steely scales on the forewings are present.

Plate 8.42. Side view of A. Female moth B. Male moth. In addition to size, sexual dimorphism is apparent in labial palpus, antenna, colour of the forewings, distribution and intensity of metallic scales, terminal black spots and intensity of white spots.

Plate 8.43. Forewing of a female moth showing terminal black spots and very few white scales preceding the spots. Individual metallic scales (relatively sparse) are present on the straw coloured wing.

Plate 8.44. Forewing of the male moth showing terminal black spots and presence of more prominent white scales preceding the terminal black spots. Presence of individual metallic scales on the straw coloured wing. Intensity of these scales are more as compared to that of female moth.

Plate 8.45. Ventral view of A. Female moth B. Male moth. The abdomen of the moths are clothed with ochreous scales. Female abdomen appears more conical.

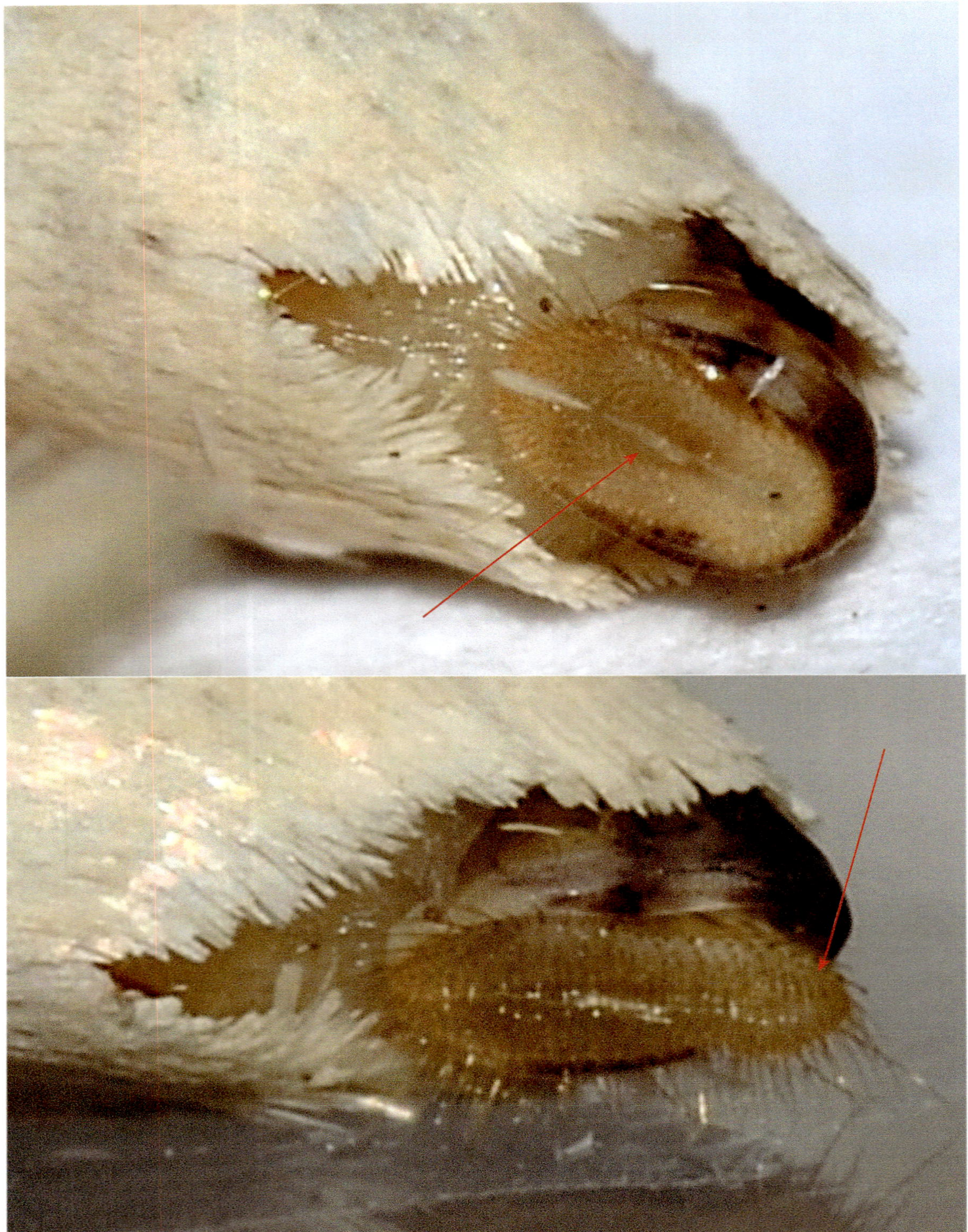

Plate 8.46. In female moth, *papillae anales* appear like two bathroom-slippers kept side by side (with median space) and papillae anales are laced with numerous sensory hairs.

Plate 8.47. Dorsal view of head of **A**. Male **B**. Female moth. Porrect labial palpi. Antennae are thicker, ribbed and smooth at the side in male. In female, antennae are *filiform*, ribbed and clothed with scales.

Plate 8.48. Dorsal view of head of a female moth. Labial palpi porrect. Face of the head ends in a broad corneous point.

Plate 8.49. Shape and colour of the compound eye. Prominent black *ocellus* is present behind an antenna.

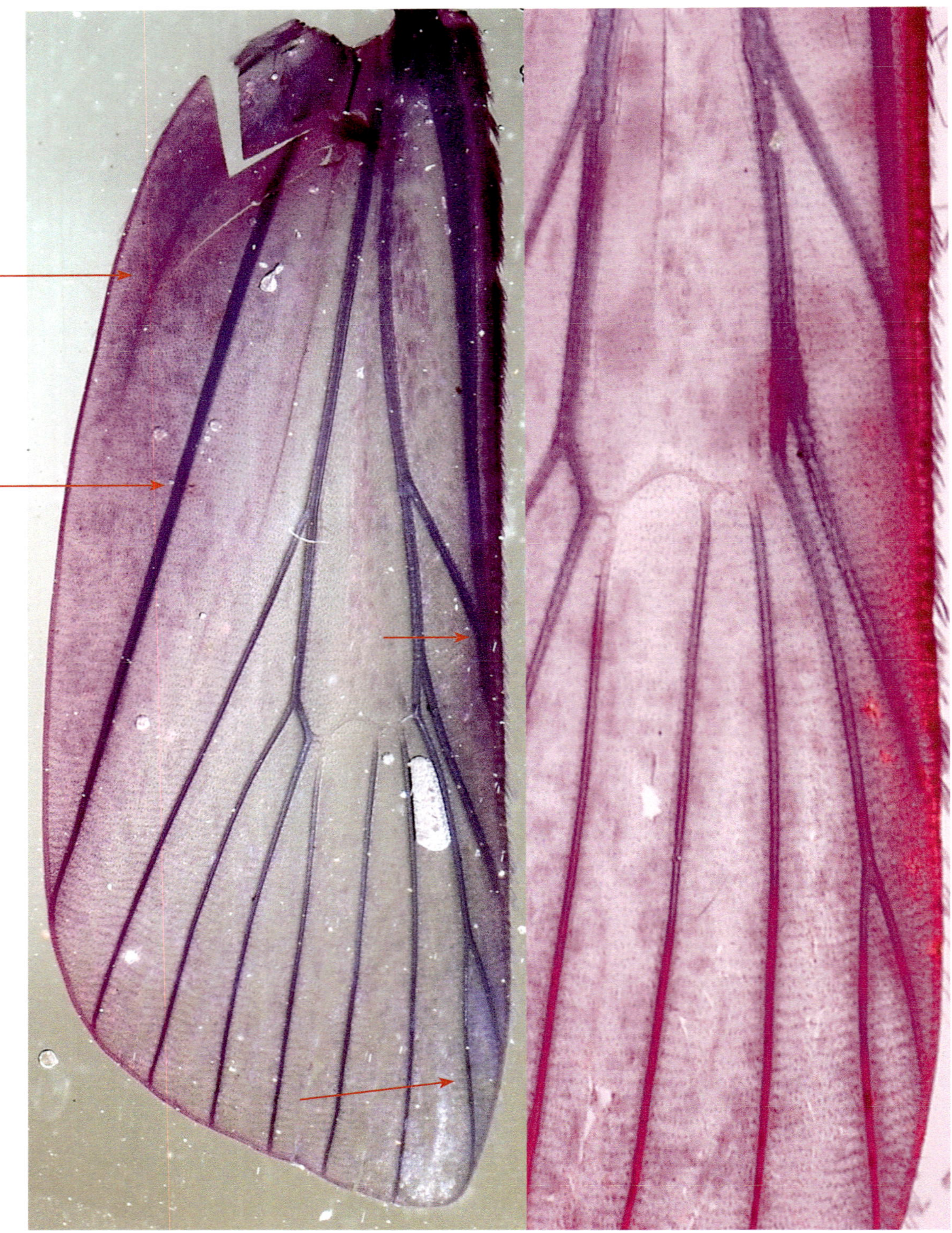

Plate 8.50. Forewing venation (female). Sc touches R_1 near the costal area. R_2 close to the stalk of R_3 and R_4(stalk curves towards R_2). R_3 is half of R_4. A_2 very prominent, A_3 rudimentary.

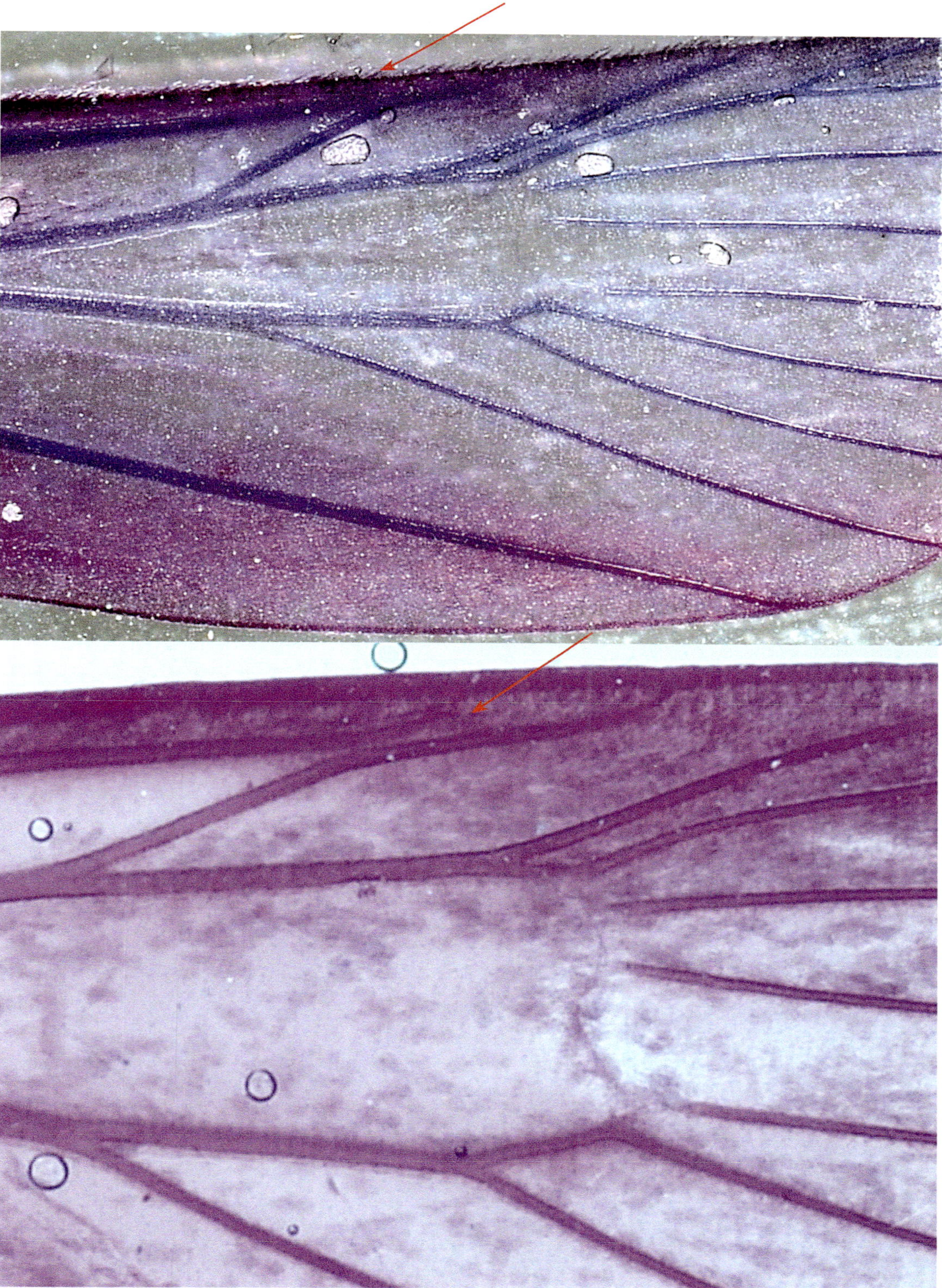

Plate 8.51. Wing venation in forewing. Sc and R_1 may remain free or Sc just touches R_1 at the terminal portion.

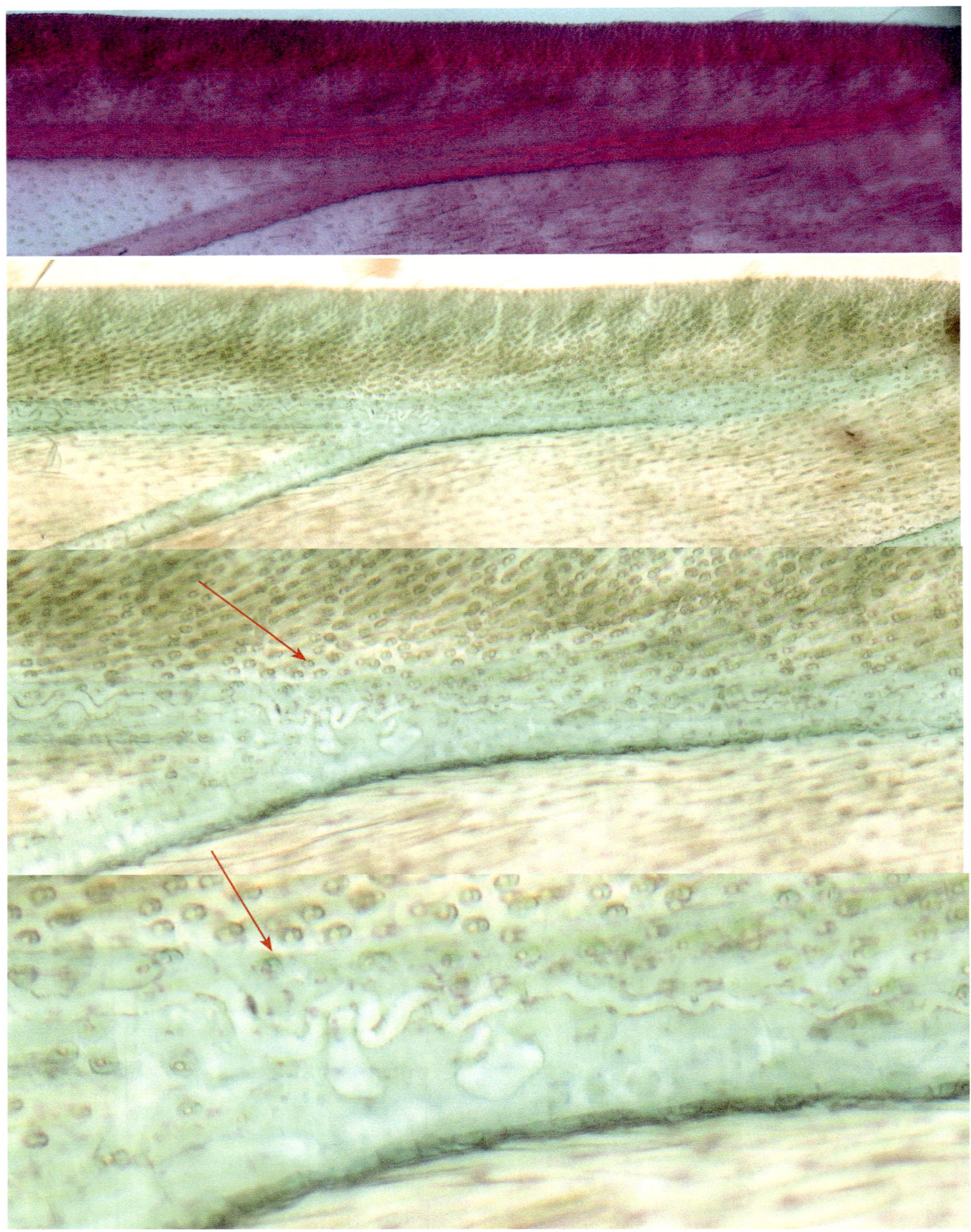

Plate 8.52. Variations in the positions of Sc and R_1. Passing of *tracheal tube* from Sc to R_1.

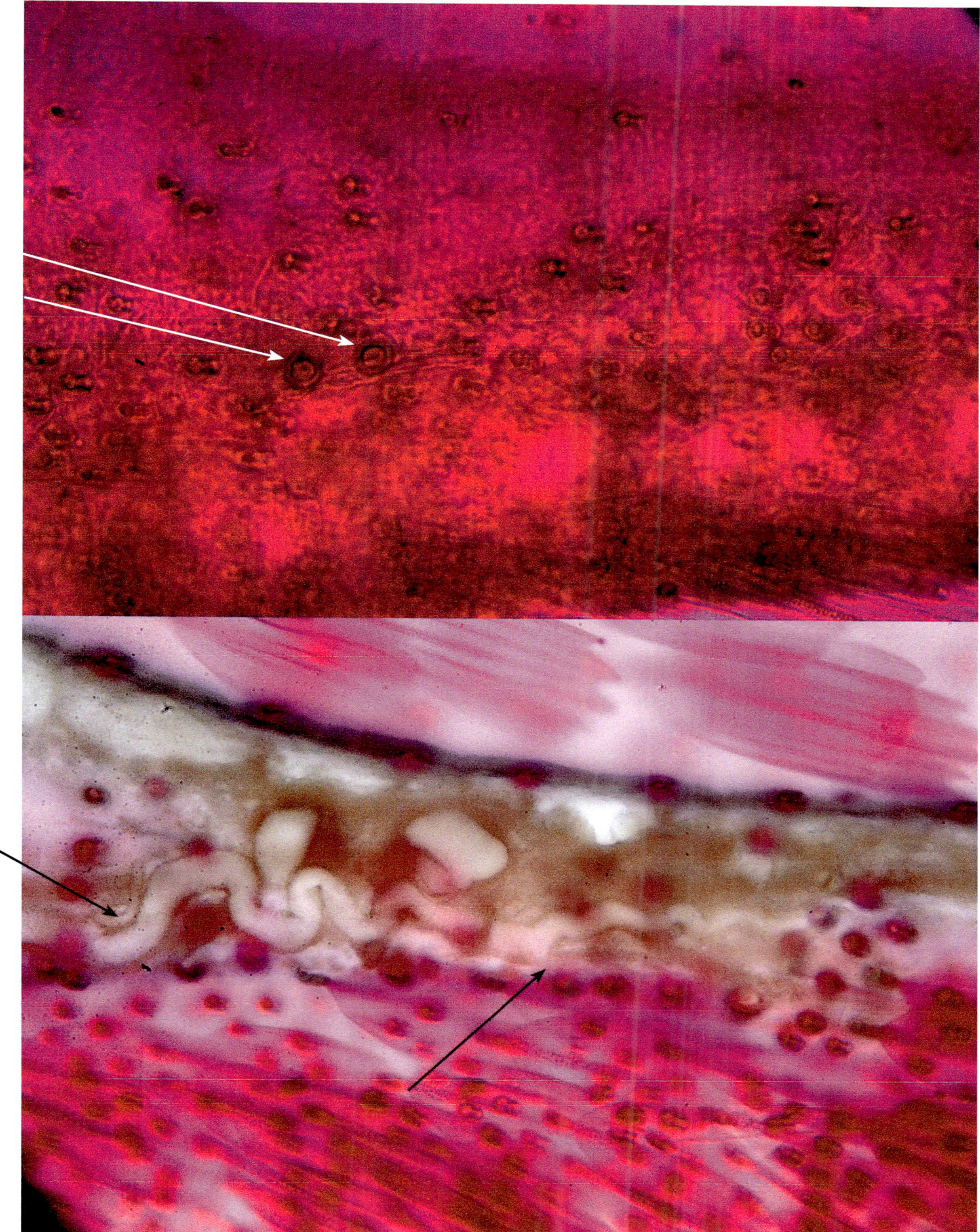

Plate 8.53. Variations in the positions of Sc and R_1. Sc touching R_1 at the end. Presence of *mechanoreceptors* indicate the termination of Sc. In the second case, Sc just touching R_1 and passing the *tracheal tube*; only a minor *tracheal branch* supports the terminal portion of Sc.

Plate 8.54 Egg laying by the female moth. Unfertilized egg masses are laid in an unorganized and scattered manner.

Plate 8.55. Close up of egg mass. Freshly laid eggs appear water-soaked white with ornamented shell. An egg appears as an up-turned saucer.

9

Stalk borer, *Chilo auricilius* Dudgeon

The Stalk borer, *Chilo auricilius* Dudgeon is a specialized borer of sugarcane and rice. In sugarcane, it extensively damages the mature cane stalk and for this reason it has got the common name *'Stalk borer'*. In rice, this borer is known as *'gold-fringed rice borer'* due to the presence of terminal scales with golden lustre on the wings of the moths. In fact, the specific epithet *auricilius* indicates this *gold-fringe* character. It is also known as '*tarai borer*" due its prevalence in the '*tarai*' region (moist as well as cold area in the subtropical belt).

Dudgeon (1905) named the moth as *Chilo auricilia* (Please refer ***Appendix II*** for original description). Fletcher and Ghosh (1919) misidentified this borer as *Diatraea* sp. (C.S. 1674 on page 389 and as The Rice *Chilo*, C.S. 1677 on page 390, Please refer ***Appendix IV*** for the details) and provided excellent drawings of the moth, larva and pupa. Later, Fletcher (1928) after comparing the moths of this species with the insect collection kept at the British Museum (Natural History), London, identified the moth of Stalk borer as *Diatraea auricilia* (Dudgeon). Kapur (1950) erected the genus *Chilotraea* and placed the Stalk borer as *Chilotraea auricilia* (Dudgeon) Kapur. However, while revising the world species of *Chilo*, Bleszynski (1970) did not agree to this further division of genus *Chilo*, into *Chilo* and *Chilotraea* by Kapur(1950) and retained this moth in the genus *Chilo* and named it *Chilo auricilius* Dudgeon.

It is an important moth borer of sugarcane in subtropical India especially in the *tarai* sugarcane belt. The larva of this moth borer remains active and continues to damage the cane even during the cold winter months (except very cold days). In contrast, during the winter months larvae of other moth borers of sugarcane proceed for hibernation (mostly remains inactive). Stalk borer rather faces a lot of hardship during the high temperature period. Its activity and population greatly diminishes during the summer months. Due its ability to tolerate cold weather, it predominates in rice crop grown in the Himalayan foothills (*tarai* belt) of Assam, Darjeeling, Himachal Pradesh, Uttrakhand, Uttar Pradesh and in Jammu and Kashmir. As per the earlier reports of its incidence, it appeared that the borer had spread to the entire subtropical belt starting from the localized area of Bihar and Bengal.

9.1 Currently accepted name

Chilo auricilius Dudgeon

9.2 Synonyms

= *Diatraea auricilia* (Dudgeon) Fletcher

= *Chilotraea auricilia* (Dudgeon) Kapur

9.3 Symptoms of infestation and brief biology

It is a specialized borer of mature stalk tissues of sugarcane and discernible symptoms develop after the formation of stalk. However, the borer can attack in any stage of the crop and depending on the stage, symptoms of infestation varies. In early stage of crop growth (shoot stage) infestation of this borer also produces *dead heart* that is quite similar to that of Early shoot borer (ESB) but without any offensive foul smell. In the later stages of the crop growth when it attacks the stalk, the damage is not easily discernible like other borers. It is only after the removal of leaf sheath of the affected cane, the damage becomes evident.

Among the sugarcane moth borers, Stalk borer larva remains active during the winter months except in very cold winter days. In this moth borer, pupation starts in the latter half of January and continues in February when most of the other moth borers remain in hibernation. Occasionally in this period, a rush in pupation takes place whenever there is a rise in the ambient temperature. The moth emergence begins by the end of January and continues through February-March. At this period, female moths mostly lay their eggs on the ratoon sprouts or in an autumn-planted crop, if available. The stalk borer larvae usually feed in the leaf sheath and occasionally venture into the central portion causing the formation of *dead hearts*. In this period, incidence of Stalk borer is generally mild (Green borer and ESB dominate in the shoot stage) and most of *dead hearts* are often counted as that of Early shoot borer attack. In fact, the Stalk borer is a cold loving species. Larvae of Stalk borer pass the summer months rather in hardship, and its population dwindles during this period. Larvae of Stalk borer are not comfortable with the high summer temperature and hardly stays in the above ground portion of the cane and usually it goes down in the stalk at the clump base (below the soil level).

Dead heart formation diminishes considerably with formation of stalk as the larva preferably goes deep inside to feed the relatively developed thicker stalk portion of the cane (mostly July onwards when ambient temperature favours its proliferation). When the cane girth is able to sustain the borer damage, hardly any *dead heart* is formed. The Stalk borer, as the name implies, is primarily a borer damaging the cane stalk. In contrast to other shoot borers, the activity of Stalk borer increases considerably during the monsoon months and overall crop damage is accentuated. For this reason, it is also grouped as '*Monsoon borers*'. During the fag end of monsoon period, Stalk borer larvae infest the water shoots and due to the infestation, a curious symptom, due to high sheath mining activity of the initial instars, appears on the leaf. A characteristic yellow-orange streak formation takes place on the lamina. One or two longitudinal orange yellow streaks running close to the mid-rib become visible and the streaks often extend from base to the tip of the lamina.

In subtropical India, with the onset of monsoon (in July) a sugarcane crop switches to grand growth phase and this phase is marked by the vertical growth of the crop due to rapid cane elongation. Stalk borer gets the cue and becomes more active and increasing number of canes in the infested field fall prey to the attack. However, the infestation of Stalk borer is not easily discernible to the normal gaze, as affected cane plants do not display any symptoms except the feeding marks of the larvae of first two instars on the leaves and leaf sheaths. In the backdrop of the vigorous cane growth, the larva surreptitiously damages the internal tissues of the cane. A heavy infestation of the crop is often noticed from August onwards. Many a times a higher incidence of Stalk borer has been recorded in a situation where lodging of cane took place during the monsoon months.

The symptom of infestation is quite deceptive, as the severity of infestation cannot be judged by observing the crop from a distance. The damage of the cane stalk becomes visible only after the removal of the leaf sheath. In the infested cane, one may find small entry hole near the bud with occasional red streaking or an elongated patch of discoloured or red area on the epidermal layer of the internode. Usually one or two holes are found near the nodal region. In certain situations, several holes may be found at different locations of an internode but these holes mostly remain indistinct except for the red streaking. In general, the affected plants show little distress on the crown leaves. However, when the infestation is heavy, the crown leaves may turn pale yellow to yellow-orange; but the affected plants hardly die. On splitting open the infested cane, reddening of internodal stalk tissues and tunnel filled with frass along with borer larva or pupa or exuviae become evident. By feeding and tunnelling the internal stalk tissues, a larva may cover several internodes of the stalk. In a pithy stalk such a multiple nodal damage is common. The feeding tunnels mostly remain filled with the frass. Due to the damage of considerable amount of internal stalk tissues, this borer is notorious for affecting both extraction of juice and its quality. The cane yield also gets affected but to a lesser degree.

Once a larva enters the stalk and establishes itself well, it usually does not come out (like the Internode borer). However, after the entry if a larva does not feel comfortable with stalk tissue, it leaves that place and moves to a new location. In such situations, little damage at the nodal region and near the bud may be observed. A Stalk borer larva, after successfully establishing itself within the stalk tissues, makes an internal feeding tunnel and pushes the frass behind. The feeding tunnel extends to several internodes, as a larva does not get restricted with the hard nodal tissues; rather it makes a tunnel through it to reach the next softer internode for feeding. In heavy infestations, the entire length of the cane stalk may be found bored through by several larvae. Due to infestations of a number of larvae in one internode, several finer holes may also be observed at different locations of the internode. In susceptible genotypes, about a dozen caterpillars may be encountered in a single cane but each larva forms its own feeding tunnel. In general, usually one to two larvae may be found in the infested cane. A larva does not push a great quantity of frass out of the feeding tunnel like the Internode borer or Plassey borer, but it pushes out a little frass to block the entry hole. Most of the frass remain inside the feeding tunnel. Before going to pupate, a prepupal larva prepares a large exit hole, which mostly remains covered with the flimsy epidermis.

An autumn-planted crop suffers more from the ravages of Stalk borer than the spring planted one. This may be attributed to better growth of the crop and early cane formation that provides preferable food to the borer larvae. In some areas of Bihar and West Bengal, the sugarcane is planted in July or August. In such cases, the crop puts up sufficient growth of the stalk by April-May, the following year. The developed stalk serves as a good shelter for the Stalk borer to tide over the critical high temperature period. As reported earlier, a heavy incidence of the pest was observed during April 1962, in a crop planted in July 1961 at the Sugarcane Research Station, Pusa (Bihar), whereas the spring planted crop in an adjoining field was practically free of the pest attack (Kalra, 1978).

9.4 Egg and neonate larva

It is very difficult to locate eggs of Stalk borer in the field even by an expert. However, occasionally an egg mass may be located in the field by carefully observing basal area of the ratoon sprouts during the rush egg laying period of February-March. A female moth hides near the base of the water shoots or ratoon sprouts, especially during February-March and lays the eggs in an egg mass. Similarly, moths also lay the eggs at the upper side of the mid rib of lower leaves, preferably near the junction of leaf lamina and leaf sheath. In captivity, female moths lay eggs in a short cluster of several imbricate rows on the underside of leaves aligning the mid-rib. Rows of eggs in an egg mass may be three, four or five and thus, an egg mass appears as an irregular shaped sheet. Freshly laid eggs are upturned saucer shaped, translucent white or water-soaked white in colour.

Larval development in the translucent eggshell is visible and development of black head appears very distinct along with the parallel stripes running through the body. The developing larva is colourless or water-soaked white. Orientation of the black head does not follow a regular pattern; it is not regular like that of Internode borer (in Internode borer developing heads of all the larva in an egg mass remain directed to the first egg). The larva emerges out from the egg by making a small slit on eggshell.

A majority of the neonate larvae uses the area of auricles/ligules to make an entry into the leaf sheath or they sneak inside the space between stalk and leaf sheath. Larva also makes a direct entry in the leaf sheath just making a fine hole on the outer epidermis. Within the inner epidermis of leaf sheath, presence of several smaller holes indicates the infestation of Stalk borer larvae. Within the leaf sheath cavity, Stalk borer larvae usually goes down to the base of the leaf sheath and feed on the internal content of the leaf sheath tissues (in contrast, internode larvae prefer to feed in the upper portion of the leaf sheath). Similarly, a few larvae may reach the central whorl from the top and feed on the immature rolled leaves. At this period, a majority of the neonate larvae perishes.

A neonate larva measures around 1.5 mm, appears dull white (water soaked white) with prominent black head, well-developed dark prothoracic shield and prominent body hairs. It bears faint stripes on the body. With maturity, five stripes on the body become prominent and the second instar larva bears all the prominent five stripes from the very beginning (even during moulting to second instar). Larvae of second instar measure around 5 mm and has very prominent white *tracheal tube* bearing dark *spiracles*. Only the tubercles become discernible due to the presence of hair (in contrast, the larvae of Internode borer develop very prominent tubercles, orange bands though become much lighter but still persist in the second instar larvae).

In the beginning of the crop season, majority of the larvae of Stalk borer pass the first two instars at the base of the ratoon sprouts as leaf sheath miner. Occasionally, neonate larvae may also reach the inner core and feed the immature stalk tissues. In such situations, several larvae of first or second instar may share the same feeding tunnel. Generally, a neonate larva wanders on the leaf sheath until it finds a suitable site to feed. Occasionally, it may also reach the rolled crown leaves, where it can conveniently hide itself and feed on the immature leaf tissues. However, at the crown portion of the cane shoot, Top borer is very fast in capturing the food area and thus, in most of infestation, Top borer always enjoys an upper hand. However, an immediate attack by the other borers at the crown portion of the shoot, in most cases seals the fate of Top borer and the larva dies of starvation due to food crunch. In the leaf sheath tissue, Stalk borer larva feeds like a leaf sheath miner; feeds the internal contents leaving both the epidermis undamaged, except for occasional entry and exit holes. However, due to internal feeding of sheath tissues, the damaged area develops some dark colouration. Presence of frass within the small outer leaf sheaths is very characteristic symptom of infestation of first and second instar larvae of Stalk borer.

At the shoot stage, neonate larvae of Stalk borer usually feed on the leaf sheath by mining into it. About a dozen young larvae may be found in a single leaf sheath (*pseudo gregarious* nature).

When it feeds on the rolled lamina inside the spindle leaves, white feeding streak, a number of feeding holes along with the frass are found on the lamina. A neonate larva feeds on the leaf tissues leaving one epidermis intact (quite similar to that of Internode borer but the feeding area always remain smaller or restricted) with occasional formation of holes in the white feeding area. Similarly, when the neonate larvae feed on the leaf sheath, they go inside the leaf sheath tissue and feed like a sheath miner, feeding the internal tissues between the two epidermal layers and also leaving a lot of frass in the tissue. The larva feeds on the leaf tissues for more than a week or so, and after mustering the requisite strength (it passes two larval instars at the feeding place, the leaf sheath) usually a third instar larva go further inside to reach developing stalk tissues near the crown. It enters into an immature internode, initially feeding on the surface tissues of immature soft stalk and finally reaches the central tissues. At this stage, usually one larva is found in the cane shoot; sometimes two to three larvae may be found sharing the same feeding hole of the stalk. Due to this feeding of the stalk tissues, Stalk borer also produces *dead heart* at shoot stage.

9.5 Larva

The larvae of *Chilo auricilius* are both agile and fragile; reddish patches often develop in the larval body. The larva is very active and moves away quickly on extrication from the cane stalk. A mature Stalk borer larva is slender, creamy yellowish in colour (colour of *ghee*), bears five prominent wavy light-pink or pinkish-violet stripes on the body. A full-grown larva measures around 20-25 mm. In the lateral view, *tracheal trunk* is clearly visible as a white shiny band with prominent black rimmed, elongated oval *spiracles* on it. The higher visibility of *tracheal trunk* helps its quick separation from the larvae of ESB. The *spiracles* are open type i.e. central portion appears white with space inside. The placement of the *spiracle* pair on the 8th abdominal segment is more subdorsal than lateral.

Five prominent wavy pinkish violet stripes are present on the larval body i.e., one median, two subdorsal and two laterals. These stripes are relatively lighter in colour in comparison to that of Early shoot borer. The median stripe partially runs through the two large dorsal tubercles (D_1 and D_1) due their close proximity in each segment. In a mature larva, the median stripe terminates at the two conjoined tubercles on the 8th abdominal segment and the stripe characteristically does not run beyond that segment (in Early shoot borer the median stripe goes beyond the conjoined tubercles of the 8th segment and continues in the 9th abdominal segment).

The two sub-dorsal stripes are prominent and just touch the dorsal large tubercles (D_1) from the lateral side and run through the smaller dorsal tubercles (D_2). In case of smaller tubercles, the lateral stripe covers only half of the tubercles, mostly up to the *seta* of the tubercle (in Early shoot borer, the corresponding lateral stripe almost fully covers the smaller dorsal tubercles). These two subdorsal stripes go beyond the 8th abdominal segment just circumventing the conjoined tubercles of the segment and the *spiracle* pair. In the 9th abdominal segment these stripes often spread out and cover a large area of the segment.

The lateral stripes are relatively thinner and present above the *tracheal trunk*. Occasionally, these stripes touch the boundary of the *tracheal trunks*. As the Stalk borer larva does not go for hibernation in winter months, the body stripes also do not fade away. The median stripe and the two subdorsal stripes always remain prominent. Occasionally, the lateral stripes become lighter in colour and nearly fade away.

As indicated earlier, the Stalk borer larva, like the larva of Early shoot borer actually bears seven stripes on its body. The last pair of lateral stripe or the ephemeral pair appears well below the *tracheal trunk*. The colour of ephemeral stripe pair is much lighter than the other prominent stripes. This stripe partly remains visible only in the limited surrounding areas around the tubercles present below the *tracheal trunk*. The presence of this pair becomes apparent when the larva is viewed from the anal end.

Head of a mature larva is light to dark brown in colour, and bears prominent *ocelli* at the lateral side. *Ocelli* are arranged in a semicircular fashion. It has a sharp white-lined inverted 'Y' shaped *epicranial suture*. *Prothoracic* shield is well-developed, light yellowish-brown in colour with light coloured median depression. Terminal area of the *prothoracic* shield is dark brown. *Mesothoracic* plate is not very prominent as it takes the body colour. *Metathoracic* plate is hardly noticeable.

Tubercles are quite similar to the body colour (translucent). The two pairs of lateral tubercles present on the *meso* and *metathoracic* segments are quite larger and give the thorax a distinctive appearance. (The corresponding tubercles are smaller in Early shoot borer). The dorsal tubercles (D_1) are square shaped, relatively larger and two large tubercles (D_1 and D_1) remain at close proximity in each segment. In the 9th abdominal segment only the two dorsal tubercles are conjoined, the other two subdorsal tubercles retain their individual identity.

The crochets or hooks on the four abdominal prolegs are arranged in a complete circle. Occasionally one or two hooks appear missing from the lateral side. The hooks are *triordinal* and follow *uniserial* arrangement. Hooks are light brown in colour, *triordinal* towards inner side (*meson*) and practically *uniordinal* at the lateral side or outer side. Hooks are larger towards the *meson* and their size gradually diminishes towards the outer side (lateral side).

9.6 Pupa

Stalk borer pupa is rather thin and slender looking, as compared to the pupae of other moth borers of sugarcane. Like the pupa of ESB, a pupa of Stalk borer also shows the capability of movement with the help of spinal processes present on the body but movement is rather slow. A freshly formed pupa is yellowish-white or creamy yellow in colour and with the passage of time colour of the pupa changes to light brown. If a pupa is present in the dark rotten stalk tissues, it also becomes very dark coloured. The pattern of wing venation of *pharate adult* on the pupal body is prominent and draws immediate attention to the wing pad area. Pupal body also retains all the five prominent stripes as observed in a larva. The sub-tracheal pair or the ephemeral pair of stripes is occasionally visible. The three dorsal stripes remain very prominent. The median stripe runs up to the 8th abdominal segment and the two subdorsal stripes go down to the 9th abdominal segment. The pupal ornamentations on the 5th, 6th and 7th abdominal segments are in the form of small but distinct blunt dark headed spines. On the 5th and 6th segments, spines mostly appear in a single clear row (followed by microscopic spines), whereas, in the 7th abdominal segment, multiple rows of spines are present (followed by microscopic spines). The ventral side is smooth throughout except for the marks of *prolegs* on 5th and 6th segments. In the 7th segment, the development of spines extends beyond the *spiracles* (on both the sides) and reach up to the imaginary place joining with straight the line markings of *prolegs* on the 5th and 6th segments. These spiny ornamentations do not completely encircle the 7th segment. The central abdominal portion always remains smooth.

Head of the pupa of *Chilo auricilius* is quite distinctive among the moth borers of sugarcane and it bears two characteristic horn-like projections pointing forward. This characteristic feature helps in its quick identification. In addition, just near the eyes, two terminal ends of *prothorax* protrude and roll over slightly in the form of rolled/curved structure. These structures protect the *mesothoracic spiracles* of the pupa. Thus, there are four prominently raised structures on the head of the pupa. The face of the pupa is convex and two setae are present at the convex area.

The ornamentation on the anal end (*cremaster*) is quite distinctive The projections appear like the petals of Dahlia flower or clove (entire petal area is broad, suddenly terminates at the peak). There are four pairs of projections at the anal end (three pairs dorsal and one pair ventral). The dorsal and ventral portions of the *cremaster* are well demarcated with prominent separation marks at both sides. The ventral pair is stouter; near the separation mark, another minor projection is also present. The anal markings are prominent. In male pupa, the genital area appears as a small, depressed furrow between two broadly raised lobes on the 9th abdominal segment. In female pupa, a long triangular area develops from the anal area up to the base of the 8th segment and at the peak of the triangular area, a prominent dark coloured depression marks the genital area. The development of *papillae anales* may also be observed in the mature female pupa (*pharate adult*).

9.7 Moth (Adult)

The female moths are dark ashy-straw coloured with smaller terminal black dots without any white scales. The *labial palpus* is porrect, appears like the tail of a squirrel and is around three to four times of the diameter of the compound eye. The face is broadly round and there is no corneous point or the point remains indistinct. *Ocellus* is present and distinct but not very prominent like that of Early shoot borer. Sexual dimorphism is evident in size, colour and antenna of the moths. Female moths are larger with one or two shades lighter coloured scales on the forewings than the male moths. Like other moth borers, the antennae are *filiform* and ribbed in females. In the males, antennae are stout, thicker, ribbed (serrated) and dorsally clothed with scales. Laterally, antennae are devoid of scales, appear smooth and light brown in colour. The *flagellum* of male antenna has about 51 segments like that of Early shoot borer but the antenna of Stalk borer are relatively thinner.

Forewings are dark straw coloured and laden with brownish or dark scales. The forewings of Stalk borer always appear darker in comparison to the forewings of Early shoot borer. In the forewings, development of collective and raised metallic spot (scales) at the *discal cell* area is very distinctive of this species. The metallic scales are more prominently raised on the forewings of male moths. In the forewings a number of terminal black spots (characteristically white scales are absent) in the furrow of the veins are present. The terminal black spots also extend even to the ventral surface of the forewing. Black spot may also be noticed on the terminal portion of the hind wings. On the forewings (mostly in female moths), a distinctive area or band or line appears and this band takes the shape of a flat-bottomed 'U'. When this is viewed on both the forewings by keeping them side by side, it forms a shape of 'W'. The terminal scales of the forewings impart a golden appearance, at the hind periphery of the moth. The corrugation pattern of veins on the forewing is more prominent. The forewing appears as triangular with broader margin area. The presence of solid terminal black spots along with the distinctive raised metallic scales on the forewings helps its identification and quick separation from similar moths of other species.

Papillae anales or ovipositor is densely clothed with sensory setae. The two halves of the *Papillae anales* are a bit elongated, bathroom slipper like with prominent space in between.

9.8 Forewing venation

The forewings of Stalk borer appear relatively shorter and much wider at the *termen* (typical triangular) in comparison to the wings of other moth borers of sugarcane. In the forewing, R_1 (*radius* one) is fully anastomosed with Sc (*subcosta*) to from a combined vein Sc + R_1. On way to *costal area*, the combined vein travels some distance but just before reaching *costal area* separates (small forking) to form Sc and R_1 again. Individually Sc and R_1 end at the *costal area*. However, the forking is not always very prominent, the presence of two round structures (*mechanoreceptors*) along with the terminal setae indicates the termination of the vein (Sc). The Sc and the Rs (*radius*) run parallel to each other for most of the distance (length). R_2 is free and terminates at *costal area*. R_3 and R_4 are stalked and the stalk is closer to R_2. Both R_3 and R_4 end at the *costal area*. R_3 is smaller (about half of R_4) in length than R_4. R_5 is free and ends in *termen*. Three *median* veins viz., M_1, M_2 and M_3 are present. These veins are well developed and equally spaced. Two *cubitus veins*, Cu1a and Cu1b are present. Among the three positions of *anal veins*, A_1 is absent, A_2 is complete and well developed and A_3 is short and partially developed.

The first peak (*radial peak*) of *discal cell* is rather flat and the bridge forms nearly a right angle with the *radius* whereas, the second peak (*cubital peak*) is rather sharp angled and positioned distinctly above the first peak.

List of plates on Stalk borer, *Chilo auricilius* Dudgeon

Description of plates

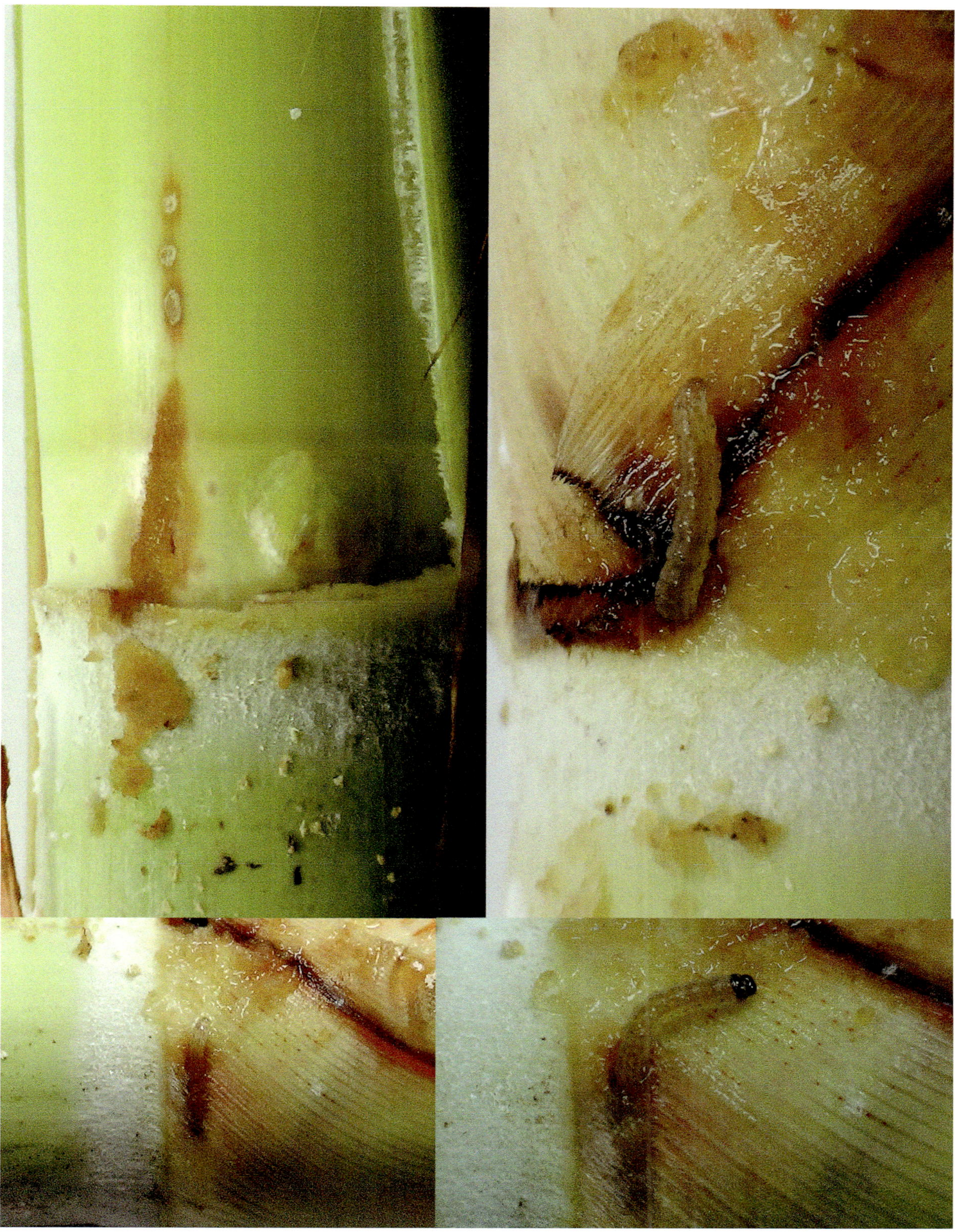

Plate 9.1. Entry of stalk borer larva in the cane stalk. A larva prefers to enter into the stalk through nodal region.

Plate 9.2. Entry and exit hole of stalk borer larva. A large hole is made for the safe exit of moth. Exit holes are usually remain blocked with dry frass.

Plate 9.3. Internal damage of the cane. Larva actively feeding inside the cane tissue and the damaged area turns muddy red in colour. A larva by feeding the hard nodal tissues reach the next soft internodal tissue.

Plate 9.4. Internal damage of the cane (close up). Larva actively feeding inside the cane tissue and the damaged area turn muddy red colour. A larva also feeds on the hard nodal tissues.

Plate 9.5. Internal damage of the cane. Multiple tunnels, tunnel filled with dry frass and pupation within the tunnel without formation of any silken chamber.

Plate 9.6. Pupation takes place inside the feeding tunnel (close up). Silken chamber is not formed. Pupal exuvium after emergence of the adult.

Plate 9.7. External and internal symptoms developed in an stalk borer infested cane. Several rounds of attack took place. Exit holes and internal damages caused by several stalk borer larvae at different period.

Plate 9.8. Neonate larvae of Stalk borer. Head and *prothoracic* shield are black. Head capsule appears appressed. Width of the *prothoracic* shield is smaller than the head. Rest of the body is water soaked white with prominent hairs. A larva bears faint stripes and measures around 1.5 mm.

Plate 9.9. Feeding behaviour and location of feeding of first instar larvae of Stalk borer. Larvae *pseudo-gregariously* feed as the *leaf sheath miner*. Larvae also can feed on the immature leaf tissue like the internode borer and produce smaller white streaks on the leaf.

Plate 9.10. Feeding behaviour and location of feeding of second instar larva. Larvae feed as the *leaf sheath miner*.

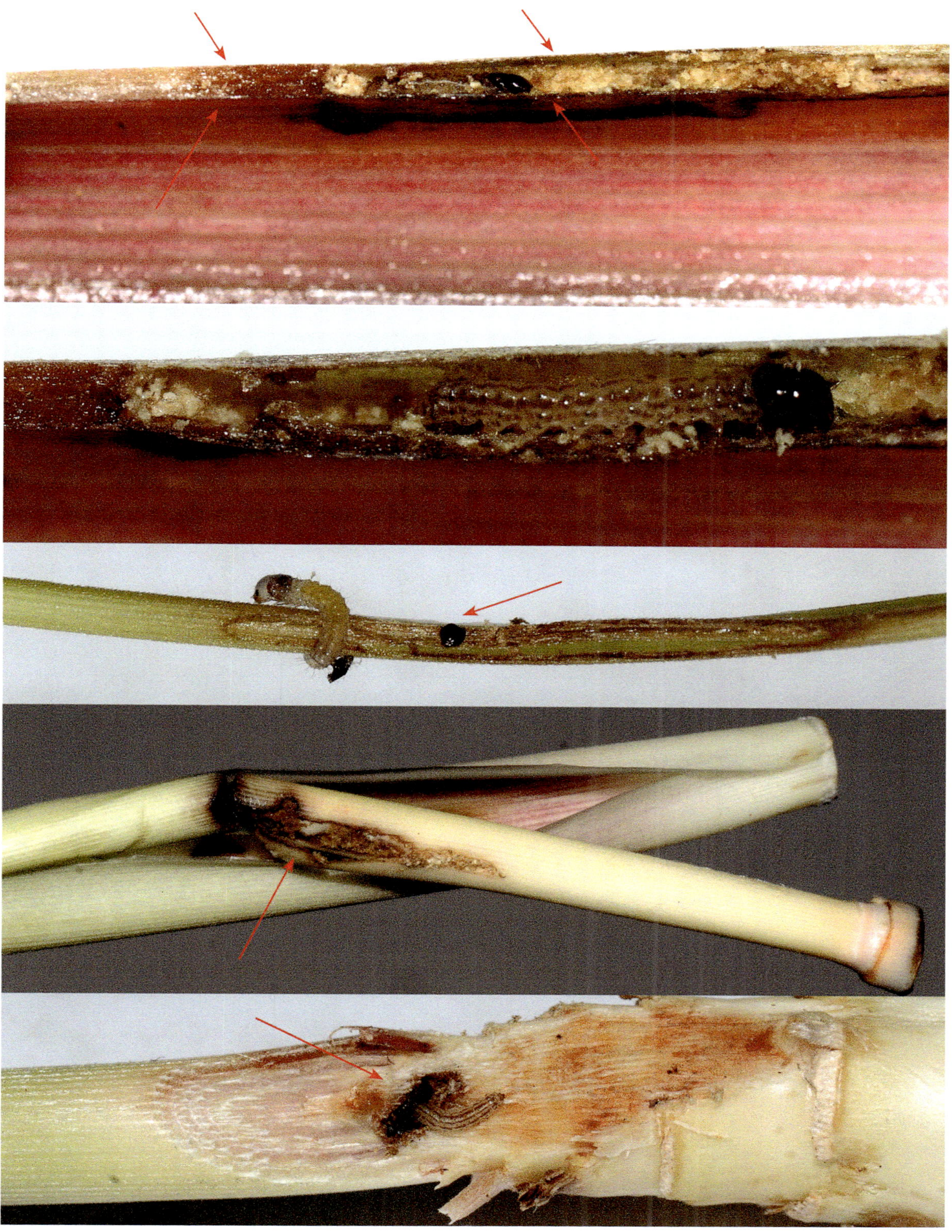

Plate 9.11. Feeding behaviour and location of feeding of second and third instar larvae. Larvae feed as the *leaf sheath miner*. Occasionally enters into the immature cane portion of the crown and causes formation of *dead heart*.

Plate 9.12. First and second instar larvae. Head of the second instar larva, immediately after moulting appears ashy-white with some pinkish colouration. *Prothoracic shield* and first thoracic segment appears as a light coloured band. *Meso* and *meta* thoracic area appear white. Hairs remain prominent on the body.

Plate 9.13. Larvae of Stalk borer (*Chilo auricilius*) in different instars.

Plate 9.14. Larvae of stalk borer (dorsal view). The median and two subdorsal stripes are wavy and prominent. Median stripe ends at the fused tubercle of the 8th abdominal segment and does not cross the segment. Body hairs are visible and are prominent at the anal region.

Plate 9.15. Larva of stalk borer (lateral view). The subdorsal stripe is wider and a shade darker than the lateral stripe. Lateral stripe is positioned just above the *tracheal trunk* but never crosses it. *Tracheal trunk* is very prominent and appear as white tube. *Spiracles* are black rimmed, elongated oval with white centre .

Plate 9.16. Dorsal view(close up). The median and two subdorsal stripes start just after the *prothoracic shield*. Large tubercles (D_1) are roundish square. The intra tubercular distance between two large dorsal tubercles (D_1-D_1) in each segment is very less and the central stripe often touches these tubercles. The tubercles take the body colour. Larva often shows some pinkish colouration at the first few abdominal segments.

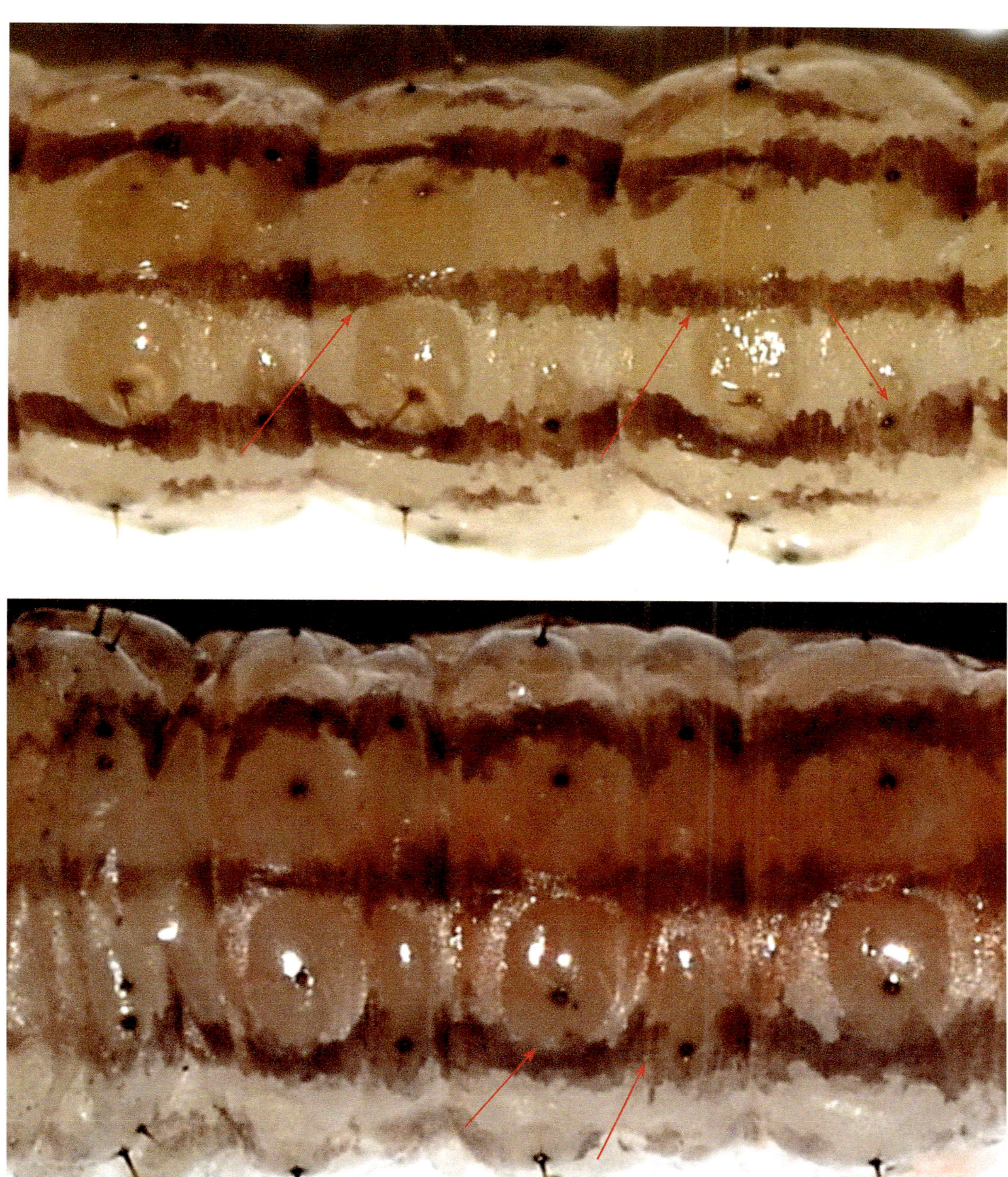

Plate 9.17. Dorsal view(close up). The intra tubercular distance between two large dorsal tubercles (D_1-D_1) in each segment is very less and median stripe often touches the large dorsal tubercles. The subdorsal stripe runs through the smaller dorsal tubercles (D_2) and reach up to the setae and also touches the large dorsal tubercles at the outer end.

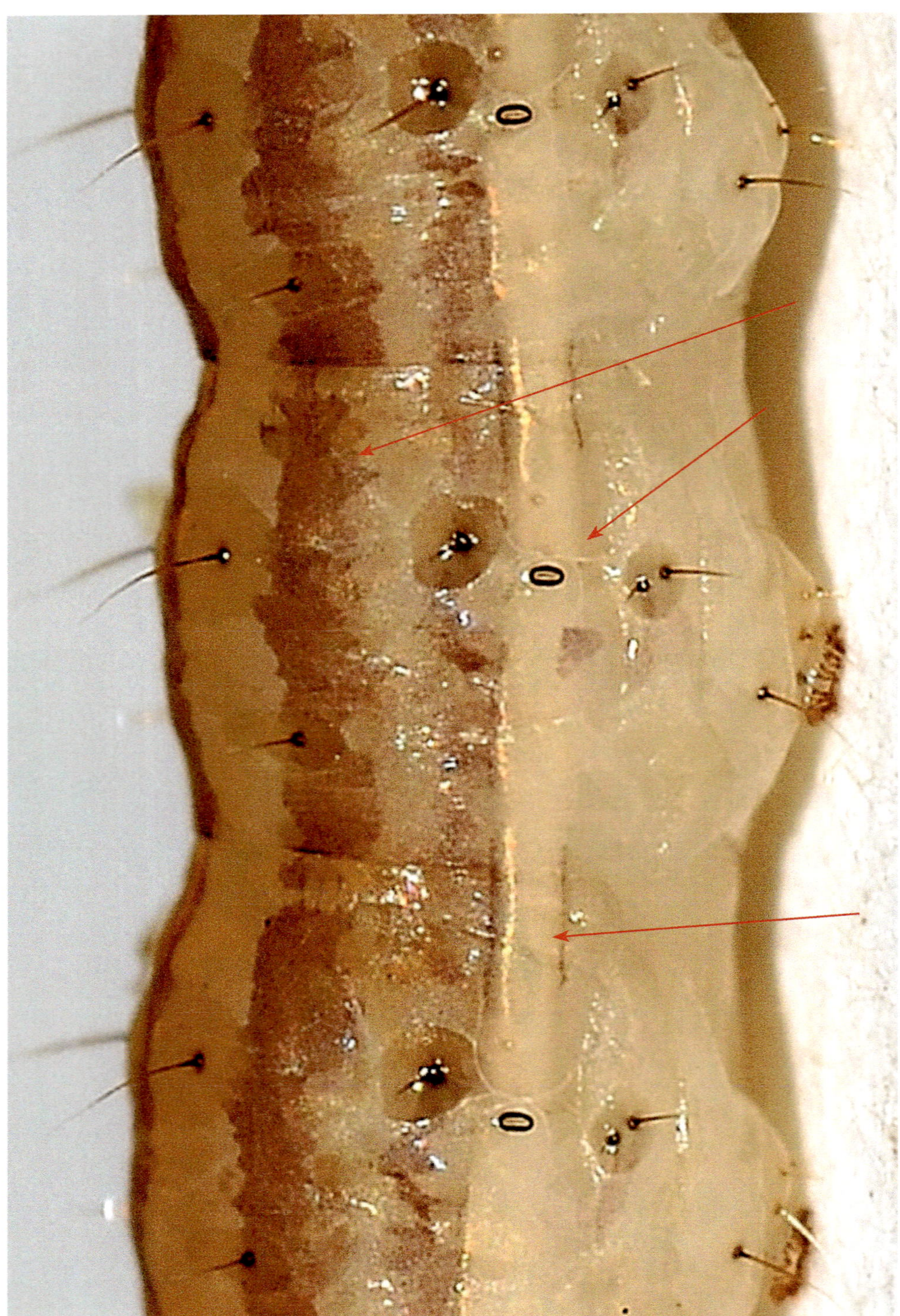

Plate 9.18. Lateral view (close up). The *tracheal trunk* is very distinct as broad white tube. *Spiracles* are black rimmed, elongated oval with white centre. The subdorsal stripe is wider and wavy. It just touches the large dorsal tubercles (D_1) at the lateral side and covers the smaller dorsal tubercles (D_2) up to the seta. Lateral stripe is narrower, sometimes becomes lighter.

Plate 9.19. Lateral view(close up) showing the presence of seven stripes on the body. The last pair of stripe is ephemeral and present below the *tracheal trunk* and is visible from the anal end. The colour of the stripes is mostly retained by the tubercles. The *spiracle* present on the 8th segment is much larger and sub-dorsal in placement. Body hairs are prominent at the anal end.

Plate 9.20. Lateral view(close up). Showing central, subdorsal, lateral and ephemeral stripes, their relative width, intensity of colour and waviness on the body. The last pair or ephemeral stripes is present below the *tracheal trunk* and runs through the tubercle present just below the *spiracle*.

Plate 9.21. View from the anal end(close up) showing the presence of seven stripes on the body. The median stripe ends at the conjoined tubercle of 8th segment. The subdorsal stripe pair cross the 8th segment and reach the 9th segment. The lateral pair also reach the 9th segment. The last or ephemeral pair also reach the 9th segment. The four dorsal tubercles on the 9th segment retain their individual identity. Anal plate is light in colour.

Plate 9.22. Anal end of larva (dorsal view). Variations in colour and spreading of stripes in the 8th and 9th abdominal segments. Presence of dark dots on the *sclerotized* area.

Plate 9.23. Head and thoracic region of the larva. The lateral tubercles present on *meso* and *meta* thoracic segment are relatively larger and easily discernible. The *epicranial suture* area is an inverted 'Y' shaped triangle is demarketed with white line. The mid-dorsal area of the prothoracic shield is also white . Prothoracic shield has patch of dark area near the periphery.

Plate 9.24. Head of a larva. The *epicranial suture* area is an inverted 'Y' shaped triangle is demarketed with white line. The mid-dorsal area of the *prothoracic* shield is also white . *Prothoracic* shield has patch of dark area near the periphery.

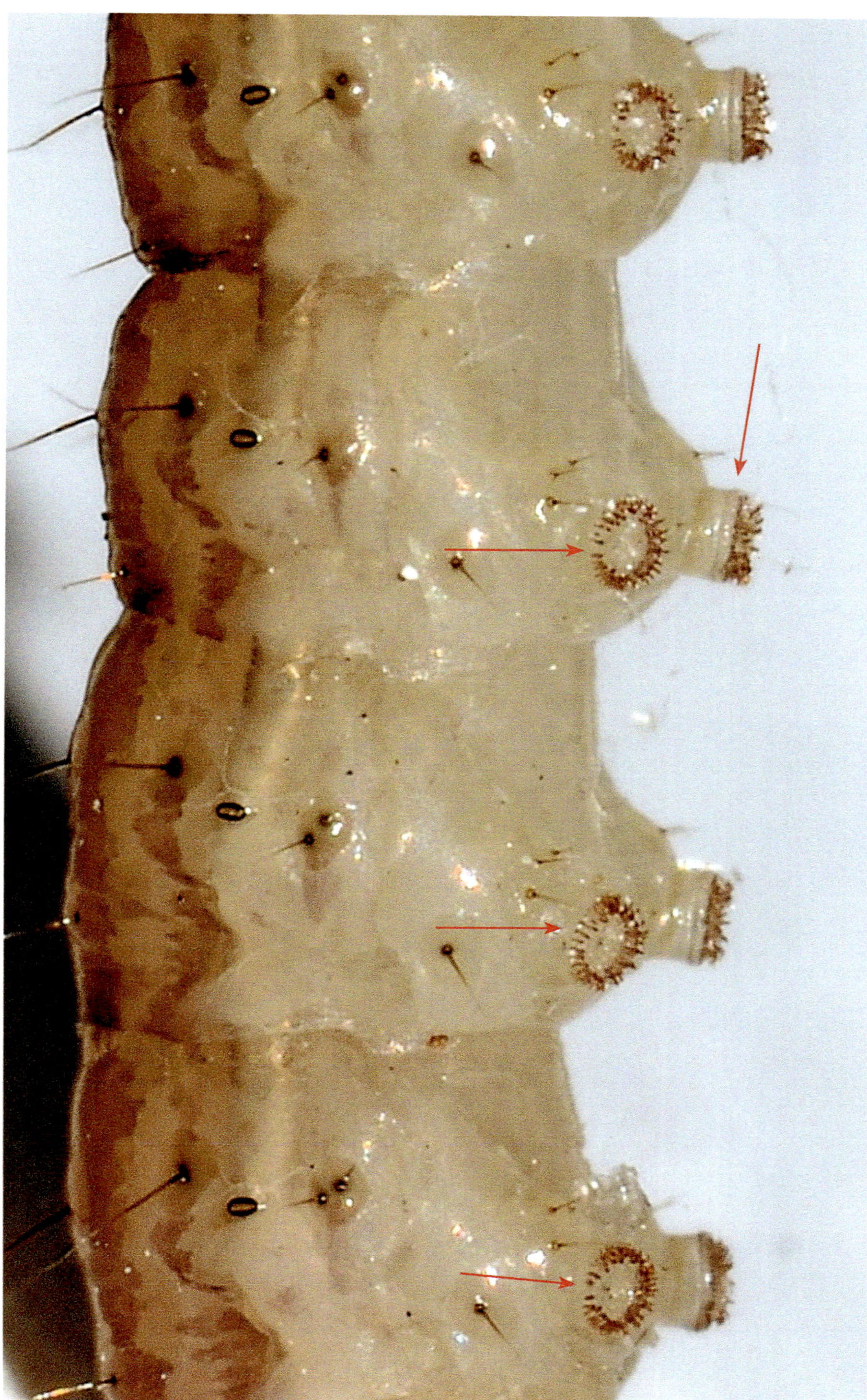

Plate 9.25. Abdominal prolegs. The crochet or hooks are arranged in a *complete circle* and follow an *uniserial* and *triordinal* pattern. Larger hooks are towards *meson*. At the lateral side one or two hooks may remain missing.

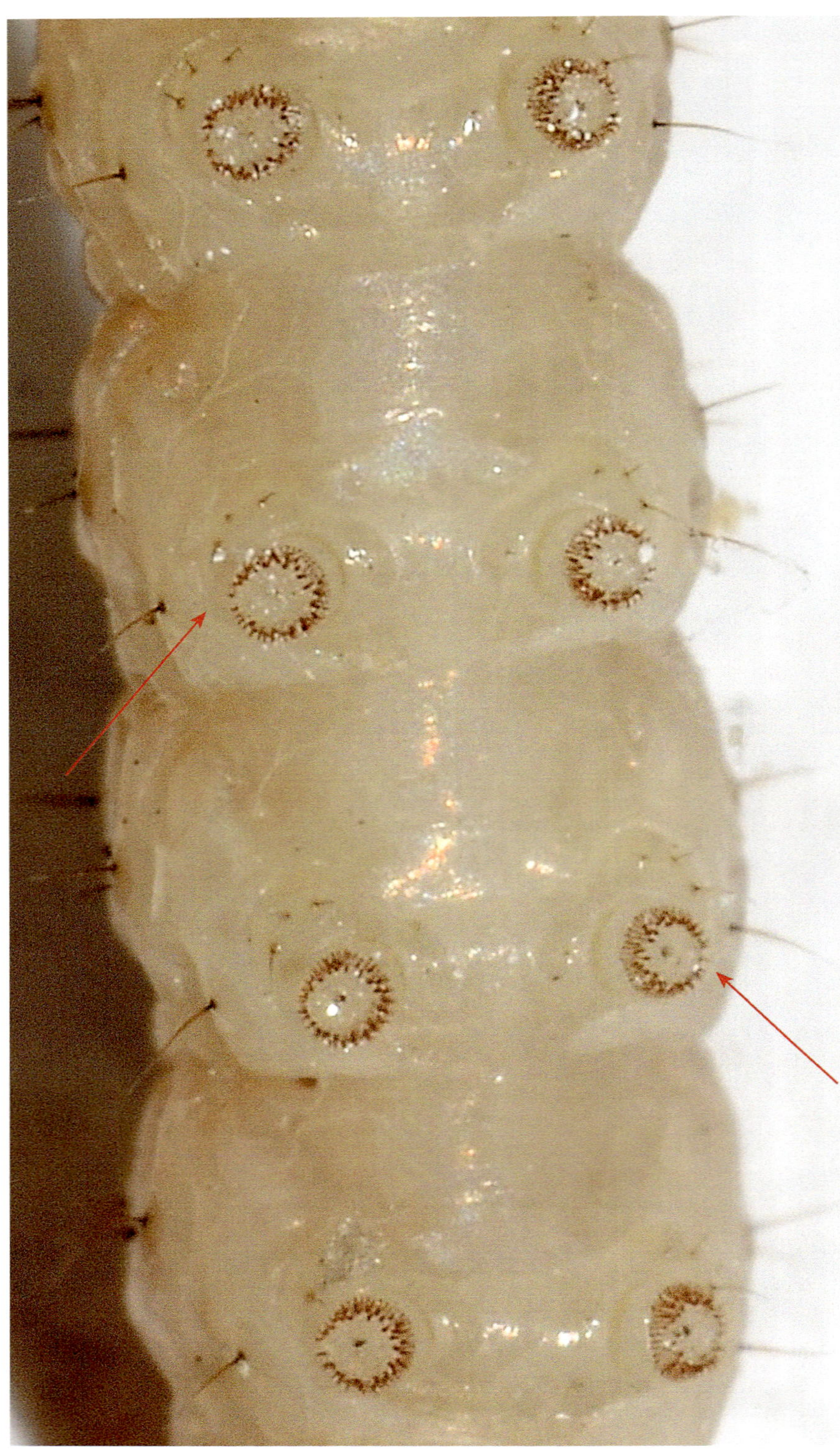

Plate 9.26. Abdominal prolegs. The crochet or hooks are arranged in a *complete circle* and follow an *uniserial* and *triordinal* pattern. Larger hooks are towards *meson*. At the lateral side one or two hooks may remain missing.

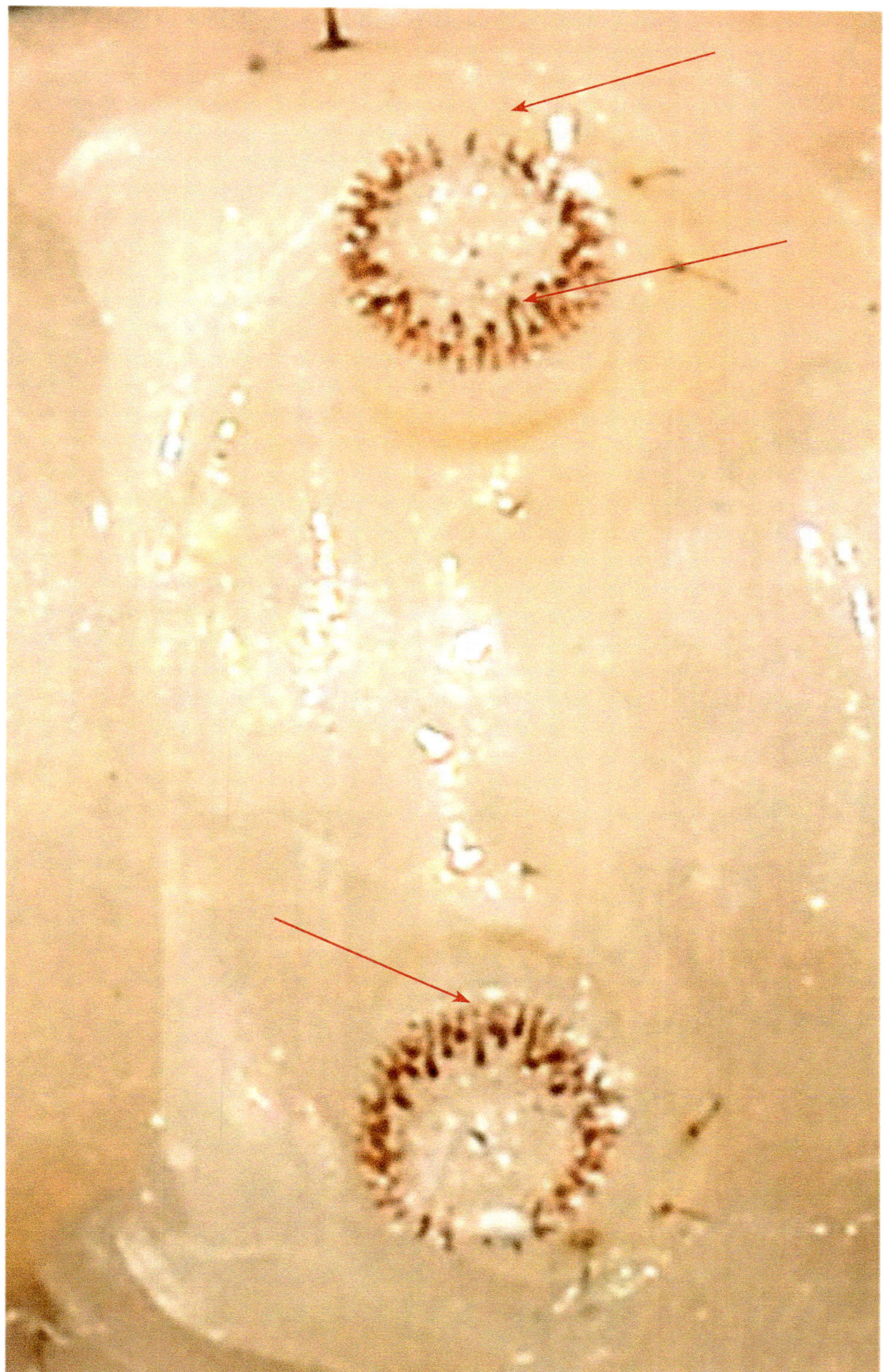

Plate 9.27. Abdominal prolegs (close up). The crochet or hooks are arranged in a *complete circle* and follow an *uniserial* and *triordinal* pattern. Larger hooks are towards *meson*.

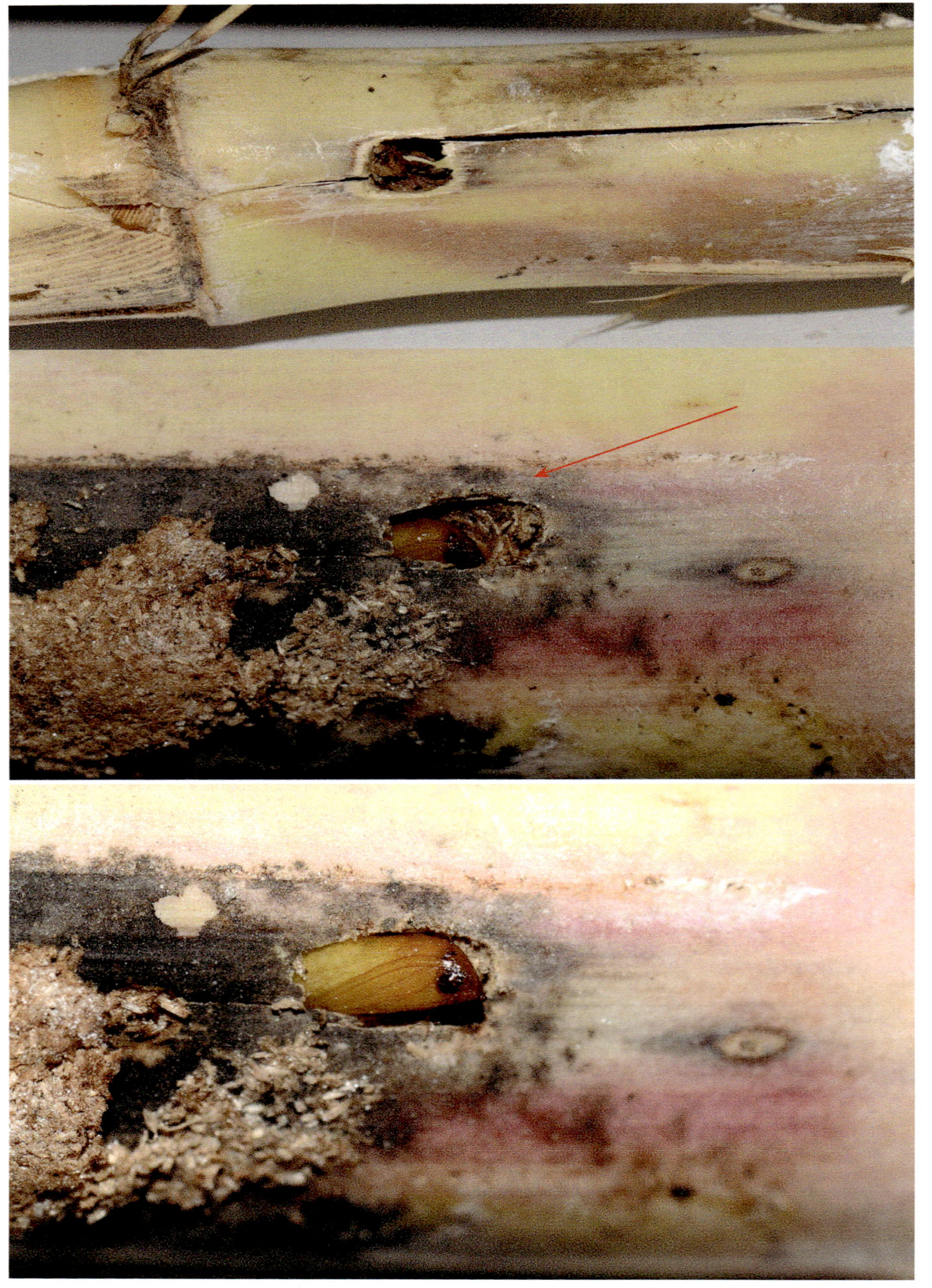

Plate 9.28. Pupation takes place inside the feeding tunnel and pupa rests very close to the exit hole. The exit hole is relatively large and irregular in shape and usually remains covered with dry frass.

Plate 9.29. Pupation takes place inside the feeding tunnel and a larva does not weaves any specialized silken chamber for the pupa to rest.

Plate 9.30. Ventral side of a light coloured female pupa. Head is marked by the presence of two horn like projections. Venation patterns are clearly visible in the wing pad area. Ventral side is smooth except the marks of prolegs. *Cremaster* with eight spines-two large ventral and six small dorsal in two groups. Dorsal and ventral side is separated with large cut mark.

Plate 9.31. Ventral side of a dark coloured male pupa. Head is marked with two horn like projections. Venation patterns are clearly visible in the wing pad area. Ventral side is smooth except the marks of prolegs. *Cremaster* with eight spines-two large ventral and six small dorsal in two groups. Dorsal and ventral side is separated with large cut mark.

Plate 9.32. Relative size of the male and female pupae of Stalk borer (dorsal view). A pupa bears five prominent stripes on the body. The median stripe terminates at the 8th abdominal segment. Stripes start form the *mesothoracic* shield. On the dorsal side on 5th , 6th and 7th abdominal segments distinct row of spines are present.

Plate 9.33. Male pupae of Stalk borer (in different views). Pupae are slender. Colour of a pupa varies widely. Depending on the environment in which a pupa rests, the colour of the pupa changes. Pupa appears elongated and relatively thin.

Plate 9.34. Head and thoracic region of the pupa (dorsal view). Head with two prominent horn like projections. *Mesothoracic spiracles* are large arch type and are very distinctive. Median ridge appears as white line and traverses all the three thoracic shields. Stripes are also present on *metathoracic* shield.

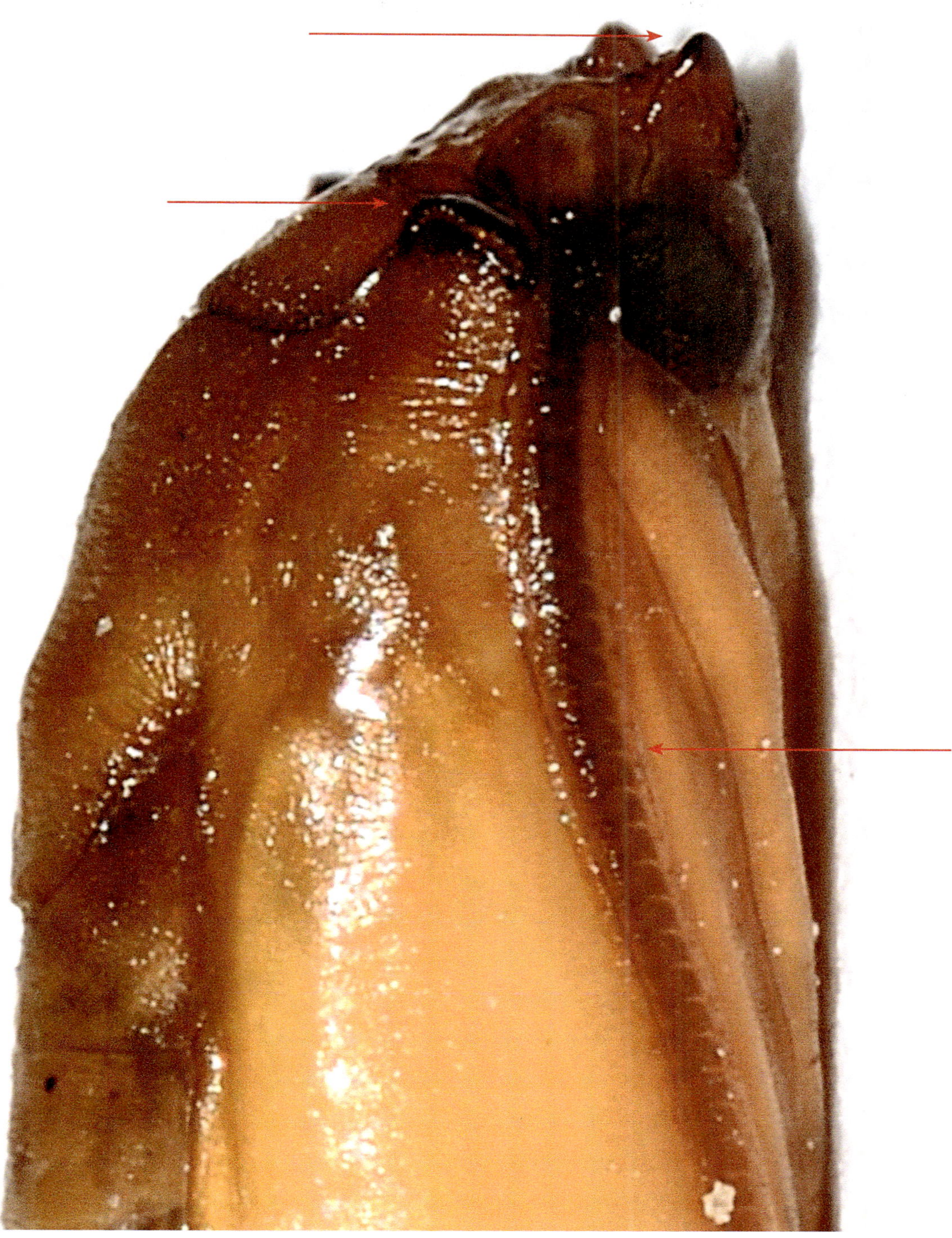

Plate 9.35. Head and thoracic region of a pupa (side view). Head with two prominent horn like projections. *Mesothoracic spiracles* are very distinctive, large and arch type. Development of serrated antennae in the *pharate* adult is visible.

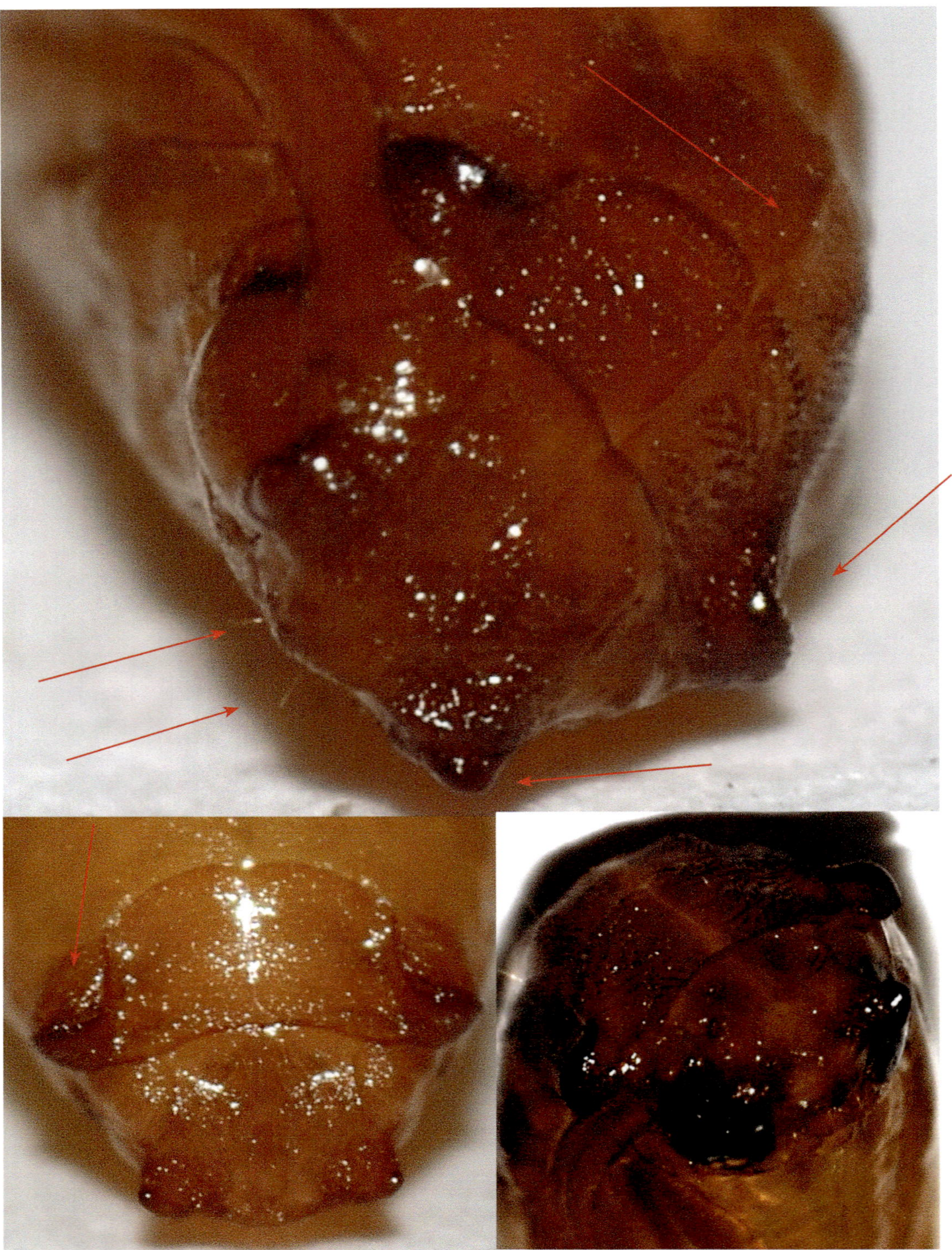

Plate 9.36. Head of the pupa (close up). Variations in the development of *mesothoracic spiracles*. Face of the pupa is slightly convex and bears two setae at the side of the convex area.

Plate 9.37. Ornamentations on the abdominal segments (dorsal view). The spines are stout and short with dark tip. One row of spine present on the 5th segment, occasional two to three rows spines are found on the 6th segment and the 7th segment supports several rows of spines . In each segment the terminal spines are larger.

Plate 9.38. Thoracic and abdominal portion (side view). Prominent arch formation above the 2nd and 3rd abdominal *spiracles*. The spines on the 5th and 6th abdominal segment reach up to the *spiracle*.

Plate 9.39. Ornamentations on the abdominal segments (side and ventral view). The spines are short with dark tip. One row of spine present on the 5th segment, occasionally two to three rows of spines are found on the 6th segment and the 7th segment supports several rows of spines. The terminal spines are larger. On the 7th segment the spine development goes beyond the *spiracle* and terminates in line with mark of prolegs. Ventral portion remains smooth except the marks of prolegs.

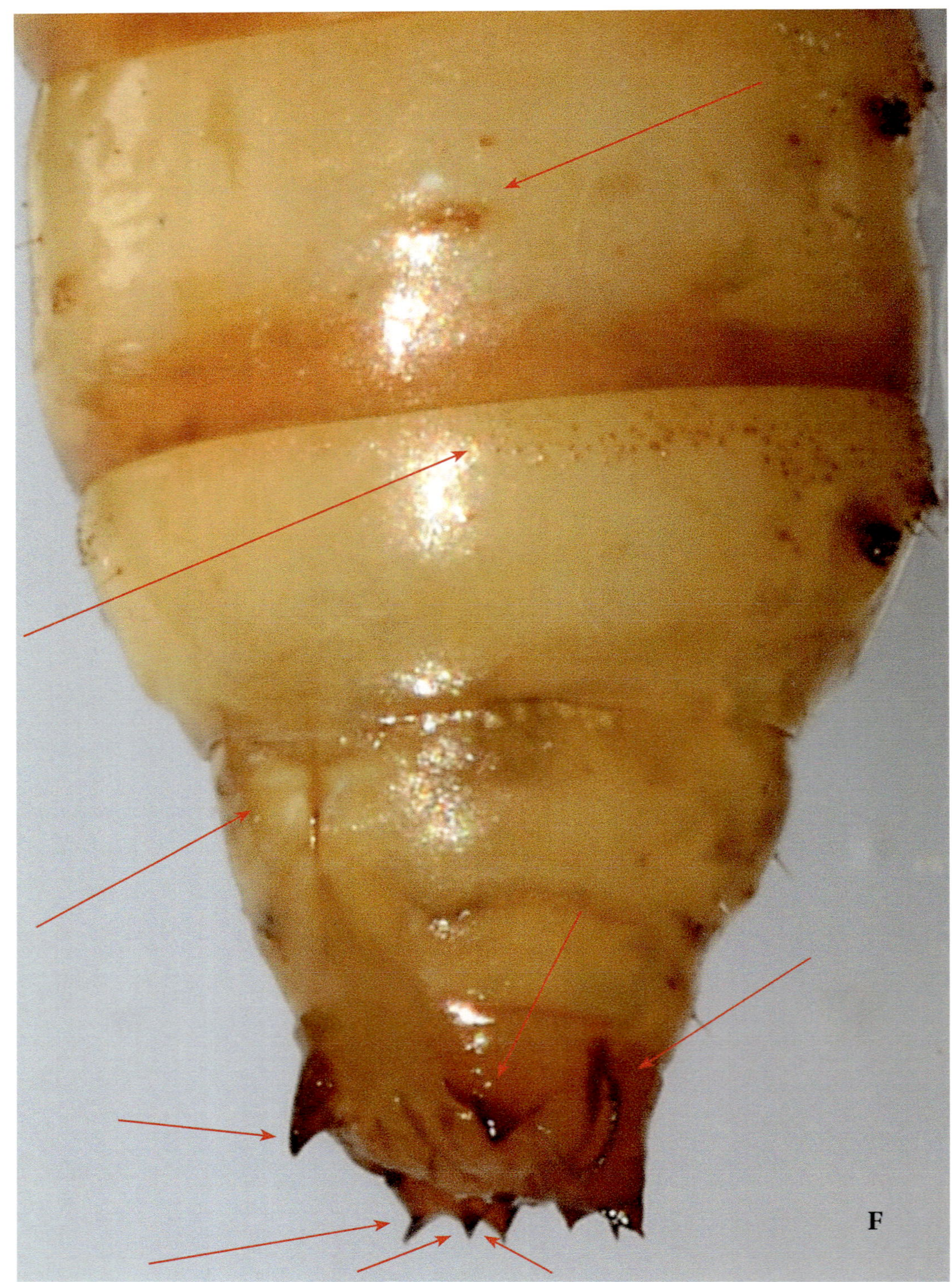

Plate 9.40. Female pupa showing genitalia and *cremaster*. The female genitalia is marked as a small depression at the middle of the 8th segment. *Cremaster* with eight spines, two large ventral and six small dorsal in two groups. Dorsal and ventral side is separated with a large cut mark. On the 7th segment, the spine development goes beyond the *spiracle* and terminates in line with mark of prolegs.

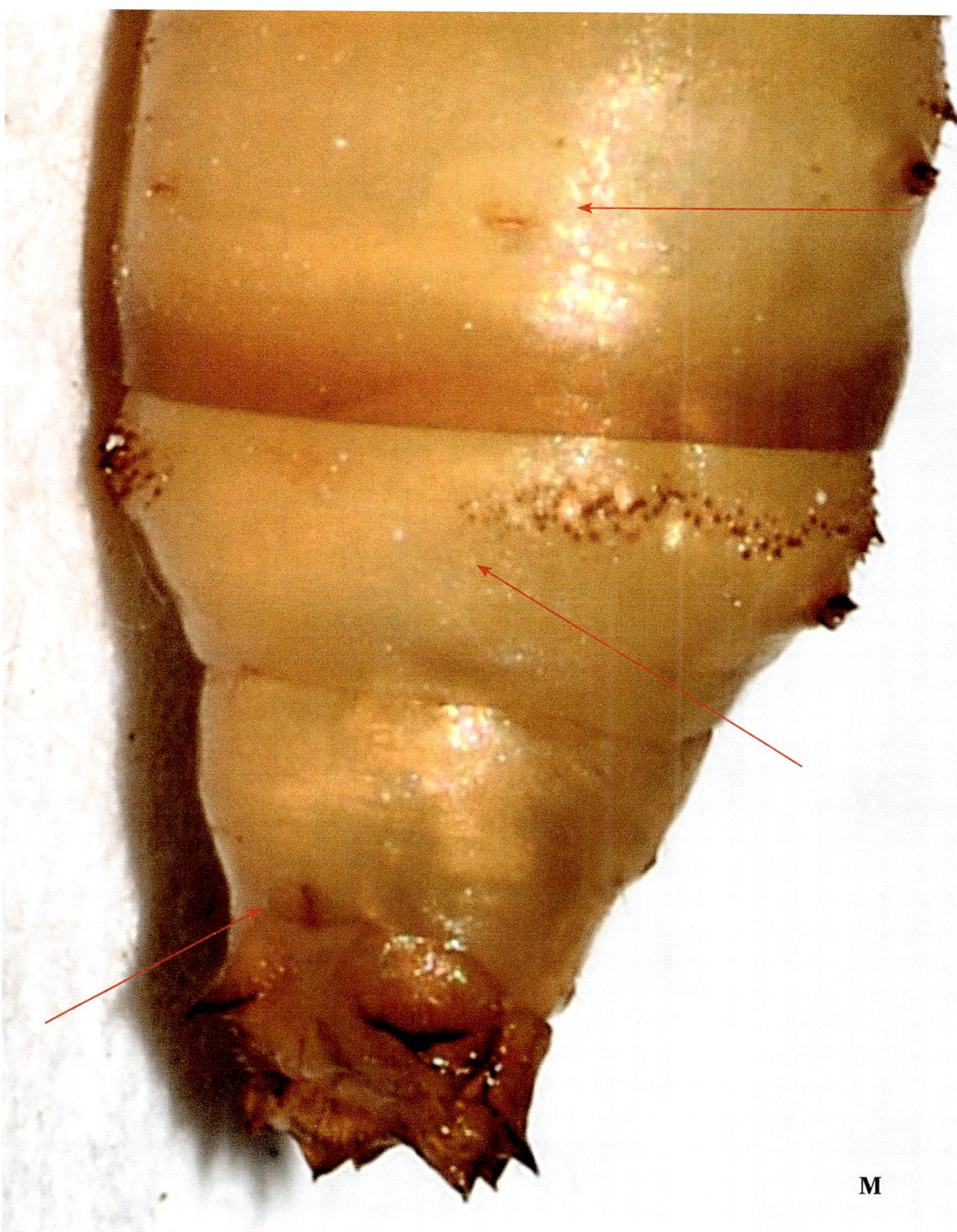

Plate 9.41. Abdominal portion of a male pupa showing genitalia and *cremaster*. Male genitalia is marked with two lobed area with central depression on 9th segment. *Cremaster* with eight spines, two large ventral and six small dorsal in two groups. Dorsal and ventral side is separated with a large cut mark. On the 7th segment, the spine development goes beyond the *spiracle* and terminates in line with mark of prolegs.

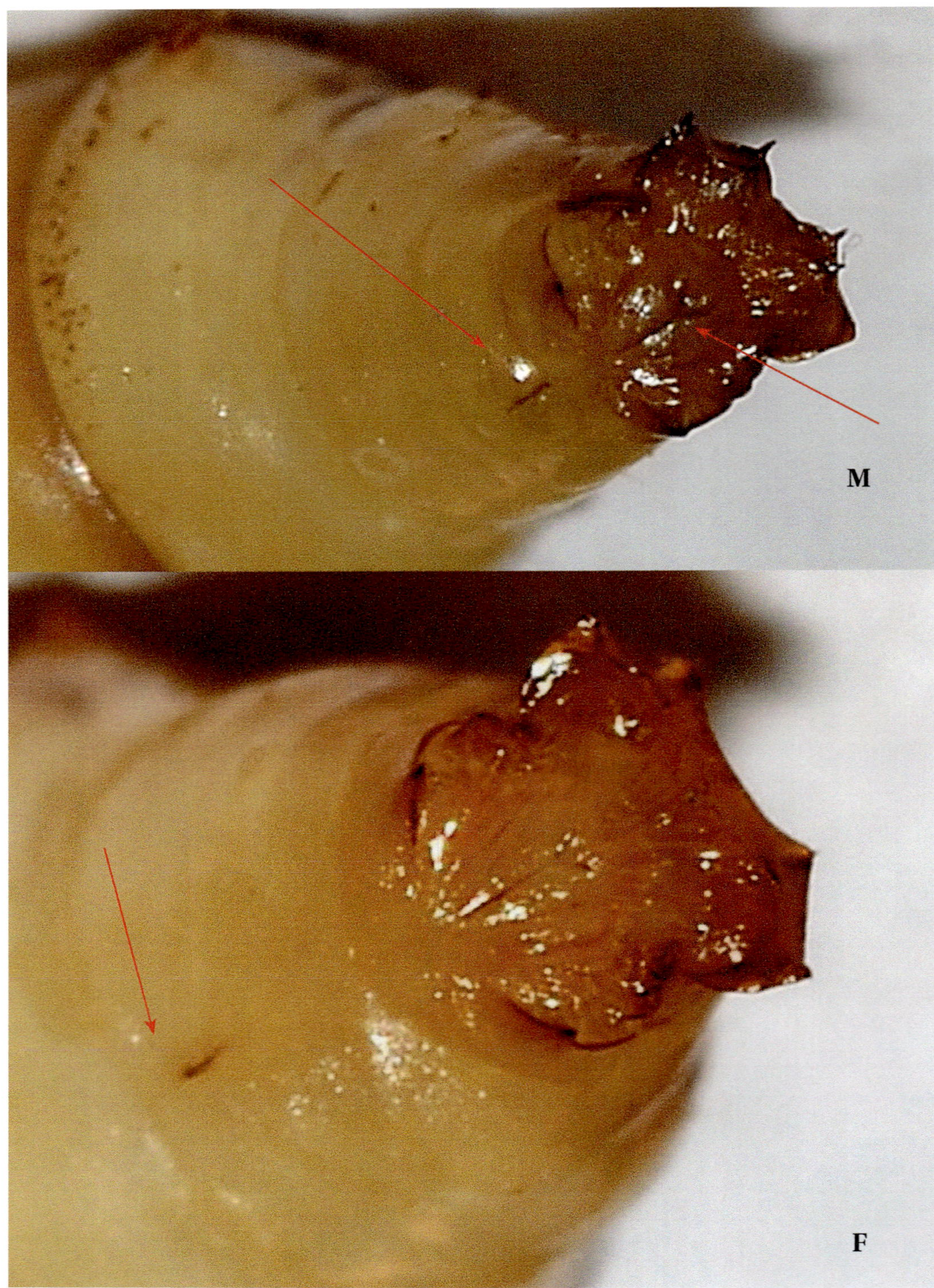

Plate 9.42. Anal segment of both male and female pupae viewed from the *cremaster* side.

Plate 9.43. Male moth of Stalk borer (dorsal view). Forewings have typical raised steely scales at five to six locations in the central area. Terminal end of the wings also bear small black dots at the furrow area, the dots have corresponding steely scales behind. Terminal end imparts a golden lustre.

Plate 9.44. Male moths of Stalk borer (side and dorsal views). Forewings have typical raised steely scales at five to six locations in the central area. Terminal end of the wings also bear small black dots at the furrow area, the dots have corresponding steely scales above. Terminal end imparts a golden lustre.

Plate 9.45. Female moth of Stalk borer. Moth is one shade lighter in colour. Forewings have raised steely scales at five to six locations in the central area. Terminal end of the wings also bear small black dots at the furrow area. Terminal end imparts a golden lustre.

Plate 9.46. Female moth of Stalk borer (side view). Forewings have raised steely scales at five to six locations in the central area but these spots are much lighter than the male moth. Terminal end of the wings also bear small black dots at the furrow area. Terminal end imparts a golden lustre. *Labial palp*i are porrect and like a tail of a squirrel.

Plate 9.47. Female moth of stalk borer feigning death. Porrect *labial palpi* resemble the tail of a squirrel. Antenna is *filiform* and serrated.

Plate 9.48. Male and female moths (ventral view).

Plate 9.49. Female and male moths on centimetre scale. Female moths are larger. There are wide variations in the size of male moths.

Plate 9.50. Head of the moth (close up). Head is round. Eyes appear greenish in photograph. *Labial palpi* are porrect and like the tail of a squirrel. Antennae are *filiform* and ribbed.

Plate 9.51. Head of the moth (close up). Face is broadly rounded. An *ocellus* is present just behind the antenna.

Plate 9.52. Antenna (female and male). Antenna is *filiform*, serrated and prominently ribbed in female. In male, the antenna is relatively broader, side of the antenna is devoid of scales and appears light to dark brown in colour. Dorsally the antenna is ribbed.

Plate 9.53. Forewing (close up) showing the terminal gold fringe, small terminal black dots on the furrow area, followed by a row of steely scales in the furrow area corresponding to each black dot and six raised patchy area with steely scales at the *discal* region.

Plate 9.54. Formation of pattern of 'W' on the forewings. A band or stripe of scales appear near the terminal end (at the steely scale area) and imparts a typical appearance to the moth.

Plate 9.55. Female moth (from the ventral side). The terminal black spotting of the wings continues to the ventral side of the forewing. Black spotting is also present on the hindwing. Female genitalia remains visible.

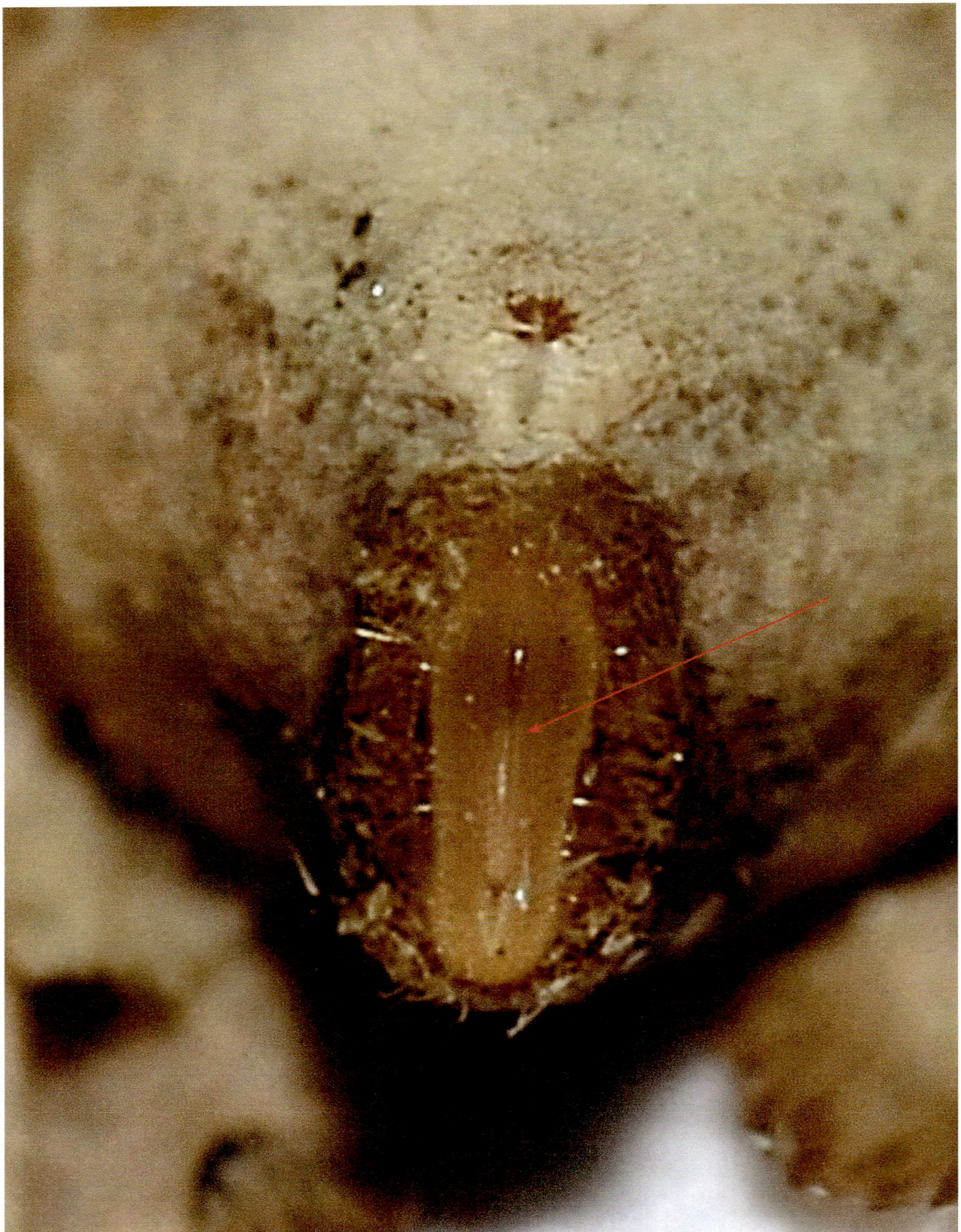

Plate 9.56. Female moth showing elongated *papillae anales* with large median depression

Plate 9.57. Forewing venations (stained and unstained). Sc is stout at the base and much thicker than the other veins. R_1 joins the Sc and forms the combined vein Sc+R_1. After running certain distance, parallel to Rs, both veins separate out. Sc separate out as a small fork and ends at costal area. R_1 continues further. R_2 is very close to the stalk of R_3+R_4. R_5 originate near the radial angle. M_2 and M_3 are separate at base and appear from the cubital angle. Cubital angle is placed higher than the radial angle. Cu1a and Cu1b are present. Cu1a and R_1 occupy same plane in the discal cell. A_1 is absent, A_2 is very prominent and A_3 is poorly developed.

Plate 9.58. Forewing venations . Sc is stout at the base and much thicker than the other veins. R_1 joins the Sc and forms the combined vein Sc+R_1. After running certain distance, parallel to Rs, both veins separate out. Sc separate out as a small fork and ends at costal area. R_1 continues further. R_2 is very close to the stalk of R_3+R_4. R_5 originate near the radial angle. M_2 and M_3 are separate at base and appear from the cubital angle. Cubital angle is placed higher than the radial angle. Cu1a and Cu1b are present. Cu1a and R_1 occupy same plane in the discal cell. A_1 is absent, A_2 is very prominent and A_3 is poorly developed.

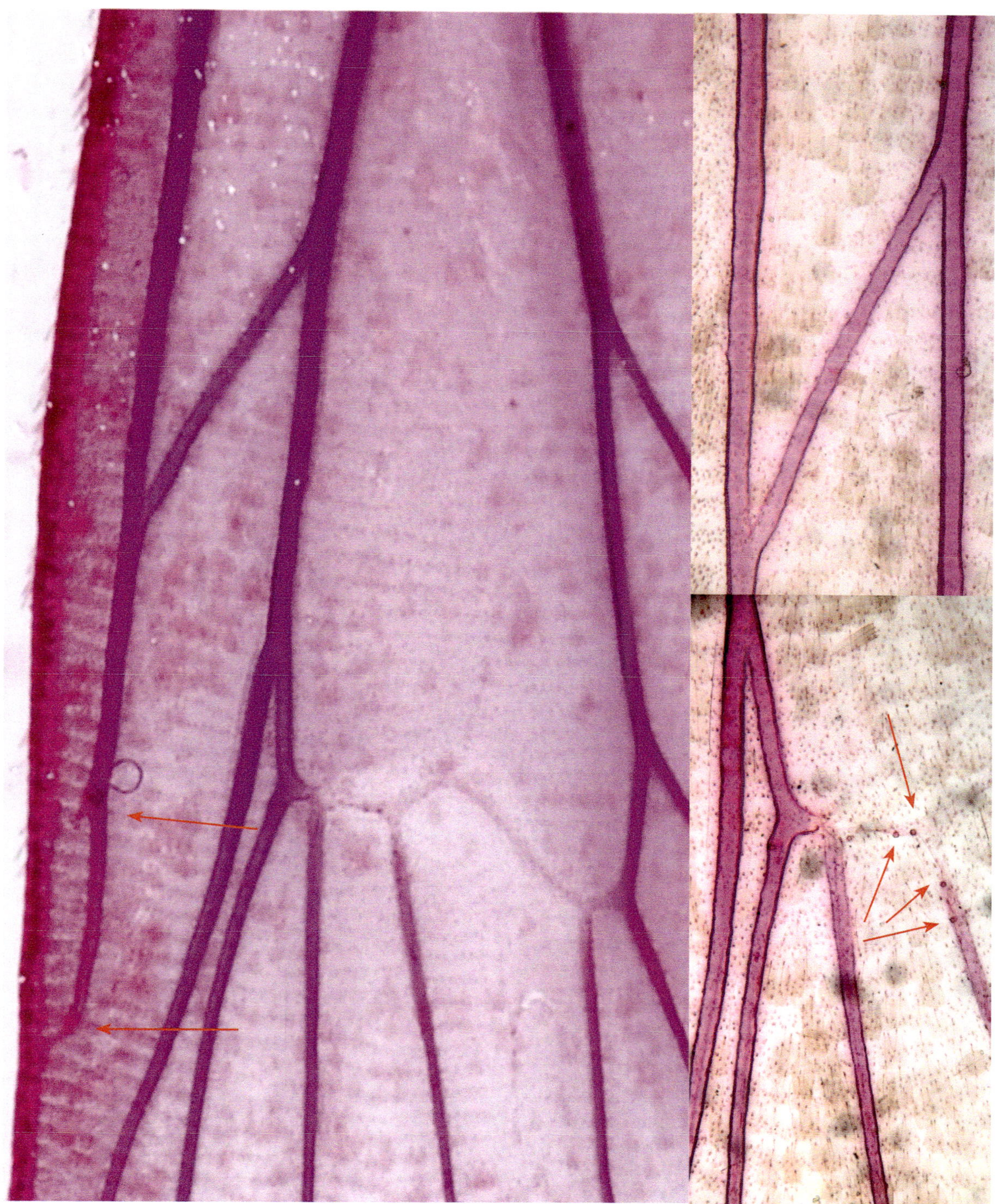

Plate 9.59. Discal region (close up). Sc runs almost parallel to Rs. Anastomosis of $Sc+R_1$ and their separation. Close proximity of R_2 (it takes a curvature) with the stalk of R_3+R_4. Origin of R_5 near the radial angle which is around 90°. M_1 appears after a good gap. Cubital peak occupies much higher position than the radial peak. M_2 and M_3 appear from the broad cubital peak. Predominance of *mechanoreceptors* at R_5 and M_1.

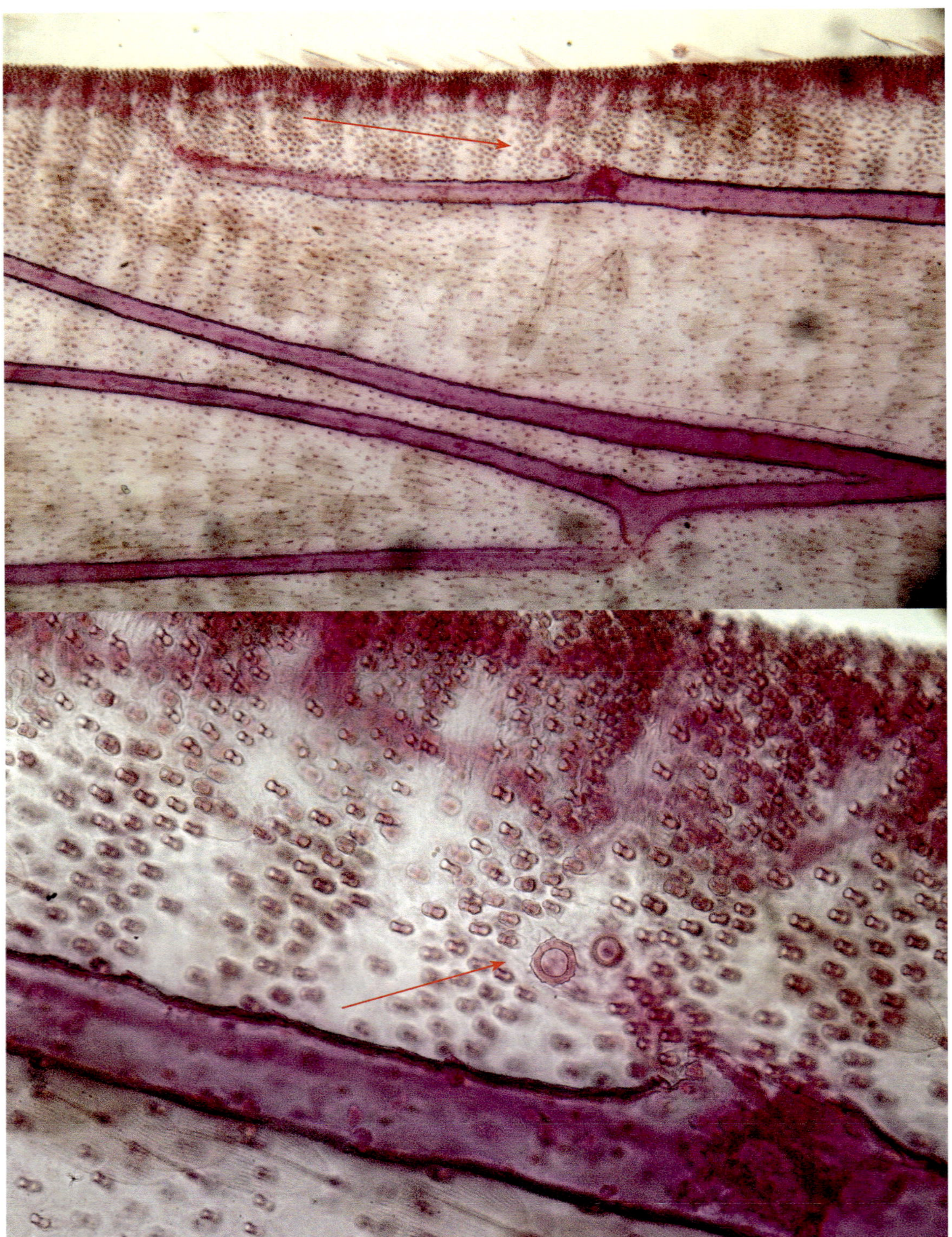

Plate 9.60. Branching of Sc and R_1 (under microscope). The branching is short. The presence of two *mechanoreceptors* indicate the end of Sc. R_1 continues further.

Plate 9.61. Egg mass of Stalk borer laid on the inner side of the mid-rib near the junction of lamina and leaf sheath in the laboratory condition, when moths were released on ratoon sprouts. Eggs were also observed on the immature leaf sheath at the soil level.

Plate 9.62. Egg mass of Stalk borer laid at the laboratory (rearing). Egg mass remain irregular in shape, as egg mass may contain 3, 4 or 5 rows of eggs. Eggs are white when freshly laid. Larval development is visible within the translucent egg shell and the stripes on the developing larval body remain visible. In black head stage, head orientation does not follow a regular pattern.

10

Internode borer, *Chilo sacchariphagus indicus* (Kapur)

The Internode borer of sugarcane, *Chilo sacchariphagus indicus* (Kapur) is an important sugarcane borer with an all India footprint. However, it is a predominant sugarcane borer in the tropical south causing considerable concern. The moth borer identified as *Chilo sacchariphagus* is currently recognised to have three subspecies viz., *Chilo sacchariphagus indicus* (Kapur) in India, *Chilo sacchariphagus sacchariphagus* (Bojer) in Indonesia, Philippines and Mauritius and *Chilo sacchariphagus stramineellus* (Caradja) in China, Japan and Formosa (Hattori, 1971). Morphologically all these three subspecies are very similar (identical) but with slight differences in the genitalia. Revising the taxonomy of genus *Chilo*, Bleszynski (1970) concluded that all the recorded populations of this borer belong either to one widely spread species or to several phylogenetically very young species. Apparently, geographical isolations of different populations resulted in slight variations in the genitalia, and these differences cannot be considered diagnostic. However, for the present treatment the name, *Chilo sacchariphagus indicus* (Kapur) is accepted for the Internode borer that infests the sugarcane crop in India.

This borer came into prominence in Mauritius after its accidental introduction to Mauritius from Java (now Indonesia) through the import of large quantities of sugarcane cuttings in mid 1800's. After that introduction in the island this moth borer established itself so well in the new environment that it turned out to be a major concern in profitable growing of sugarcane in Mauritius. Accordingly, Mr. Bojer was deputed to look into the problem and to find a suitable solution to get rid of this menace. Mr. W. Bojer identified this moth borer pest as *Proceras sacchariphagus* Bojer, in 1856. Later on, in 1863 Walker identified this species as *Chilo venosatus* Walker (from the collection of Mr. Saunder from Sarawak, Borneo; described on page 144). One gets an excellent description and colour drawings of this species (*Diatraea striatalis* Sn.) in the publication of Deventer (1906) from Java.

In India, it was probably studied first by Lefroy (1906) from western India but he described the borer species involved as *Chilo simplex* (from the description in the text and furnished drawings, fig. 4 plate XL, forewing of moth it becomes apparent). Fletcher(1914) in his book "*Some South Indian Insect*" on page 422 remarked that

> "***Remarks*** - The insect has hitherto been confused with *Chilo simplex*, but is distinct in structure, habits and main foodplants. It is perhaps *Diatraea venosata*, Wlk., of which *striatalis*, Snellen well known as a serious pest of cane in Java, is a synonym."

For a long time in the Indian literature this moth borer was identified as *Diatraea venosata* (Walker) Hampson. Fletcher and Ghosh (1919), provided excellent drawings of both the larva and adult of this species (page 388, plates 50 & 51, Please refer **Appendix IV** for the details). Although this species is now grouped in *Chilo*, characteristically this moth species does not possess *ocellus*. (As presence of *ocellus* constitute an important character in the identification and separation of *Chilo* moths. Similarly, the characteristic absence of *ocellus* constitutes an important identifying character of *Diatraea* moths including *Diatraea saccharalis* (F.), the American sugarcane borer.)

10.1 Currently accepted name

Chilo sacchariphagus indicus (Kapur)

10.2 Synonyms

= *Proceras sacchariphagus* Bojer

= *Borer saccharellus* Guenee

= *Chilo mauriciellus* Walker

= *Chilo venosatus* Walker

= *Diatraea striatalis* Snellen

= *Diatraea venosata* (Walker) Hampson

= *Diatraea mauriciella* (Walker) Hampson

= *Proceras venosatus* (Bojer)

= *Proceras indicus* Kapur

= *Chilo sacchariphagus* (Bojer) Bleszynski

10.3 Symptoms of infestation and brief biology

The Internode borer, *Chilo sacchariphagus indicus* (Kapur) is a major borer pest of sugarcane in the tropical sugarcane belt of India causing substantial damage to the crop. Over the years, Internode borer has increased its presence in the subtropical sugarcane belt also, occasionally causing major concern in sugarcane cultivation. This moth borer remains active throughout the year in tropical sugarcane belt whereas, in the subtropical belt, it hibernates during the cold winter months (in larval stage). The larva of this moth borer mostly feeds in the immature internode of a cane stalk and rarely damages mature portion of the cane. For his high affinity to the intermodal tissues, it was named Internode borer. In contrast to the Stalk borer, where more than one larva stay in one cane stalk, in the Internode borer, usually only one larva is encountered in a mature cane stalk. However, despite its regular occurrence, this pest seldom assumes a very high population in a field. Probably, the bio-agents collectively contain the development of a large population of this moth borer.

The neonate larvae prefer the immature crown leaves for their early feeding. Soon after hatching, most of the larvae of one egg mass pass through a near *gregarious phase* or *pseudogregarious phase*. A number of the neonate larvae may congregate at one location and feed the spindle leaf tissues by scrapping and leaving substantial amount of wet frass. The partial or loose folds of the emerging spindle leaf provide the requisite hiding space. Due to the feeding of the young larvae, characteristic white streaks appear on leaf lamina on unfolding. In scrapping, the lower epidermal layer are usually left untouched and thus the eaten up portion appears white. Several such elongated white areas are formed on the young leaves. In this way, larvae of Internode borer pass first two instars. Third instar larvae usually disperse from the *pseudogregarious phase* and reach the adjoining cane plants. Reaching an uninfested plant, a larva bores at the tender portion of cane at the crown and produces a large *dead heart*. However, no offensive smell emanates on pulling out the *dead heart*. At shoot stage, a *dead heart* is formed from its infestation.

In a developed cane, a larva may go down further in the stalk and make an entry into the internode. In that situation the water supply is not snapped fully due to the feeding of the larva and the crown remains green (no *dead heart* is formed). Later on, the larva moves down further in the immature stalk and feeds in the internode in a very characteristic manner. In contrast to other moth borers, a larva of Internode borer usually feeds the cane in a horizontal fashion or in a semi-circular fashion (initial feeding is quite similar to that of Gurdaspur borer). If a larva makes straight vertical feeding tunnel in the stalk, then the tunnel will also have occasional horizontal feeding areas (labyrinths). Labyrinth formation in the stalk is also a characteristic feeding behaviour of this moth borer. In some cases it restricts its feeding at one side of the internode and in that situation, several irregular holes appear on the epidermis (borer holes remain restricted to one patch or portion of the stalk).

Larva of Internode borer has a habit of pushing out the wet frass outside to keep the feeding tunnel clean. During monsoon period, fresh frass is always observed sticking outside the leaf sheath. Such frass may stick much longer on the leaf sheath and the dried mass of frass may be seen hanging from the feeding hole. At later stages, in an immature internode the tunnel wall becomes dark coloured or turns black. On the contrary, in the mature cane, the tunnel area remains dry and gets a light colour. The typical feeding habit of the larva produces a partially circular damaged or cut area on the epidermis externally and a semicircular damaged horizontal area in the internal tissues of the internode leaving only a little portion undamaged. Occasionally, feeding at the growth ring area (just above the node) results in the formation of galleries. The crown portion of affected cane plant often survives receiving partial supply of water and nutrients though the undamaged portion of the stalk. From this semicircular affected intermodal area, the affected cane often breaks or topples down due to blowing of wind at high velocity and or due to the movement of an animal or during any field operations. Thus, the Internode borer infested field shows toppling down effect of the crown. Such toppling down of the crowns is mostly observed during post monsoon period.

The attack of the pest in young shoots (plant crop and ratoon sprouts) results in the formation of *dead heart*. However, in the grown up canes, *dead heart* formation diminishes considerably. The larval damage in the cane stalk may be observed after removal of the leaf sheath. The larva of Internode borer is also migratory in habit and after feeding at one location for some time, it moves to a new uninfested cane stalk or to a new location in the same cane stalk. Thus, one larva may damage several canes before it attains maturity and proceeds for pupation. Usually, only one larva is found in one infested cane.

Pupation may take place within the stalk (in the feeding tunnel) or outside the stalk. In the season, pupation mostly takes place outside the stalk, whereas the hibernating larval population pupates within the damaged stalk where it takes shelter during the winter months. A prepupal larva prepares a loosely woven silken pupal chamber. The pupal period lasts around a week to ten days. When pupation takes place within the stalk, a prepupal larva makes a larger exit hole. A larva does not fully cut the exit hole, rather it carefully leaves the thin epidermal layer intact. In any case, if the cover of the exit hole is breached, the larva immediately seals it, weaving silken threads. After completion of the pupal period, an adult emerges out pushing the thin epidermal covering of the exit hole. When the pupation takes place outside the stalk, it usually uses the space between stalk and a mature leaf sheath. Before pupation, a fully grown larva comes out of the feeding tunnel and takes shelter underneath a tight fitting leaf sheath (in the space between stalk and leaf sheath) of the cane. It constructs a loose silken cocoon to pass the period of reclusion. This pest is carried over to the succeeding crop as hibernating larvae or pupae along with the seed cane. It also gets carried on through the clasping leaf sheaths with attached cocoons. After harvest of the crop, these cocoons are quite often left in the field along with the trash. Moths emerging from these pupae fly away to the neighbouring sugarcane fields to lay eggs on a fresh crop.

It has been observed that sugarcane varieties with soft rind generally have a higher incidence of Internode borer than those with a hard rind. Young larvae prefer to feed on the tender portion of the stalk rather than the well-formed hard internodes. Since the newly hatched borer larvae generally remain inside the leaf sheaths before feeding their way slowly into the cane, varieties with loose fitting leaf sheaths gets a preference from this moth borer.

10.4 Egg and neonate larva

There is no distinct preference among the sugarcane varieties for laying eggs. However, in a given situation more number of egg masses is spotted on a variety having broad green leaf. Female moths of Internode borer lay eggs in an imbricate (overlapping) row of egg mass parallel to the mid-rib. Usually only two rows of eggs are laid on the leaf and an egg mass contains around 20-28 eggs. Eggs are mostly laid on the upper surface of the sugarcane leaf, however there is no hard and fast rule regarding the place for laying eggs. Egg masses can be found in any suitable place that a female moth prefers for laying the eggs. Freshly laid eggs are scaly, elongated oval, flat and white in colour. The eggs initially appear creamy white and gradually the colour changes with the development of larva inside. Within the translucent eggshell, the developing larva with prominent orange bands representing each segment of the larval body remains visible. The head and *prohoracic shield* of the developing larva are black and the head of all the developing larvae remain directed to the first egg (layed). The orange coloured larva comes out from the eggshell by making a fine slit. Neonate larva measures around 1.5 to 2.0 mm. The larval body has characteristic orange band in each abdominal segment and numerous dark setae appear on the orange band. The orange bands are very light on the thoracic segments (*meso* and *metathroracic* segments). The presence of specific orange bands helps to differentiate a larva of Internode borer from the neonate larvae of other moth borers of sugarcane.

10.5 Larva

The larva of Internode borer is also very active like that of Stalk borer or Early shoot borer and on extrication from the cane, a larva quickly moves away to hide itself under a cover. A full-grown larva measures around 30 mm. In a mature larva, head and *prothoracic shield* are yellowish brown. Head appears relatively smaller and the well-developed *Prothoracic shield* appears broader. In the season, presence of very dark or black tubercles on the body immediately distinguishes it from the larva of other moth borers of sugarcane. Four pinkish-violet broad stripes run through the body. Two broad stripes are placed dorsally and two narrow stripes are placed laterally. Dorsal stripes are wider, more or less straight and run through both the larger (D_1) and smaller (D_2) dorsal tubercles of each segment. The lateral stripes run through the *tracheal trunks* and partially cover the lateral tubercles. However, the surrounding areas of a *spiracle* retain the body colour (a halo area around the *spiracle*). The lateral stripes after crossing the *spiracle* pairs end at the 8th abdominal segment, whereas the two dorsal stripes continue further and reach the 9th abdominal segment. This pattern become more apparent in hibernating larvae, especially when colour of the tubercles fades away and in turn renders the stripes more distinct on the body. In each segment, these dorsal stripes characteristically bear a light coloured narrow patch or *cross stripe* or *horizontal cut mark* in front of the larger tubercles (D_1). This cut mark appears very characteristic of this species for identification especially when colour of the tubercles fades away. Depending on the season and the instar a larva is passing through, wide variations in the colour of tubercles are observed. During the monsoon months, the tubercles turn distinctively black (just like a few burnt grains in a fired cobb of a maize). However, the black coloured tubercles gradually become lighter before the larva proceeds for pupation. The tubercles becomes light creamy (black tubercles take body colour) and the stripes become faint or fade away. During hibernation also, the colour of the stripes as well as tubercles usually fade away.

In prepupal or hibernating larva when the colour of both the tubercles and stripes become very light, horizontal light cut marks or cross stripes along with the presence of the two dorsal stripes on the 9th segment often become handy in identification. In addition, presence of large tubercles at the lateral side *meso* and *metathoraic* segments (like that of Stalk borer) further helps in the identification. In majority of the mature larvae, the two larger tubercles (D_1) on the 8th segment only remain fused at the base and thus impart a 'V' shaped appearance; the smaller tubercles (D_2) following behind mostly remain free. On the 9th segment, the two dorsal and two subdorsal tubercles join to form a larger *sclerotized* plate. The colour of this plate remains lighter than the colour of tubercles present on the other abdominal segments.

Spiracles are distinctly black rimmed and prominent. However, the appearance of the *spiracles* varies depending on the season and the instar of a larva. Generally, inside of a *spiracle* appears white and central area appears open; however, in certain situation central area appears black. The *tracheal trunk* usually remains faintly visible on the larval body. Occasionally, when it becomes visible, it appears as a dark coloured tube passing through the body. The colour of the larval body also varies depending on the season and the instar it is passing through and in certain situations (in many hibernating larvae), it becomes light cream to water soaked white in colour.

The crochet or hooks of Internode borer appear fine and sharp. Hooks are placed at the peripheral area of *prolegs* and curved backward i.e., directed outside from the circle. The arrangement of crochets in this species shows variability. In the season (during monsoon months), the crochets on the abdominal *prolegs* form a complete circle and are *multiordinal* (more than three sizes). There is a *biserial* arrangement of hooks towards *meson* and a *uniserial* arrangement in rest of the area. The hooks on the inner side of the *proleg* (towards *meson*) are larger and dense, whereas the hooks at the outer side (lateral side) are smaller and sparse. Sometimes at the outer side, two or three smaller crochets remain absent. This imparts an appearance of an incomplete circle. Often there develops a dark area at the centre of the *proleg* indicating the centre of the imaginary circle of crochets. However, in the hibernating larvae, the arrangement of crochets is distinctly an incomplete one. Towards the lateral side, many of the hooks remain absent, arrangement hooks become *uniserial* but it remain *triordinal* (three sizes). This *triordinal* crochets help its differentiation from the larva of Gurdaspur borer (prolegs are always arranged in a *uniserial* and *uniordinal* manner).

10.6 Pupa

A freshly formed pupa is light reddish brown in colour and with maturity, its colour changes to dark reddish brown. Pupa also bears four distinct stripes on the body like the larva. In the four abdominal segments of the pupa, roughness (i.e. a rough or velvety area without any definitive spines) is present on the dorsal surface (roughness is very mild in the subdorsal

List of plates on Internode borer, *Chilo sacchariphagus indicus* (Kapur)

Description of plates

Plate 10.1. Feeding symptoms of neonate Internode borer larvae on leaf lamina. On immature crown leaves, several small elongated white streaks develop due to the feeding of neonate larvae on laminar tissues.

Plate 10.2. Feeding symptoms on leaf. Larvae of Internode borer feeding on the immature leaves of the crown (staying within the rolled leaves). Larvae do not eat away the lower epidermis which later appears white on drying. Presence of a lot of frass indicates the larval feeding.

Plate 10.3. Feeding symptoms on leaf. Larvae of Internode borer feeding on the immature leaves of the crown and remain in hiding even in the partial leaf fold. Close up of the white streaks; usually streaks are devoid of holes.

Plate 10.4. Feeding of Internode borer larvae in the immature rolled leaves of the crown (in shoot stage). Occasionally neonate larvae also travel down the immature rolled leaves of the crown and feed in that area and finally result in the formation of a *dead heart*..

Plate 10.5. Neonate larvae also go down in the space between leaf sheath and *pseudostem*, especially during the shoot stage and feed on the leaf sheath tissues like that of Stalk borer larvae. The feeding behaviour of Internode borer larvae is haphazard. Larvae prefer to feed at the upper half portion of leaf sheath. Rotting of leaf sheath is less evident.

Plate 10.6. Feeding symptoms on immature stalk portion of the growing cane. An Internode borer larvae pushes out a lot of wet frass from the feeding tunnel. The frass mostly remains visible.

Plate 10.7. Feeding symptoms on stalk at grand growth stage of the crop. An Internode borer larva pushes out a lot of wet frass from the feeding tunnel. Frass may accumulate within the leaf sheath or may comes out, if there is feeding hole in the leaf sheath.

Plate 10.8. Feeding symptoms on stalk at the fag end of grand growth stage of the crop. An Internode borer larva pushes out a lot of wet frass from the feeding tunnel. Dry mallable frass may remain hanging from the feeding hole.

Plate 10.9. Typical feeding symptoms of Internode borer larva on immature stalk. Development of a horizontal feeding tunnel in the cane with a larger hole. The horizontal feeding appears as a half cut mark bearing small dark punctures on the epidermis.

Plate 10.10. Variations in the feeding pattern of the larva of Internode borer. Sometimes a patchy area with multiple feeding holes may develop. Patchy area may show rotting. Rooting and sprouting are also common.

Plate 10.11. Variations in the feeding pattern of the larva of Internode borer. Formation of *dead heart* in mature plant due to feeding at the crown area. Profuse rootings take place from the nodes near the affected area.

Plate 10.12. Variations in the feeding pattern. Larva may feed at the immature cane area, may feed just above the node and may form irregular tunnels. Sometimes feeding of the larva may results in more than one spiral cut marks on the internode. Stalks often break or get dislodged from the affected area.

Plate 10.13. Profuse rootings from the node both above and below the afftected internode take place. Sprouting of the buds below the affected portion may also be observed during the rainy season.

Plate 10.14. Toppling of the crown from the horizontal cut area of the stalk. This weakness developed due to the feeding of an Internode borer larva in semicircular fashion. A larva may stay at the upper or lower portion of the damaged stalk.

Plate 10.15. An affected clump showing toppling of the upper portion of the stalk from the damaged area.

Plate 10.16. Feeding symptoms as observed when the affected stalk is split open. The feeding tunnel remains messy with wet frass and the wall of the feeding tunnel mostly turns dark or black.

Plate 10.17. Feeding symptoms as observed when the affected stalk is split open. The feeding tunnel in the mature cane internode may remain light coloured. Often mature canes show formation of multiple cross tunnels or labyrinths.

Plate 10.18. Neonate larvae of Internode borer. Larvae appear orange coloured due to the presence of orange coloured band in each abdominal segment of the body. The orange colour bands remain visible in the developing larva within the translucent egg shell.

Plate 10.19. Neonate larvae of Internode borer. Emergence of larva after cutting a slit on the egg shell. Larvae are orange coloured with prominent orange bands in each of the abdominal segments. Head and prothorax are dark brown. Larvae come out in batches.

Plate 10.20. Neonate larvae of Internode borer (Close up). Showing presence of setae, orange bands on the abdominal segments, appressed head capsule and well developed dark brown *prothoracic shield*.

Plate 10.21. *Second instar* larva feeding within the rolled crown leaves (A). Larva with prominent tubercles but stripes are yet to develop. Head and *prothoracic* shield are well developed and black. With further growth, stripes appear on the larval body. *Third instar* larvae (C) have prominent stripes and tubercles on the body. Head capsule is light orange red and *prothoracic* shield is lighter in colour. The larval body is essentially white as non-striped area remains white.

Plate 10.22. Variations in the larva: Variations are evident in colour of the head, *prothoracic shield*, dorsal tubercles and the colour of the stripes.

Plate 10.23. An Internode borer larva bears four broad stripes. The dorsal stripes are much wider, straight and run through the both larger (D_1) and smaller (D_2) dorsal tubercles. Stripes fade away during hibernation and fading of colour is faster in the tubercles. In each abdominal segment there is a *horizontal cut mark* on the dorsal stripes.

Plate 10.24. A mature larva of Internode borer measures around 30 mm. Colour of the head and *prothoracic shield* becomes much lighter. In some situation, colour of the stripes fades away, while the colour of the tubercles remain lighter.

Plate 10.25. Larvae of Internode borer (close up, dorsal view). The two dorsal stripes are wider and run through the large (D_1) and small (D_2) tubercles on the abdominal segments and extend to the 9th segment. The dorsal and subdorsal tubercles on the 9th segment remain separate. The *horizontal cut mark* in front of D_1 is very characteristic.

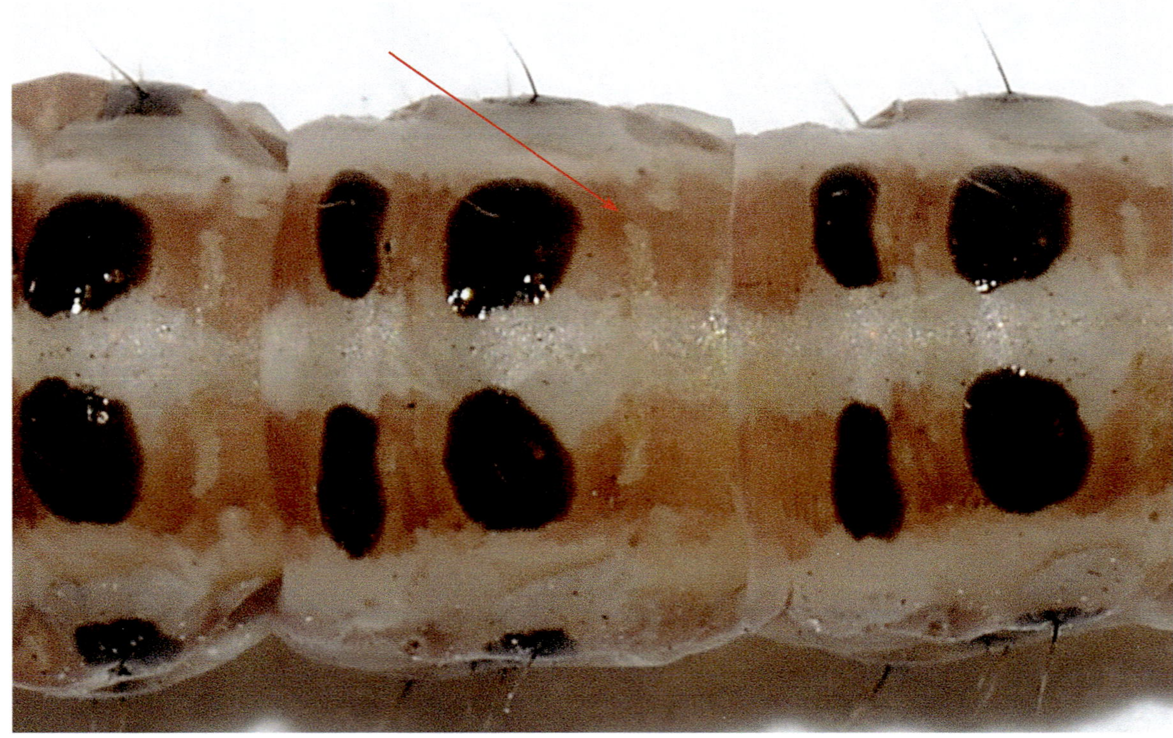

Plate 10.26. Larva of Internode borer (close up, dorsal view). The two dorsal stripes are wider and run through the large tubercles (D_1) and small tubercles (D_2). In an abdominal segment the presence of horizontal cut mark in front of D_1 is very characteristic.

Plate 10.27. Larva of Internode borer (side view). Lateral stripe runs through the *tracheal tube. Spiracles* are prominent elongated and dark rimmed and usually surrounded by a halo. Many a times *tracheal tube* is not visible as the stripe runs through it. When the *tracheal tube* is visible, it appears as dark coloured tube.

Plate 10.28. Larvae side view (close up). *Spiracles* are prominent, elongated and black with surrounding lighter area (halo around *spiracle*). Lateral stripe runs through the lateral tubercle, *tracheal tube* and *spiracles*.

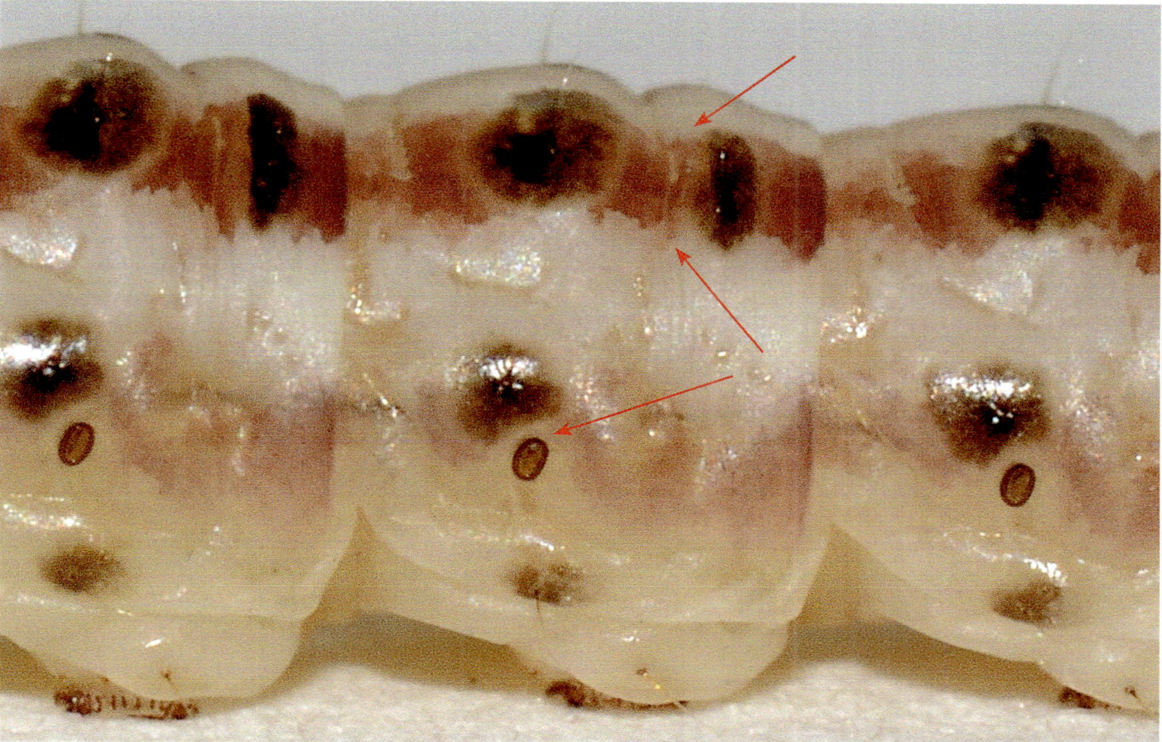

Plate 10.29. Larvae side view (close up). Lateral stripe runs through the both lateral tubercles and *tracheal trunk*. Variation in colour of the *spiracles* (centrally may be black or light and rim may be black or light brown). Lateral tubercles is truncated above a *spiracle*.

Plate 10.30. Head and thoracic region (close up). Head and *prothoracic* plate have prominent median line. *Epicranial suture* is prominent with sharp white line. *Prothoracic shield* is well developed and has dark colouration all round, more darker at side and towards *mesothoracic plate*. *Mesothoracic plate* is similar to tubercle. The two pairs of lateral thoracic tubercles are large.

Plate 10.31. Head and thoracic region. Head and *prothoracic plate* has prominent median line. *Prothoracic plate* is well developed and has dark colouration all around, more darker at side and towards *metathoracic plate*. Shape of thoracic *spiracle*. *Ocelli* are present and follow a 1+4+1 pattern.

Plate 10.32. Prepual larva in season and during hibernation. In season, tubercles retain colour and during hibernation tubercles become colourless.

Plate 10.33. Anal end of the larva. The dorsal stripes cover the 9th segment, whereas the lateral stripes terminate at the 8th segment. *Spiracle* is brown rimmed with black inside. The *horizontal cut mark* on the dorsal stripe is visible. On the 9th segment only the two dorsal tubercles fuse together.

Plate 10.34. Stripes and cut marks remain visible even in larva going through the process of *ecdysis*. Immediately after moulting the head, thorax and tubercles remain white. In the head only the mandibular area remains dark.

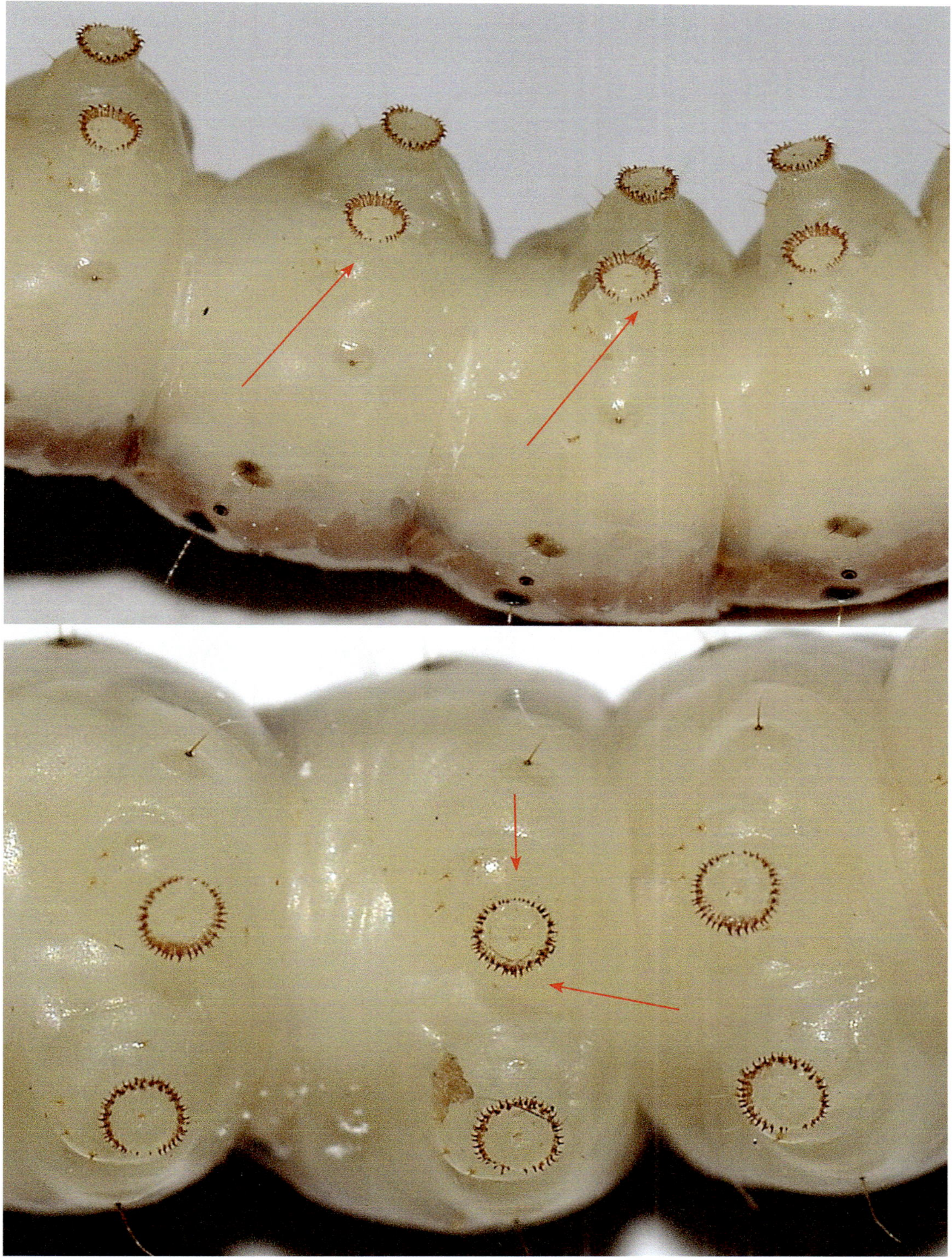

Plate 10.35. The crochets on the abdominal *prolegs* are sharp, fine and form a *complete circle*; crochets are larger towards *meson* and smaller towards the lateral side. Occasionally a few crochets remain missing from the lateral side.

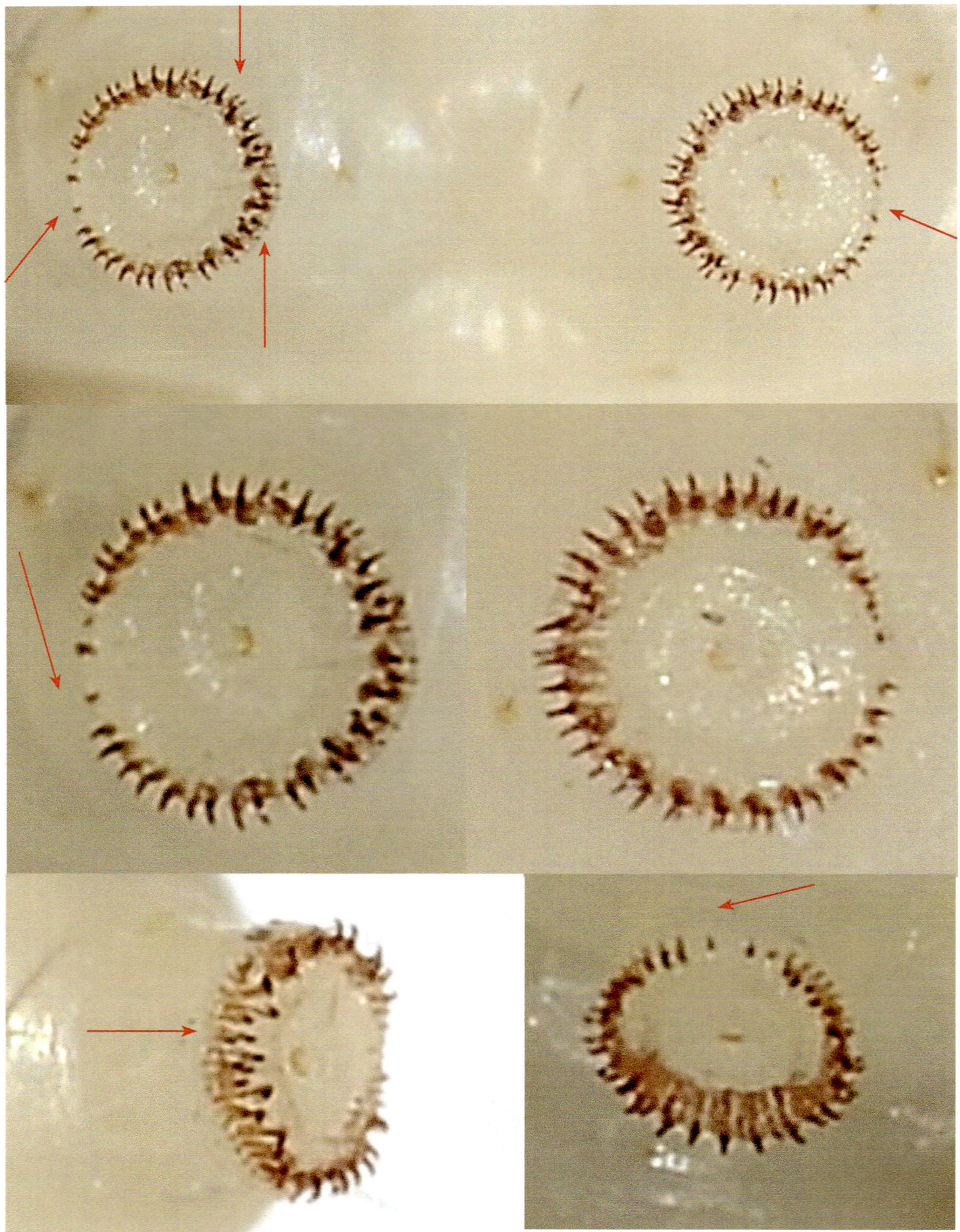

Plate 10.36. Crochets are *biserial* towards the *meson*. Crochets are *triordinal*. Size of the crochets gradually diminish towards the lateral side.

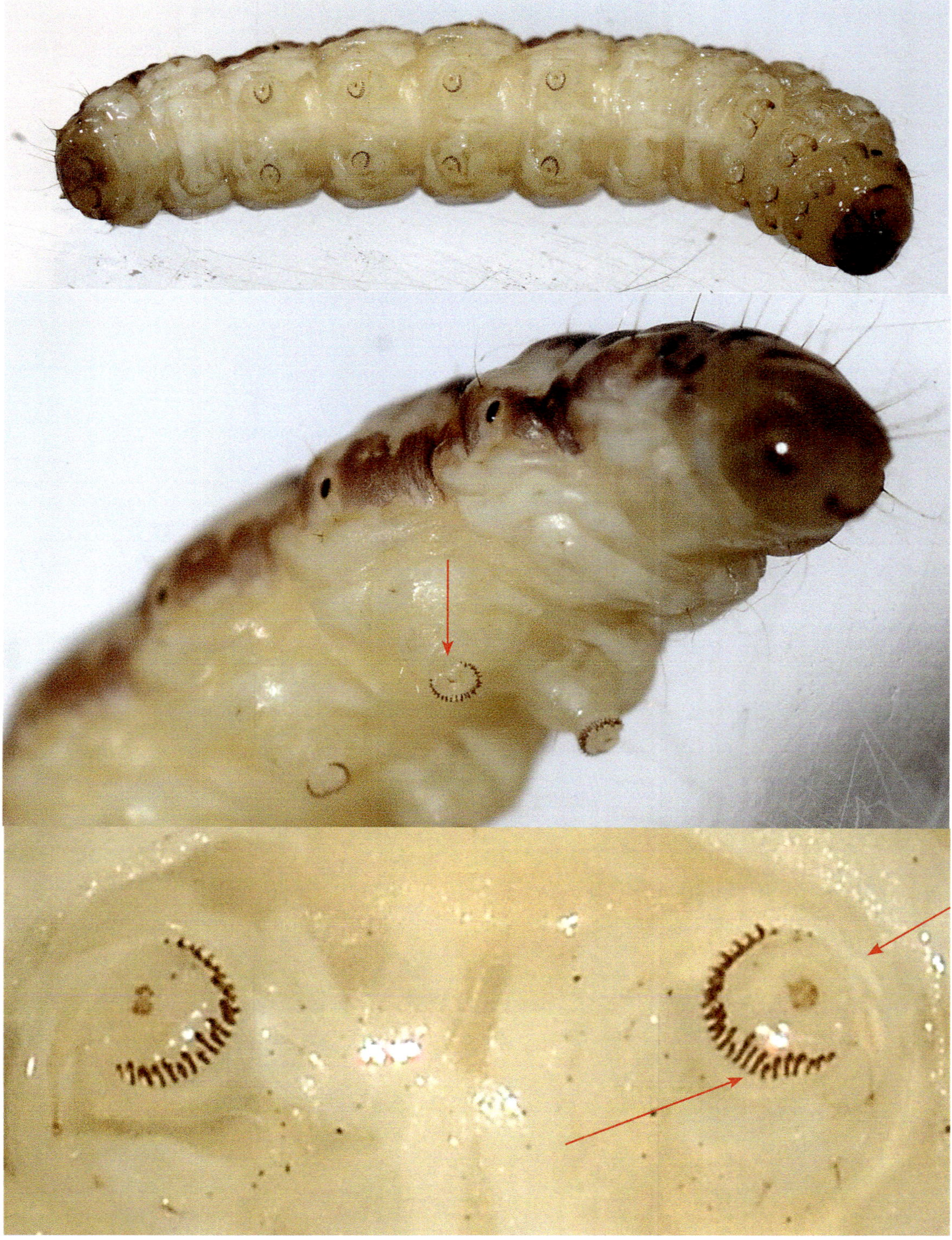

Plate 10.37. In certain situation the crochets on the prolegs remain incomplete. In that situation, the crochets appear *uniserial* but it continues to be *triordinal*.

Plate 10.38. Pupation takes place within the feeding tunnel. Prepupal larvae prepares an exit hole which remains covered by with the thin epidermal layer of the host (larva carefully keeps the epidermal layer intact). Larva prepares a loose silken chamber and pupates within it.

Plate 10.39. Pupae of Internode borer. Colour of the pupae vary from yellowish to reddish brown. A pupa appears stouter and bears four prominent stripes (like the larva) on the body.

Plate 10.40. Dorsal view of male and female pupae. The four stripes are present on the pupal body. The median *prothoracic* ridge is prominent.

Plate 10.41. Ventral side of male pupa. There is slight variations in wing pad area. Ventral abdominal side is smooth except for the marks of prolegs on 5th and 6th segments. The anal segments impart a solid appearance.

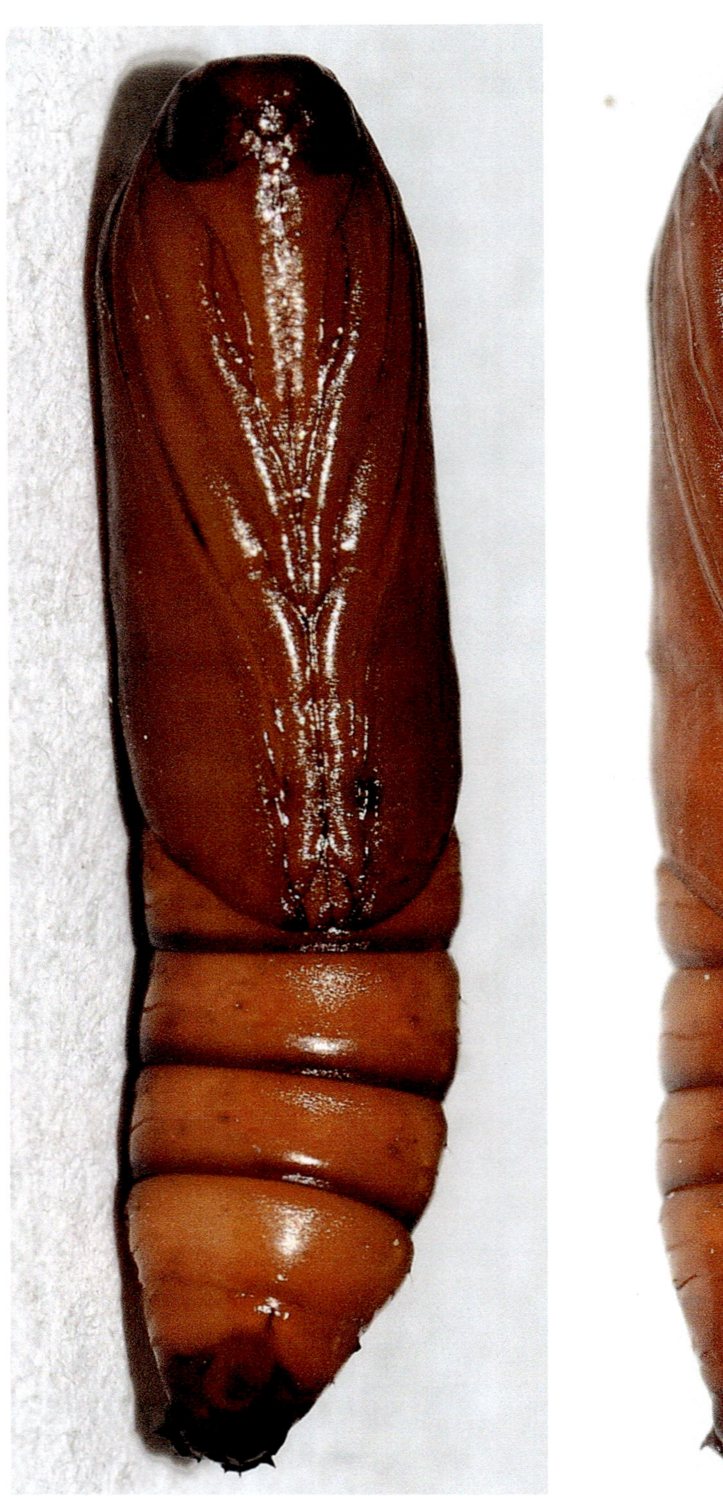

Plate 10.42. Ventral side of female pupa. Wing pad area forms a smooth curvature. Ventral abdominal side is smooth except for the marks of prolegs on 5th and 6th segments. The anal segments impart a solid appearance.

Plate 10.43. Side view of the female and male pupae. The velvety rough ornamentations on the 5th ,6th , and 7th abdominal segments extend a little to the subdorsal area.

Plate 10.44. Side view of a pupa (close up). The velvety rough ornamentations present on the dorsal side of a pupa on the 5th, 6th, and 7th abdominal segment extend a little to the subdorsal area.

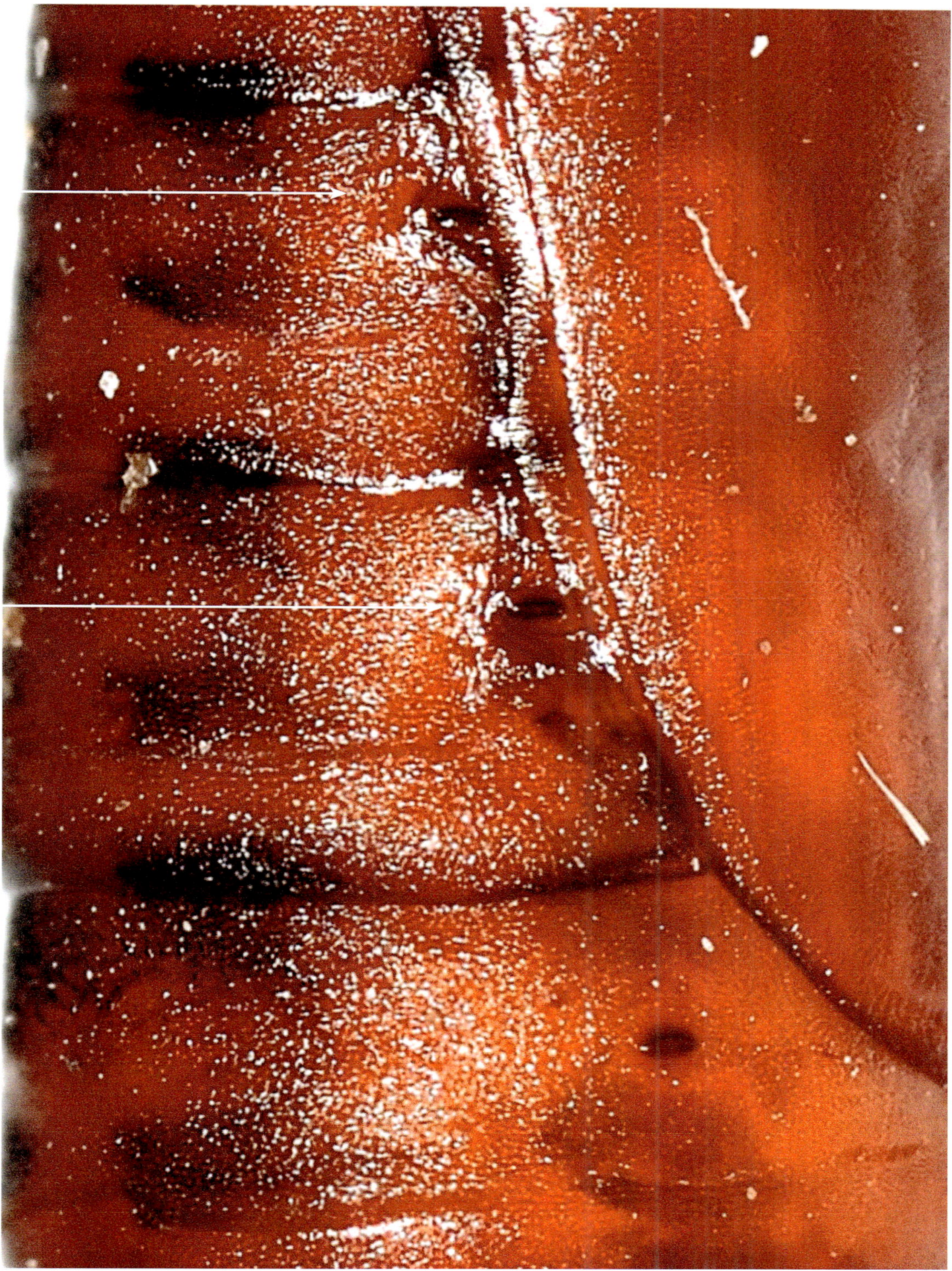

Plate 10.45. The two abdominal *spiracles* in the wing pad area support the development of arches above them. However, these arches are not very prominent.

Plate 10.46. Spine arrangements at the *cremaster* (close up) from dorsal side. There are six spines (in three sizes). Lateral pairs are larger, curved and with broad base. The subdorsal pair are medium and straight. The central pair or *meson* pair are tiny and often missed.

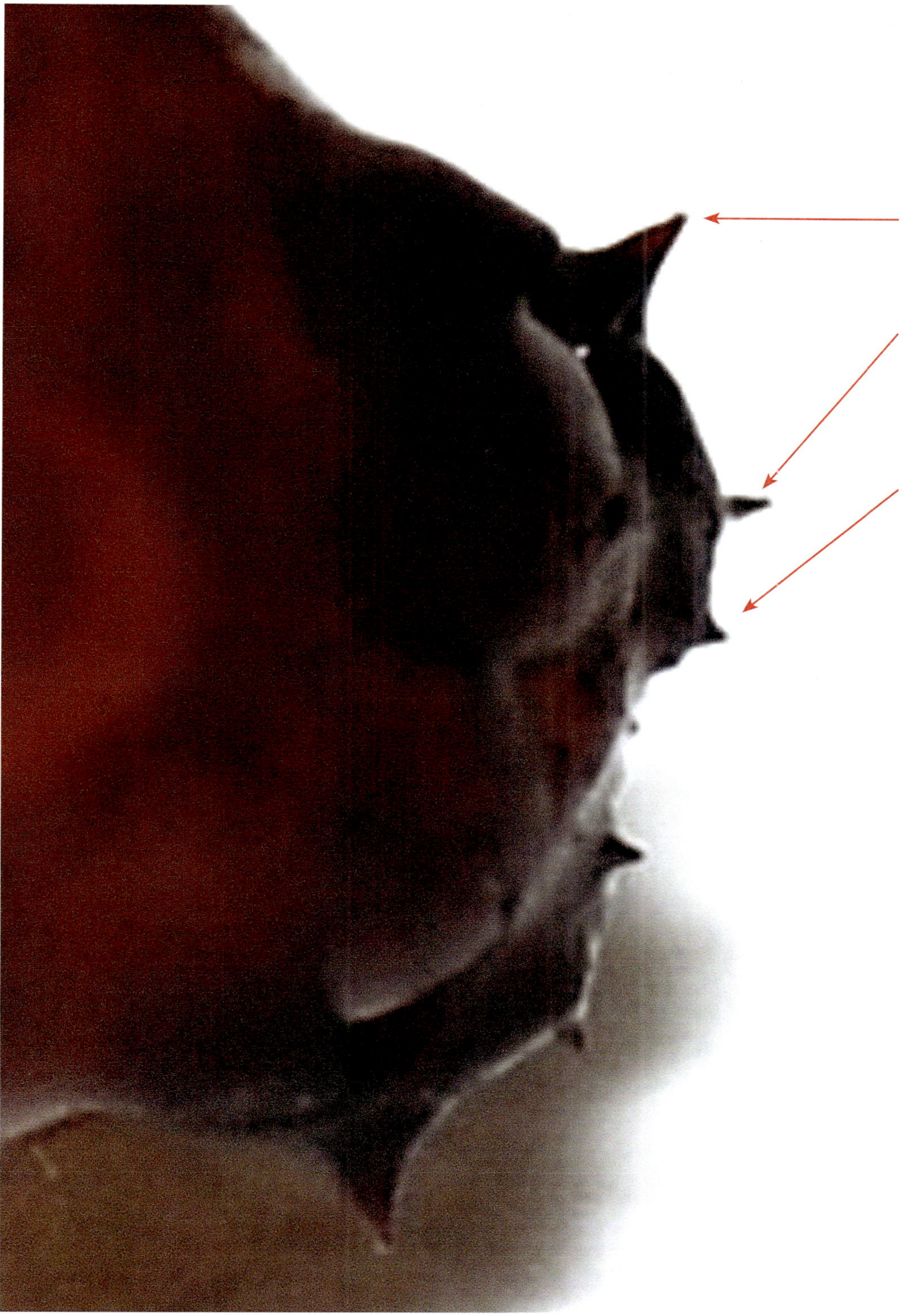

Plate 10.47. Spine arrangements at the *cremaster* (close up) from ventral side. On the dorsal side each half supports the development of three spines; large lateral spine, medium subdorsal and small central (*meson*) spine.

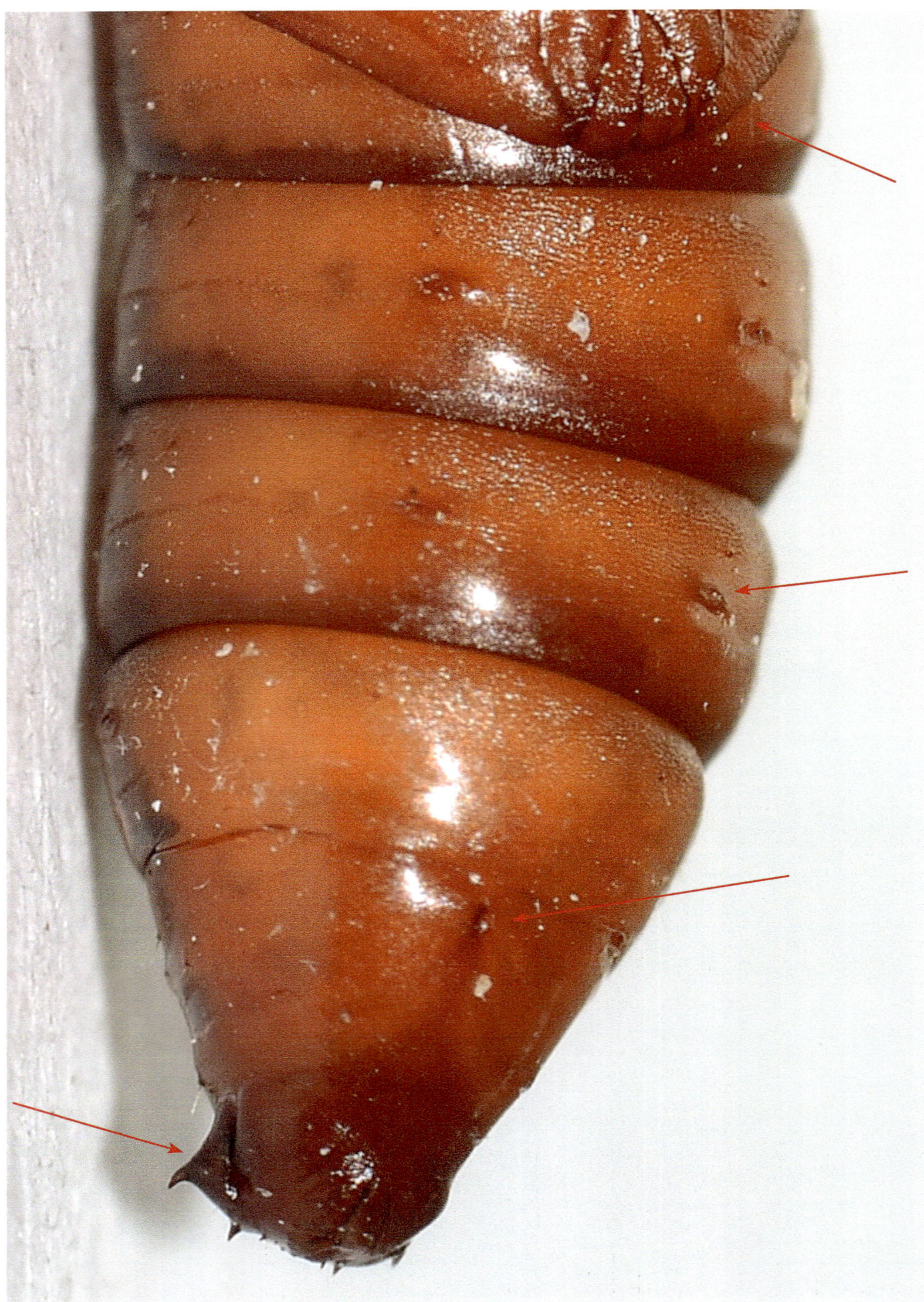

Plate 10.48. Ventral side of female pupa (close up). The wingpad area appear smooth and round and at periphery almost touches the end of the 4th abdominal segment. Marks of prolegs on 4th and 5th abdominal segments are prominent. Ventral side is smooth. Anal segment appears solid.

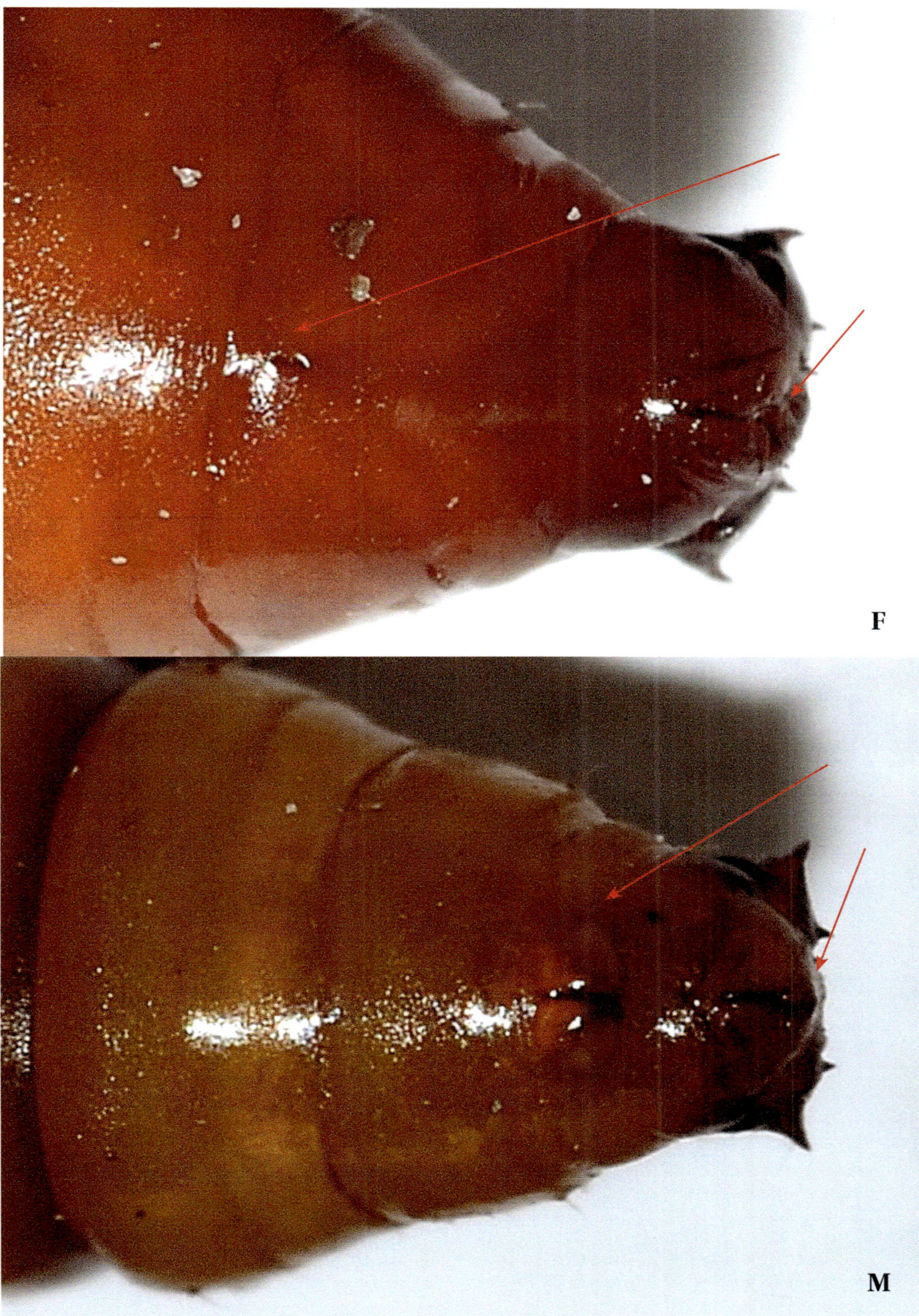

Plate 10.49. Genitalia of female and male pupae. In female it is represented by a small depression on the 8th segment. In male it is represented by two broadly raised lobes with slight median depression present on the 9th segment.

Plate 10.50. Female moth (dorsal view). Forewings with striations. Coppery scales run in the crest area, parallel to the costal end. A brownish spot at the *discal* region. Terminal brownish spots are present but spots are not prominent.

Plate 10.51. Female moth (dorsal and side views). Sitting habit, criss- cross antennae at sitting position. Relative size of eyes and porrect *labial palpi*. Presence of coppery scales in furrow area and *cubital discal* spot on the forewing. *Apex* is conical.

Plate 10.52. Male moth (side view). Sitting habit, broader antennae. Sides remain devoid of scales and appear light brown in colour . Relative size of eyes and porrect *labial palpi*. Presence of dark coppery scales (darker than female) in furrow area and also darker *cubital discal* spot on the forewing. *Apex* is conical.

Plate 10.53. Moth showing criss- cross antennae at resting position. A tuft of characteristic scales is present behind the eyes and continues to the thoracic region (in both female and male moths).

Plate 10.54. Head area of female moth (close up). Dorsal and ventral views showing relative size of eyes and porrect *labial palpi* (4 times of eye dia), scaled *proboscis,* pattern of scales on head, face is round and *filiform* antennae. Presence of characteristic scales behind the eyes.

Plate 10.55. Head area of male and female moths. Antenna is stout and doubly serrated in male moth, whereas in female moth it is *filiform*, finely serrated and fully clothed with scales. Antennae appear ribbed at the dorsal side. In male , lateral side of antenna is devoid of scales and appears light brown in colour. Characteristic tuft of scales behind the eyes are present in both male and female moths.

Plate 10.56. *Papillae anales*: It is broad bathroom slipper shaped (ventral side is broader) with numerous long sensory setae. A long gap is present between the two halves with a median depression.

Plate 10.57. Terminal black spots are prominent at the ventral side of the fore wings. Black spots are also present on the terminal area of the hind wings . Coppery scales are present at the costal area of hind wings.

Plate 10.58. Wing venation (forewing, female). *Apex* is extended like a cone. A_1 is rudimentary, A_2 is robust and takes slight upward curvature near the terminal area of *dorsum* and ends at *termen*. A_3 is feeble and reaches up to the middle of the wing but does not end at the *dorsum*.

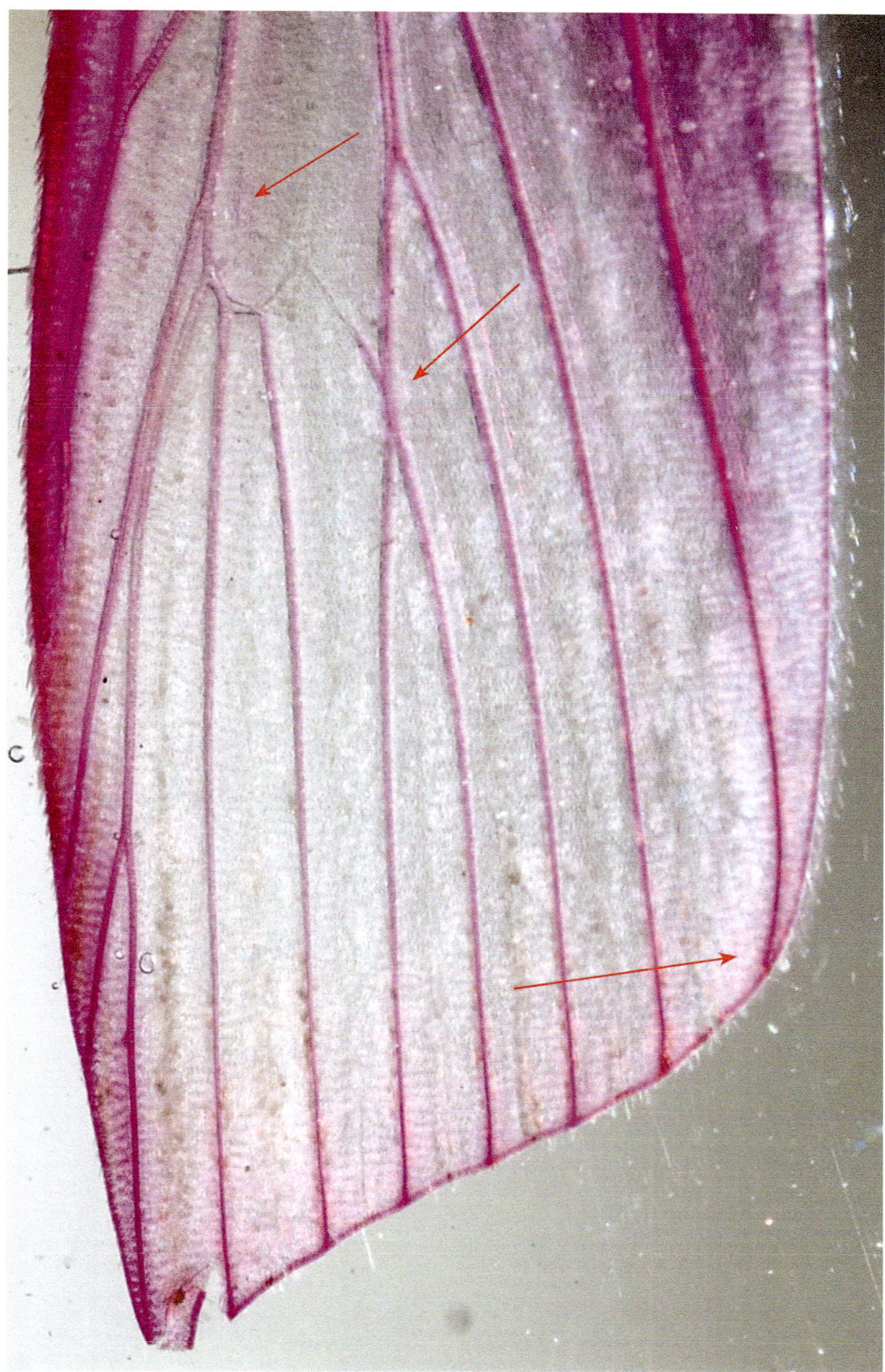

Plate 10.59. Forewing venation (close up). Atypical shape of *discal* cell. Second peak (*cubital* peak) is sharp and extended much above the first peak (*radial* peak) formed by the M_2 and M_3. Sc anastomoses with R_1 and form the combine vein $Sc+R_1$; it separates again to form Sc and R_1 at the terminal portion. R_2 is very close to the stalk of R_3 and R_4.

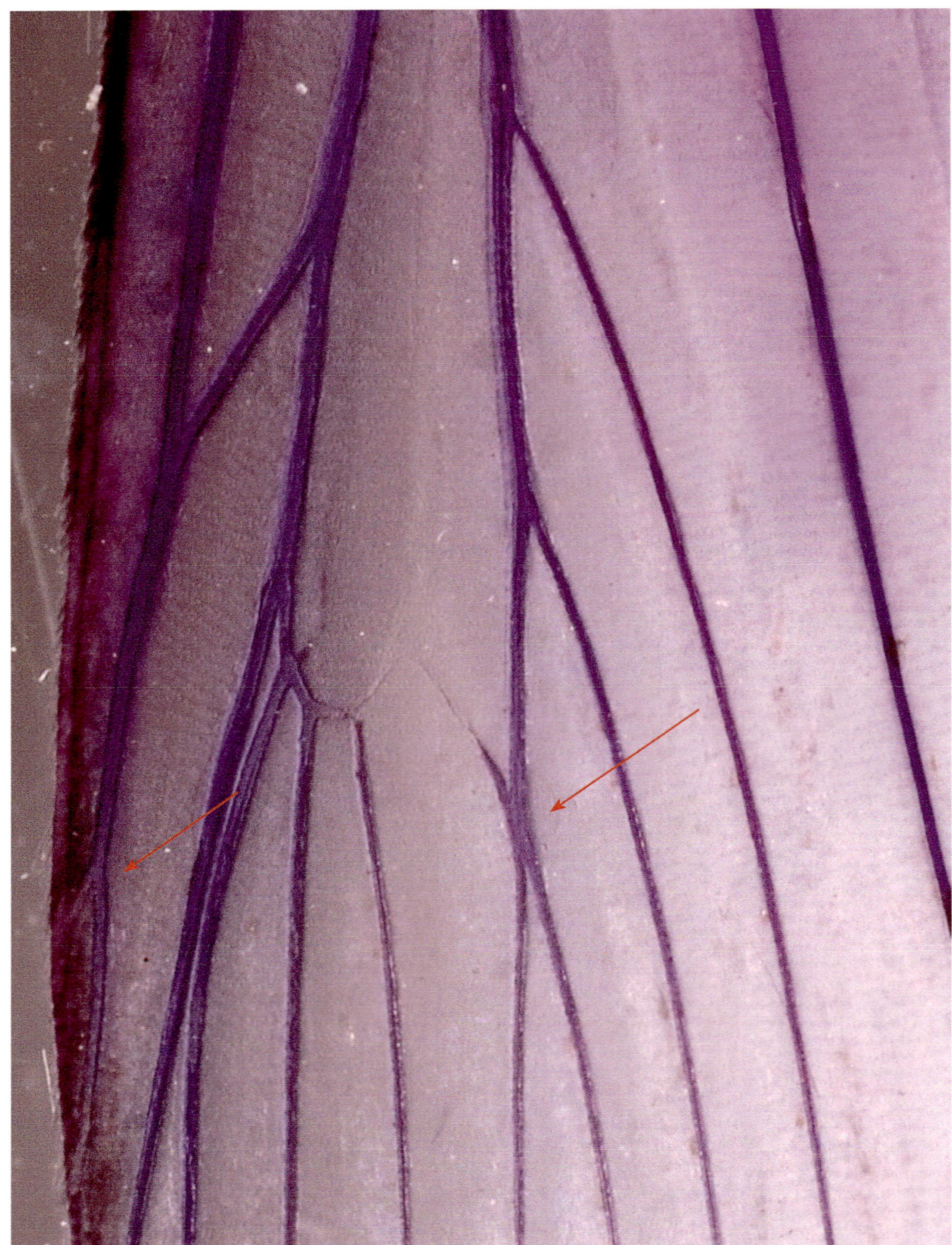

Plate 10.60. Forewing venation (close up). Atypical shape of *discal* cell. Second peak (*cubital* peak) is sharp and extended much above the first peak (*radial* peak). M_2 tangentially touches and M_3. Bifurcation of Sc+R_1 and both veins end at costal margin. Stalk of R_3+R_4 takes a curvature and appear very close to R_2. M_2 and M_3 just tangentially touches to form the *cubital* angle.

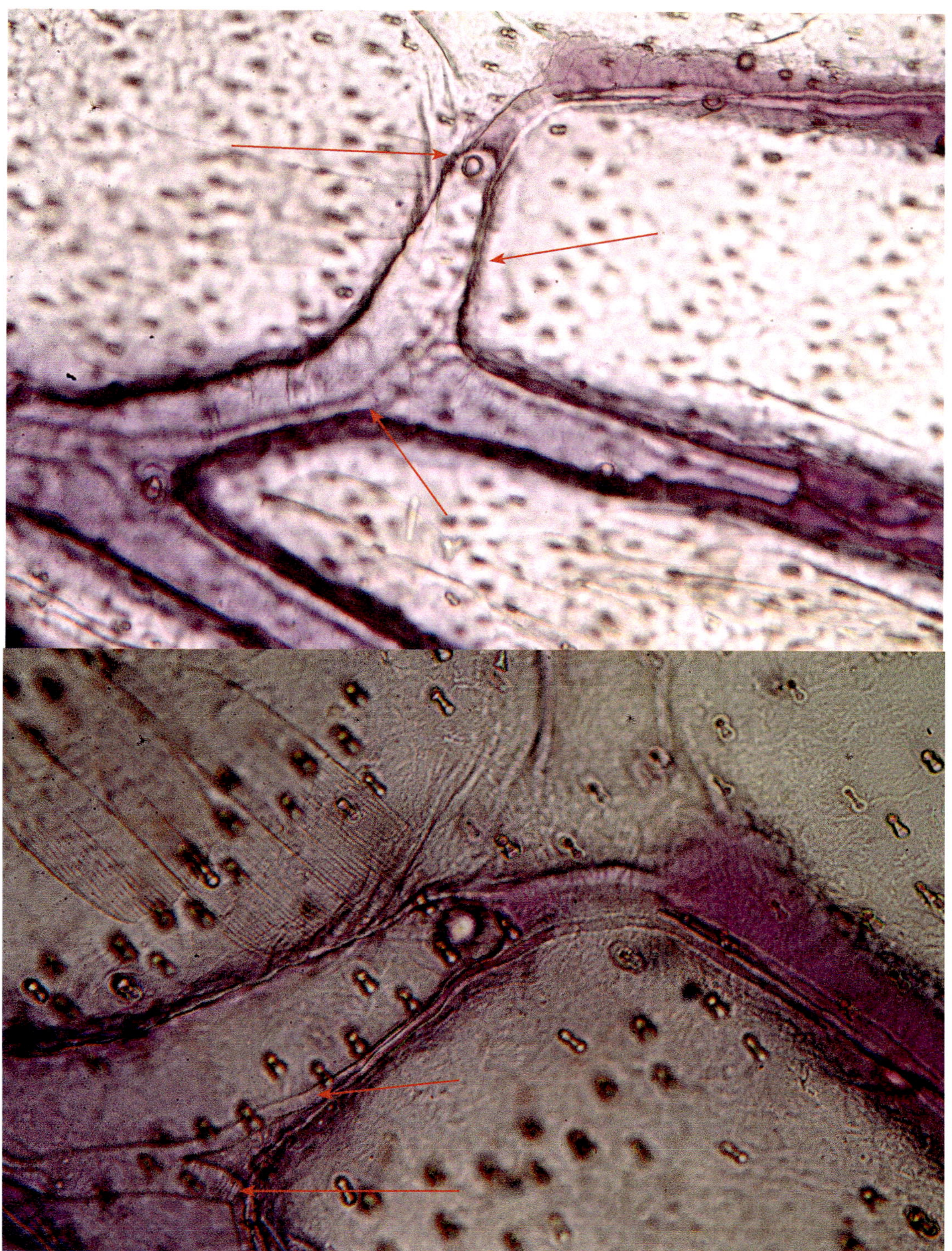

Plate 10.61. Bifurcation of *tracheal tube* from R_5 to reach the M_1. Presence of round *mechanoreceptors* on the veins.

Plate 10.62. Egg mass of Internode borer. Freshly laid eggs are translucent white, gradually change colour (due to the development of larva). At black head stage prominent orange girdling become visible. Egg mass may be found on both upper or lower surface of the leaf lamina. Egg mass may be located even on the mid-rib.

Plate 10.63. Size and two rows arrangement egg mass. Eggs are layed in an imbricate manner of two rows of eggs. Heads of the developing larvae remains oriented to the first laid egg of the egg mass.

Plate 10.64. Emergence of a larva from the egg shell. A larva cuts a small slit on the egg shell and comes out. Larvae emerges out in batches.

Plate 10.65. Left over egg shells of egg mass after the emergence of larva. A larva cuts a small slit on the egg shell and comes out.

Plate 10.66. Parasitized egg mass turns black or dark. Adult parasitic wasp emerges by making round holes on the egg shell.

11

Gurdaspur borer, *Acigona steniellus* (Hampson)

Gurdaspur borer, *Acigona steniellus* (Hampson) is also a destructive pest of sugarcane but localized in its occurrence. This species of moth was first recorded by Hampson (1899) from a collection received from Tezpur, Assam (in the collection of moths by Mr. Rothschild) and he described it as *Chilo steniellus* Hampson. In the same year, he also named a similar moth as *Donacoscaptes steniellus* Hampson.

4202*a*. Chilo steniellus, n. sp.

♂. Head and thorax pale grey-brown; abdomen whitish, dorsally ochreous towards base. Forewing long and narrow; pale grey-brown irrorated with fuscous and suffused with fuscous in the interspaces, especially the cell and discal fold, leaving the veins slightly paler; traces of two very oblique lines from below apex to middle of inner margin; a marginal series of black points. Hindwing pure white.

Habitat.—Tezpore, Assam. *Exp.* 30 mm. *Type*—In coll. Rothschild.

However, after the first record of Hampson (1899), this moth was never reported from the eastern part of India (Assam), rather around two decades later, a "*New pyralid borer*" was observed damaging the sugarcane crop in the western parts of India (Sialkot, now in Pakistan). At a later date, this "*New pyralid borer*" was identified and described as *Chilo trypetes* by Mr. G. A. Bisset. Later on this moth was identified as the same moth that had Hampson (1899) identified earlier and finally it was named as *Acigona steniellus* (Hampson) by Bleszynski (1970).

As a pest of sugarcane, it was first recorded in 1923 from a rain-fed sugarcane crop from Sialkot district, Punjab province (now in Pakistan) by Mr. M. Afzal Hussain. He called this borer a "*New pyralid borer*", as the borer inflicted altogether a different kind of damage to the sugarcane stalk. Moreover, the larva was not very similar to the known larvae of moth borers that were described by Fletcher and Ghosh (1919) in the Proceedings of Third Entomological Meetings at Pusa, Bihar.

However, a good description of an unidentified moth borer from *Saccharum spontaneum* as furnished on page 393 of Fletcher and Ghosh (1919) indicates a vivid description of the Gurdaspur borer. The first hand account is furnished below.

PROCEEDINGS OF THE THIRD ENTOMOLOGICAL MEETING 393

? Pyralid borer.

Plate 64, fig. 1.

Foodplant—*Rarhi* and *Batri* (*Saccharum spontaneum*).

These caterpillars, which we have not yet been successful in rearing, occur in very large numbers. The hard stems of grown-up plants are converted into hollow tubes throughout their length, a single larva usually occurring in individual plants. The heart shoot dies and dries only when the larva bores near the top.

The insect has only one generation in the year. Young larvæ are found about July and they feed up to about the early part of November, by which time they become fullgrown. Then they rest through the winter and the hot weather, pupating and emerging as moths about June-July. The larvæ have been observed to rest in the Insectary up to the middle of August and then die.

The fullgrown larva is about 35 mm. long. The shape is cylindrical or rather semicylindrical being slightly compressed dorsoventrally. The head is glossy-yellow. The segments of the body have an elongated and slightly chitinized and glossy appearance. The hairs are small and black. The spiracles are elongated oval, brown with a dark narrow rim. The five pairs of prolegs are short, equally developed and have their hooklets in complete circles.

Hereafter, this moth borer pest was recorded affecting large areas of sugarcane crop in the Gurdaspur district of Punjab, India. This "*New pyralid borer*", got the common name "***Gurdaspur borer***" due to its prevalence in the district of Gurdaspur (in local language it was popularly known "*Gurdaspuri garuan*"). Moths collected from the borer infested sugarcane field of Punjab were sent to Mr. G. A. Bisset for identification (at the British Museum of Natural History, London). He described this "*New pyralid moth*" as *Chilo trypetes* Bisset (Bisset, 1939).

Morphologically, the *Chilo* and *Acigona* adults as well as the larvae are very similar. Revising the world species of the genus *Chilo*, Bleszynski (1970) transferred this species to *Acigona* Hubner. Since then the name, *Acigona steniellus* (Hampson) is in vogue for this species of moth borer. In Uttar Pradesh, its incidence was first observed in 1938 in some localities of Dehradun district of Uttar Pradesh (now in Uttarakhand) and thus, it was earlier called "*Doon borer*".

This borer was recorded in almost all the cane-growing areas of Punjab, Himachal Pradesh,

Haryana, and Rajasthan. Gurdaspur borer has also spread to the adjoining districts of U.P., like Saharanpur, Meerut, Muzaffarnagar and cane growing areas of Uttarakhand like Pantnagar, Dehradun, etc. However, during recent years incidence of Gurdaspur borer has gone down considerably.

11.1 Currently accepted name

Acigona steniellus (Hampson)

11.2 Synonyms

= *Chilo steniellus* Hampson

= *Donacoscaptes steniellus* Hampson

= *Chilo trypetes* Bisset

= *Bissetia steniella* Kapur

11.3 Symptoms of infestation and brief biology

The infestation of Gurdaspur borer takes place only after the summer months have elapsed and ambient temperature has gone down with the arrival of monsoon rains. This period mostly coincides with onset of grand growth phase in the sugarcane crop. At this stage, due to the rapid growth of the crop, the infestation of first brood of Gurdaspur borer mostly remains unnoticed. It is in the second and third brood that its infestation becomes apparent in the crop; the affected cane plants turn pale yellow. Later on, the affected plants die.

In a brood, Gurdaspur borer has two definite phases of infestation of the sugarcane crop viz., *gregarious phase* and *solitary phase*. In the *gregarious phase* of infestation, congregation of neonate larvae takes place; usually all the neonate larvae of one egg mass assemble at one place and move like a small battalion to attack. The leader of the larval group first makes a hole for entry into the cane at a place of its choice. Mostly the hole is made in the upper immature internodal portion of the stalk and the leader larva makes an entry into the cane. The other larvae follow suit, and usually all the follower larvae wait for their turn and use the same entry point for making entry into the cane. The entry-hole gradually broadens due to feeding of the larvae following the leader. In the *gregarious phase*, around 20-25 larvae are usually found feeding the internal tissues of an infested immature internode of the cane. The *gregarious phase* of Gurdaspur borer is mostly limited to first and second instar larvae; only immature larvae share the same feeding space. (In contrast, in Plassey borer the *gregarious phase* continues further and many a times a number of mature larvae share the same space and feed at one place; one may find several pupae at the common feeding tunnel). From third instar onwards, when the larvae have attained sufficient size and strength, larvae of the *gregarious phase* try to lead an independent life and move out singly to find a new place to feed. From the third instar, larvae enter into the *solitary phase.* A larva usually migrates to the adjoining cane plant; generally only one larva is found in most of the affected cane (presence of two or three larvae in a cane is not uncommon). This *solitary phase* is considered the most damaging phase of this pest as it suddenly engulfs the entire sugarcane crop from almost negligible incidence.

A third instar larva of Gurdaspur borer, after reaching the new location searches for a suitable spot in the upper portion of the immature cane stalk for making an entry hole. Usually a larva feeds on the cane tissues at the nodal region; first at the bud or adjacent to the bud. Many a times, a larva enters directly into the stalk by boring through the bud and then it proceeds further inside the internode by feeding the internal stalk tissues. In the beginning, a larva feeds only the peripheral tissues (just below the epidermal layer) of the internode and produces a shallow feeding tunnel just beneath the epidermis. By feeding the intermodal tissues, a larva gradually moves upward in the internode in a characteristic manner; the upward spiral journey of larva may be *clockwise* or *anticlockwise*. During its upward feeding, a larva makes a series of characteristic small holes in the epidermal layer, making its path of feeding journey visible externally. Once the leaf sheath is removed, this passage mark appears as a dark spiral streak with tiny and dark punctured areas in the epidermis. When around half to two third of the internode gets damaged in this manner, the caterpillar moves to the central portion or the pith area of the internode. Hereafter, the larva eats away the internodal tissues vertically upward and thus, makes a long straight and relatively wider tunnel in the central portion of the cane. This type of initial spiral streak formation on the cane, followed by a straight and wide upward central feeding tunnel is the very characteristic of the attack of Gurdaspur borer. (In contrast, the Internode borer produces a semicircular streak on the rind and usually it is not followed by a central feeding tunnel. If at all a feeding tunnel develops, it never remains straight, rather it shows characteristic horizontal feeding areas / labyrinth formation throughout the length of the tunnel.)

In the *solitary phase*, crown of the infested plant shows distinctive yellowish-orange leaves (from a distance, appearance of the crown leaves is quite similar to that of red rot affected cane. In the borer infestation, only top few leaves turn yellow, the lower leaves always remain green. As the infestation progresses further, the entire affected crown dries up (quite similar to that of Plassey borer infestation). The spirally damaged stalk becomes weak and breaks off easily with the gush of wind or with the movement of an animal shaking the cane, (similar type of breakage of stalk is also observed in the infestation of Internode borer).

Gurdaspur borer is truly a cane borer in the sense that its attack commences when the shoot stage of the crop is over and sufficient stalk formation has taken place. This *Monsoon borer* emerges in the scene only after the summer period is over, and the sugarcane crop is in the grand growth phase. Gurdaspur borer appears late and disappears early from the scene. However, during its short presence (only two effective broods), it can damage the sugarcane crop beyond recognition. Due to its preference for stalk, early cane forming varieties always bear the burnt during the appearance of the first brood, as the borer larvae prefentially attack the available cane stalks in the field.

Larvae of Gurdaspur borer hibernate during winter months in the stubble portion of the cane and remain in an inactive or quiescent state. The hibernation is unusually prolonged and it continues up to summer months. Moths of Gurdaspur borer start emerging during middle of the summer and continue until late summer (probably pre-monsoon shower provides the requisite cue for emergence). The newly emerged female moths, by laying eggs initiate the first generation in late summer period. As the larvae hibernate in the stubble, the stubble crop or the ratoon crop always receives higher incidence of the borer infestation in comparison to the corresponding plant crop.

With the lowering of ambient temperature in October-November, the caterpillars feeding at the top portion of the cane leave the feeding site and move out. These larvae (different instars) migrate to the lower portion of the same cane or to the adjoining canes. After reaching the lower internodes of the stalk, a larva bores into the cane internode. Entering in the lower portion of the stalk, a larva goes further down internally by boring through the central portion or the pith region of the cane (usually a larva goes below the soil level) where the larva subsequently hibernates. Once the period of hibernation is over, a larva prepares an exit hole and pupates in the tunnel. Moths begin to emerge in a week or so and in about a fortnight, the infestation symptoms of Gurdaspur borer are noticed in the crop (mostly in late June-July).

The Gurdaspur borer remains active only for a brief period i.e., around five months (June-July to October-November) of a calendar year and it remains inactive for the major period i.e. from November to May next year (almost seven months, spanning two calendar years). Depending on the prevailing weather, a minor variation may be observed. In a calendar year, the Gurdaspur borer only passes through two to three generations. From June-July to October, all life stages of the borer may be observed in an infested cane field. Due to the presence of different life stages in a staggered manner, overlapping broods are also common in this borer.

11.4 Egg and neonate larva

Eggs are laid in clusters in and along the groove of the mid-rib on the upper surface of the leaf near the leaf sheath. Freshly laid eggs are white to pale cream in colour and gradually change to dark grey. Newly opened leaves of the crown are preferred by the female moths for egg laying. One egg cluster usually contains around 20 to 50 eggs. After a week of incubation, hatching takes place. The neonate larvae of one egg mass congregate at one place and feed for a certain period on the surface tissues of the crown leaves. After certain growth, the leader larva ventures out and makes a feeding hole in the immature portion of a cane stalk. Rest of the members follow the leader and enter the cane through the same entry hole.

11.5 Larva

Among the striped borers of sugarcane, mature larva of Gurdaspur borer and that of Internode borer are usually a bit lengthier and stouter than the larvae of other moth borers. A mature Gurdaspur larva measures around 30-35 mm in length. Larva is slender, slightly flattened dorsoventrally and rather blunt at the anal end. Larval head is light orange to light brown in colour. *Prothoracic shield* is well developed and light brown in colour with patches of dark brown area.

In the larval stage, like Plassey borer, Gurdaspur borer also exhibits variations in the formation of coloured stripes and coloured tubercles on the body. However, in a typical larva, the presence of four pinkish brown stripes (two dorsal to sub-dorsal and two very dark laterals) and an incomplete-circle arrangement of *uniserial* and *uniordinal* crochets in the four abdominal prolegs help in its correct identification. The larva of Gurdaspur borer is basically white in colour and four prominent stripes run through the larval body. The mid-dorsal area (between to sub-dorsal stripes) and the area between the two lateral stripes appear distinctly white.

The sub-dorsal stripes are narrow and light pinkish brown in colour, whereas the lateral stripes are broader, very prominent and pinkish brown to black in colour. The lateral stripes are much wider, cover wider area of the lower lateral portion and extend well below the *tracheal trunk*. When present, these dark lateral stripes help in quick identification of the species in the field. The sub-dorsal stripes along with the characteristic dark coloured lateral skin folds sometimes impart a flat square appearance (dorsal view) to the larval body rather than the usual cylindrical appearance of larvae of other moth borers of sugarcane. *Spiracles* are elongated and distinctly black but mostly remain masked by the overwhelming presence of the dark lateral stripes that run through the *tracheal trunks*. These stripes usually become lighter and may completely fade away during hibernation. The dorsal tubercles (D_1) are large but not prominent due to their light grey colour. Smaller tubercles (D_2) that follow the larger tubercles are relatively larger in Gurdaspur borer in comparison to the larvae of other moth borers. Presence of setae on the larval body is also prominent.

Crochets or hooks on the abdominal *prolegs* of a larva characteristically form an incomplete circle; hooks are present at the inner side (*meson*) (in about 40% circumference of the imaginary circle outer side or lateral side remains without hooks). Hooks are dark coloured, stout, *uniordinal* and *uniserial*. In resting position crochets appear as if arranged in a straight line (larva uses only few crochets of the interior region of *proleg* for the movement). In the season, depending on the prevailing weather, a larva may take about five to six weeks to attain maturity.

11.6 Pupa

Like the pupae of other moth borers of sugarcane, a pupa of Gurdaspur borer also appears stout and slender. Freshly formed pupa is creamy white in colour and gradually changes to yellowish brown. A pupa of Gurdaspur borer measures around 14-19 mm and it also retains the dark lateral stripes running through the *spiracles*. Front of the pupal head is distinctly conical in shape. Characteristically, the pupa of Gurdaspur borer does not bear any specialized ornamentations on 5th, 6th and 7th abdominal segments. However, numerous microscopic spines are present in the abdominal segments. The dorsal side of the anal segment (*cremaster*) has two prominent upwardly curved projections. At the ventral side, there are three smaller spines in each half.

11.7 Moth (Adult)

The moth of Gurdaspur borer is light straw (ashy straw) coloured. Forewings are creamy yellowish white (ashy white) or pale grey with pale yellow base, numerous greyish spots (due to development of dusky brown terminal area in many scales) appear on the entire forewings. The seven terminal black spots on each of the forewings are also prominent.

Labial palpi are porrect, prominent, bushy and relatively longer and are more than three times of the diameter of the compound eye. The compound eyes appear white in the photograph. The face of the moth is distinctly conical with a very prominent corneous point that is sharp like a needle head. (The maize borer, *Chilo partellus* has also a conical face with a sharp corneous point). Like the adults of Internode borer, *Ocellus* is characteristically absent in the adults of Gurdaspur borer. In male, the antennae are serrated as well as finely ciliated. The joints of the male antenna are very broad and widely located - a character that differentiates this species from rest of the moth borers. In the female moth, antennae are simple and *filiform*.

11.8 Forewing venation

Usually the Sc and R_1 remain free and both end at the costal area. R_1 takes a curvature towards Sc. However, in certain cases Sc and R_1 may cross each other. R_2 is free. R_3 and R_4 are stalked. R_3 and R_4 end at the costal area. R_3 is relatively smaller than the other moth borers. R_5 is free and ends at *termen*. Cu1a and Cu1b are present. M_1 is free. M_2 and M_3 arise from the same point from the lower cell angle (*cubital* angle) and both end separately at the *termen*. A_1 is absent. A_2 is well developed and ends at *dorsum*. A_3 is rudimentary.

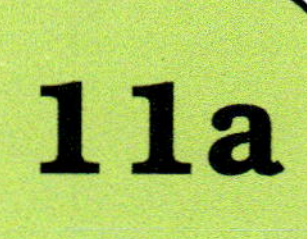

List of plates on Gurdaspur borer, *Acigona steniellus* (Hampson)

Description of plates

Plate 11.1. External symptoms of infestation of Gurdaspur borer. Typical spiral girdling of the cane in the internode. The girdling and the hole appear after removal of the leaf sheaths. Borer enters above a nodal area and continues feeding in the upper portion of the internode.

Plate 11.2. Internal symptoms. Initial cross wise tunnelling (mostly remains filled with frass) and formation of peripheral holes. At mature larval stage, a larva forms broader clean central feeding tunnel near the immature portion of the stalk.

Plate 11.3. A larva of Gurdaspur borer is basically creamy white with four prominent stripes. Lateral stripes are broader and darker and hide the visibility of the *spiracles*. Dorsal stripes are placed more subdorsally and in dorsal view impart squarely appearance to the body.

Plate 11.4. Larvae of Gurdaspur borer (dorsal view). Dorsal tubercles are apparent on the stripe and almost similar in colour with the stripe. Larva bears prominent setae on the body. Anal end is broader and larva appears squarely throughout its length.

Plate 11.5. Larvae of Gurdaspur borer (Side view). Due to typical placement of sub-dorsal stripes, the body appears squarish. The creamy white colour of the body is prominent. Lateral stripes are wider and darker. Lateral skin folds are large and distinct. Head capsule is yellowish-orange.

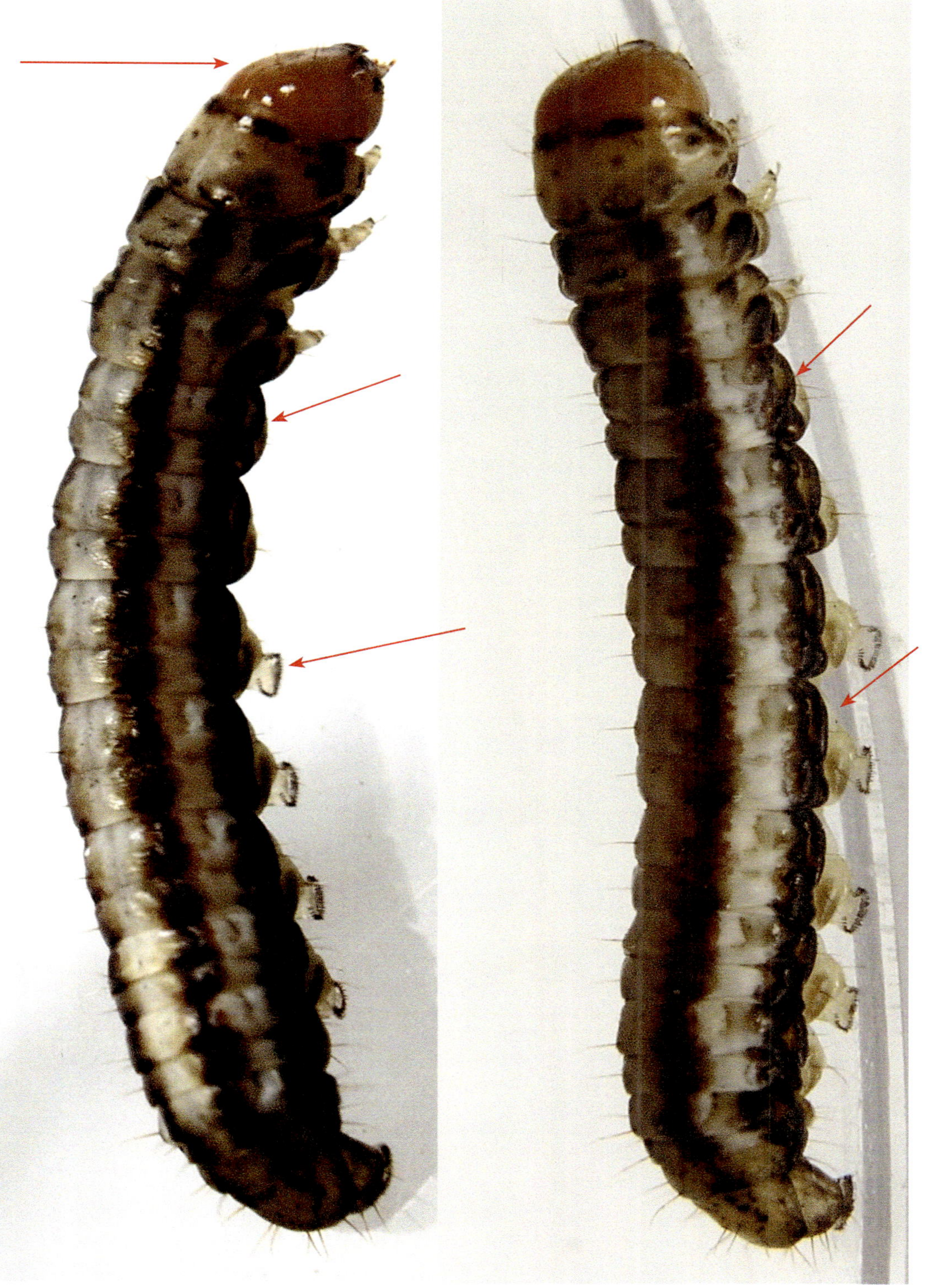

Plate 11.6. Larvae of Gurdaspur borer (side view). Lateral stripe is wider and darker and hides the *spiracles*. The creamy white colour of the body is prominent. Lateral skin folds are large and distinct. Head capsule is yellowish-orange.

Plate 11.7. Larvae of Gurdaspur borer. (close up, side view). Lateral stripe is wider and dark, and skin folds are very prominent. Crochets follow *uniserial* and *uniordinal* arrangements on the abdominal *prolegs*. Hooks form characteristic *incomplete circle*; about 1/4th of the hook line of a proleg remains vacant.

Plate 11.8. Lateral stripe is wider and dark and skin folds are very prominent and the colour retain even after preservation in alcohol. Crochets are *uniserial, uniordinal* and *incomple.*

Plate 11.9. Female moth of Gurdaspur borer. Antennae are *filiform*. *Labial palpi* porrect and about 3.5 times of the diameter of compound eye that appear white in photographs. Terminal portion of many scattered scales on the forewings are dark, and thus impart a distinctive appearance to the moth.

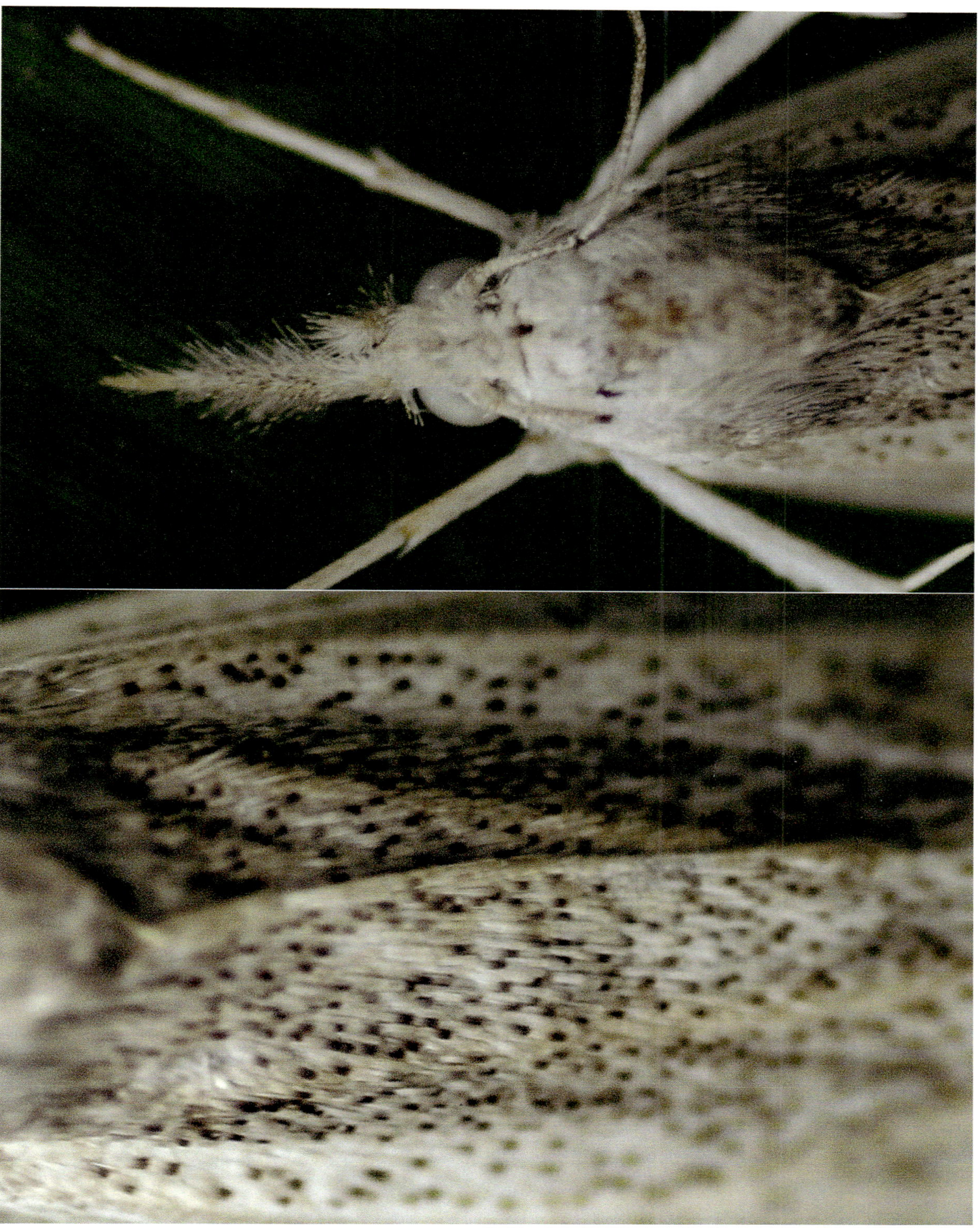

Plate 11.10. Female moth of Gurdaspur borer. *Labial palpi* porrect and conical. Antennae *filiform*. Small dark scales are present through out the forewings.

12

Plassey borer, *Chilo tumidicostalis* (Hampson)

Plassey borer, *Chilo tumidicostalis* (Hampson) is another important pest of sugarcane causing considerable concern. This pest is endemic in the eastern part of India, starting from eastern Uttar Pradesh and extending up to Assam. Of late, this moth borer is spreading towards the central districts of Uttar Pradesh. The author has noticed the presence of this moth borer in the districts of Hardoi, Shahjahanpur and Lakhimpur-Kheri. However, the intensity of infestation was always limited. In 1901, Mr. E. P. Stebbing (of the Indian Forest Service who was then stewarding the Indian Museum) visited some places of north Bihar and published his account in the *Indian Museum Notes Vol. 5, page 164-182* with heading "*VII - Insect pests of the sugarcane*" (please refer ***Appendix III*** for the details). The description clearly indicates the presence of Plassey borer in the district Champaran, Bihar (though Stebbing identified the borer as *Chilo simplex*).

However, it was Hampson (1919) who first described the adult of this moth borer as *Argyria tumidicostalis* from Bengal. Fletcher and Ghosh (1919) recorded this borer from the districts of Pabna and Dhaka (now in Bangladesh) and placed it under the genus *Diatraea* (C.S. 1610 with the footnote that Sir George Hampson had named it *Argyria tumidicostalis*, Hampson). However, in the plates provided in that publication, Fletcher and Ghosh (1919) wrote the name *Argyria tumidicostalis*, Hampson (Plate 52, 53, 54. Please refer *Appendix IV* for the details). Kapur (1950) transferred this borer in the genus *Chilo* as *Chilo tumidicostalis* (Hampson). It got the specific epithet due to the very prominent costal area of the forewing.

12.1 Currently accepted name

Chilo tumidicostalis (Hampson)

12.2 Synonyms

= *Argyria tumidicostalis* Hampson

= *Chilo gemininotalis* Hampson

12.3 Symptoms of infestation and brief biology

This borer, though recorded way back in the beginning of the 20th century, took another four decades to realise its destructive potential. Probably, during this period, due to very similar appearance of maize borer larva, sugarcane workers could not realise that the larva obtained from sugarcane is of a different kind of borer. However, the borer came into prominence in 1956, when a large sugarcane area under Ramnuggar Cane and Sugar Co., Ltd. at Plassey, District Nadia (West Bengal) was infested by this moth borer species. In the factory farm, an incidence of about 95% was recorded during August 1956 in several sugarcane varieties viz., Co 419, Co 453, Co 527 and BO 11. Consequently, the sugar factory had to bear heavy losses in both the cane tonnage and sugar recovery. An SOS was received by the IISR, Lucknow and consequently Mr. B. D. Gupta, the Central Sugarcane Entomologist and his Senior Research Assistant Mr. P. N. Awasthy were entrusted to look into the matter. They immediately visited the borer-ravaged sugarcane fields at Plassey, West Bengal. This team stayed there and made a detailed study of this moth borer. Mr. B. D. Gupta named the moth borer as "***Plassey borer***" due to the epidemic it caused at Plassey, West Bengal.

In a brood, like that of Gurdaspur borer, Plassey borer infests sugarcane crop in two phases viz., *gregarious phase* or primary infestation and *solitary phase* or *pseudosolitary phase* or secondary infestation. In the *gregarious phase* of infestation, larval congregation takes place. Usually, all the freshly hatched larvae of an egg mass congregate at one place (*gregarious* in nature). Usually around 20-30 neonate larvae form a group and enter into an immature internode and collectively feed inside the cane tissues. In the beginning, the neonate larvae of Plassey borer appear sluggish. They feed by scrapping the leaf surface and gradually move towards the base of lamina. Finally, these larvae make an entry in the space between leaf sheath and the growing stalk. Once the larvae find the suitable portion of the immature cane stalk, they make feeding holes and enter within the stalk tissue. Collectively, all these larvae eat away the internal tissues of the internode and form a big central hollow filled with dark messy frass. The surroundings of the feeding central hollow of the internal stalk also turn dark brown or black.

In case of primary infestation, more than 100 caterpillars may be found feeding in an intermingled manner in the internodal cavity staying in the midst of frass. Due to the presence of a large number of larvae at one internode and their active feeding on the surrounding stalk tissues, many relatively large feeding holes are also formed on the epidermis. From these multitude of feeding holes fresh shining red frass usually oozes out. All these actively feeding larvae collectively devour up almost the entire portion of the internal contents of the young stalk. Due to this type of feeding, the water and nutrient transport to the crown portion of the cane is greatly impaired and the entire crown turns yellow and eventually dries up. The clasping leaf sheaths help to retain the weight of drying crown. Such affected crowns often topple down with slight jerk. Presence of multitude of irregular holes (as if '*hazar dwari* of *Mushidabad*' a famous building with multitude of doors and is located near Plassey, is quite an indicative of the formation of a multitude of exit holes) on the upper portion of the stalk is the very characteristic of the infestation of Plassey borer. Borer holes may be observed even without removal of the leaf sheath. In addition, profuse rooting as well as sprouting of buds in the upper unaffected portion of the cane (just below the damaged area) takes place. Drying up of the entire crown of sugarcane (in the eastern states of the country) is very characteristic of the primary infestation of Plassey borer. (In the western states similar drying is observed due to infestation of Gurdaspur borer).

Like most of the other moth borers of sugarcane, Plassey borer also starts emerging with the advent of favourable temperature in February-March (in the eastern part of the country) and cause the formation of a large *dead heart* (having more than two prominent dry leaves). However, damage to the sugarcane crop by the first brood is not pronounced. This borer is a typical *Monsoon borer*. The ravage of this borer is pronounced during the rainy season (in August-September) when the crop has sufficiently grown up and formed adequate stalk tissues for the preferred feeding of the larvae. The larvae of Plassey borer usually prefer the immature upper half of the stalk for feeding.

The secondary infestation or the infestation of *pseudosolitary phase* results from the dispersal and establishment of larvae of the *gregarious phase* in the adjoining cane stalks of the same or neighbouring clumps. (Larvae of Plassey borer stay longer in the *gregarious phase* and several larvae of advance instars generally share the same feeding tunnel and pupate in the same tunnel). The grown up caterpillars disperse to the adjoining cane plants or migrate to the lower healthy portion of the same cane stalk (showing primary infestation). In secondary infestation, the attack spans several internodes of the cane and larvae mostly form a long central tunnel and cross the barrier of nodal tissues like that of Stalk borer. In this infestation, as it takes place in middle to lower portion of the cane, crown leaves usually do not dry up; infested cane plants rarely look distressed and sickly. The number of larvae in secondary infestation of cane may vary from one to a dozen (*pseudosolitary phase*). Pupation takes place within the cane and a number of pupae may be found at one large feeding tunnel. The borer also goes in hibernation during the winter months. At this period, a larva makes a central tunnel in the cane and stays near the node. Usually a larva does not breach the barrier of protective nodal tissues. One cane stalk may support several hibernating larvae.

12.4 Eggs and neonate larva

An average egg mass of Plassey borer contains more eggs (usually 80-100 eggs in an egg mass) in comparison to that of other *Chilo* moth borers of sugarcane. However, the number of eggs in an egg mass varies widely. Eggs are laid in a typical Pyralid fashion i.e., overlapping rows of eggs (in imbricate rows) on the underside of the first three opened leaves of the crown. Usually three to four rows of eggs constitute an egg mass. Eggs are flattened and dirty white in colour when fresh. The neonate larva has a reddish brown head with prominent dark coloured *prothorax*. A neonate larva measures around 1.5 mm, very hairy with a deep grey body. Four indistinct rows of blackish spots (tubercles) are present on the larval body. Tubercles are prominently setose.

12.5 Larva

A fully-grown larva measures around 25-30 mm. The general colour of the body is pale yellowish white. *Prothoracic shield* is large (appears bulging sideways) and dark brown in colour. In a typical larva, the large dorsal tubercles (D_1) are prominent, appear like a square blotch (like fur markings on a *cheetah*), and bear prominent setae. The smaller tubercles (D_2) always remain aligned with larger tubercles at the central or mid-dorsal area but remain significantly extended towards the lateral side. A tubercle above the *spiracle* always forms a concave curvature downward. On the 8th segment, larger tubercle pair is fused, similarly the following smaller tubercle pair is also fused. On the 9th segment, all the four tubercles (two dorsal and two subdorsal) are fused together to form a very large *sclerotized* area (four setae present in a line on the fused tubercles). The four longitudinal stripes (two dorsal and two laterals) are pinkish and intensity of pink colour varies widely. The lateral stripes run through the lateral tubercles and always remain above the *spiracles*. (In contrast, lateral stripes run through the *spiracles* in the Internode borer). *Tracheal trunk* is clearly visible. *Spiracles* are black rimmed, elongated oval and appear open inside. The *spiracle* pair on the 8th abdominal segment is placed subdorsally and is relatively larger with much thicker black rim. This *spiracle* pair draws quick attention due to their relative large size on the larval body. The crochets on the abdominal *prolegs* are stout and are arranged in a complete circle. The hooks are *multiordinal*, light brown to dark brown (black) and follow a *biserial* or *multiserial* pattern (towards *meson*); hooks appear *triordinal* at the outer or lateral side. Crochets are larger at the inner side (towards *meson*) and the length gradually diminishes towards the outer or lateral side. Larvae of Plassey borer hibernate in the feeding tunnel. During the hibernation the colour of the stripes fades away and the larvae appear creamy white or dirty white in colour.

Like Gurdaspur borer, in the larval stage, this pest also exhibits variations in the colour of stripes and colour and development of tubercles. The variations are so striking that many a times it was described as *polymorphism* in the past (it is not appropriate to term these variations as *polymorphism*). Four distinct larval types may be observed. (i) All the four stripes and tubercles are prominent. (ii) Stripes are lighter but the tubercles are distinct. This type is most commonly encountered in the fields. (iii) Stripes are indistinct and tubercles are very light in colour, almost similar to the body colour. (iv) Dorsal stripes are broader but discontinuous, as if forming brackets around large dorsal tubercles from the lateral side. Due to these wide variations, larvae of Plassey borer often appear quite similar to that of maize borer, *Chilo partellus*. However, the presence of asetose tubercles in line with the *tracheal trunk* in the *Chilo partellus* immediately helps in its identification and differentiation.

12.6 Pupa

The pupa of Plassey borer is rather stout, appears strong and armoured with strong spines on the 5th, 6th and the 7th abdominal segments. The freshly formed pupa is yellowish white in colour with four prominent pinkish stripes. Colour of the pupa gradually changes to dark brown with maturity. Head of a pupa appears slightly projected. On the head, two stout horn-like projections are present. Similarly, two smaller projections are also present above the eyes. The two *mesothoracic spiracles* appear closed. Prominent arches are present above the *spiracle* on the 2nd and 3rd abdominal segments. In the wing of a *pharate adult*, venation may be observed and in a male, *pharate adult*, development of terminal black spots on the wings become prominent and may be observed easily.

The spiny abdominal ornamentations are quite distinctive and help in quick identification of this species from the pupae of other moth borers of sugarcane. The spines present on the three abdominal segments (5th, 6th and 7th) are prominent, stout and sharp. Usually the spines may be arranged in one or two rows. Many of these sharp and dark spines are branched. In the 5th and 6th abdominal segments, spines are present only on the dorsal area and slightly extend to the subdorsal area but never reach the *spiracle*. The ventral area of these two segments is smooth except the marks of prolegs. In the 7th abdominal segment, growth of the spines continues further and completely encircles the segment. In the process of encircling the segment, smaller spines form a characteristic curvature line above the *spiracle* (towards head) and avoid the *spiracles* of the 7th segment. The anal end projections or the *cremaster* are not very distinctive in the pupa of Plassey borer like pupae of other sugarcane moth borers. On the dorsal side, two raised half-circular area or spoon like depression with outer projection are present. On each half of the projection, three spines are present. However, variations in the number of spines may be observed; sometimes spines may remain absent. The separation mark demarcating the dorsal and ventral portion of *cremaster* is visible but not very prominent. Ventral portion of the anal end has two smaller well-separated projections at each half. In the male pupa, anal end imparts a distinctive conical shape and the genital area appears as a smaller depression between the two broad prominent lobes on the 9th segment. In female pupa, genital opening is marked by a smaller depression bordering the segmental boundary of the 8th segment.

12.7 Moth (Adult)

The moths are drab straw coloured (dark scales predominate). Forewings have terminal black spots with central white spot. Due to the presence of darker scales, the wings are usually two shades darker than the usual straw colour. The distinctive white markings on the terminal black spots on the forewings help in its quick identification. The area of a white spot starts from the centre of a black spot, continues up to the *termen* and thus, imparts a characteristic '*cut -mark*' appearance to the naked eye; (it appears as if the wing ends have been cut through the outer black spots). Individual white spot area is relatively larger and more prominent in the forewings of male moths. Moreover, the white spots on the forewings are larger and more distinct than those found on the wings of other moth borers of sugarcane.

Head of the moth is round, face is normal with broadly convex corneous point. *Ocellus* is present and prominent. *Labial palpi* are porrect, prominent and almost three times of the diameter of its compound eye. *Labial palpi* are bushy in male. Like other moth borers of sugarcane, the antennae are *filiform* in females and appear ribbed on the dorsal side. Similarly, antennae of male moths are stout and thicker. Dorsally the antennae have scales and appear strongly ribbed; laterally they are devoid of scales, appear smooth and light brown.

Papillae anales or the ovipositor is densely clothed with sensory setae. Ovipositor appears like a bathroom slipper with wider toe area (ventral side). The two halves of the *Papillae anales* are with prominent space in between and the toe area appears separated.

Moths are positively *phototactic*, and may be conveniently trapped using a light trap.

12.8 Forewing venation

The *costal* area of forewings appears darker and thicker than that of the other cane borers. Due to this distinctive appearance of the *costal* area, the species has got its specific epithet '*tumidicostalis*' meaning prominent costal area.

The *subcosta* (Sc) is much thicker at the base and is free and ends at the *costal area*. The *radius* (Rs) has five branches. R_1 moves very close to Sc and runs parallel to it for a good distance. Sc passes *tracheal tubes* around three locations of R_1 but without touching the vein. *Tracheal tubes* directly pass through the wing membrane. R_2 is free and ends at *costal area*. R_3 and R_4 are stalked; R_3 is shorter than R_4 and both end at *costal area* before the *apex*. R_5 is free and ends at *termen*. The three median veins M_1, M_2 and M_3 are present and distinct. M_2 and M_3 are very close at the base but have different origin (never form a stalk). Cu1a and Cu1b are present. Out of the three anal veins, only two are present. A_1 is absent, A_2 is well developed and A_3 is rudimentary. A_3 is often missed out, and thus considered as absent. Two round structures (*mechanoreceptors*) are present at the end of the each vein. *Mechanoreceptors* are also present on the body of the veins and these are congregated at the area above the bridge of the *discal cell*. Occasionally, M_1 supports three *mechanoreceptors*.

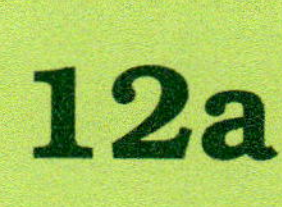

List of plates on Plassey borer, *Chilo tumidicostalis* (Hampson)

Description of plates

Plate 12.1. Symptoms of infestation of Plassey borer. Presence of multitude of holes (in all directions) on the upper portion of the cane stalk. Fresh glistening reddish frass oozes out of the holes. Rooting from the nodes. Large exit hole for the adult emergence.

Plate 12.2. Presence of multitude of holes on the upper portion of the cane stalk. Fresh glistening reddish frass oozes out of the holes. Rooting from the nodes. Large exit hole for the adult emergence.

Plate 12.3. Occasionally, Plassey borer larvae also produce cut mark like symptoms on the epidermis. Many larvae feed inside the cane and form a much larger central hollow full of frass.

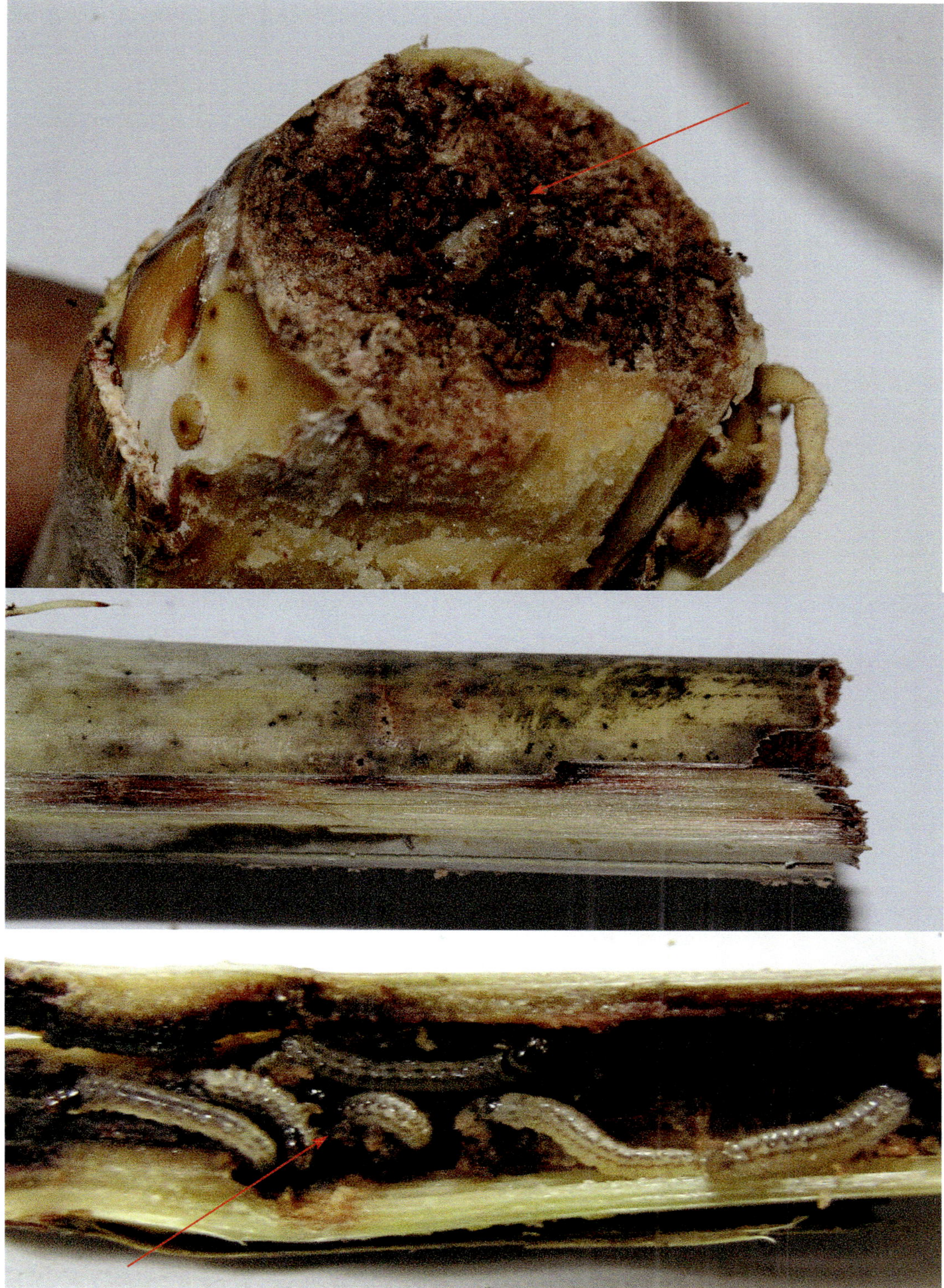

Plate 12.4. Presence of cut mark and breakage of canes. Gregarious feeding of the entire internal content of the internode by the larvae.

Plate 12.5. Symptoms of infestation. Many larvae may stay together even up to the maturity. Larvae may come out from the feeding tunnel in search of new location (solitary life) or after sometime may even re-enter the gregarious feeding area.

Plate 12.6. A prepupal larva is making an exit hole for the safe exit of adult moth from the feeding tunnel. It doest not prepare any silken pupal chamber.

Plate 12.7. A larva during hibernation. A larva makes a central tunnel in the internode and stay at the area just near the node. The colour of the larva becomes creamy yellowish white. Many larvae may stay in one cane but one in each internode.

Plate 12.8. A prepupal larva makes an exit hole for the adult to emerge from the hibernating tunnel. It doest not prepare any silken pupal chamber.

Plate 12.9. Larvae of Plassey borer are gregarious in habit. Larvae obtained from one damaged internode (top). Larvae show variations in colour pattern and colour of the tubercles.

Plate 12.10. Larvae of Plassey borer. The tubercles and the setae are very prominent. *Prothoracic shield* is very prominent, *meso* and *metathoracic* shields are visible.

Plate 12.11. Larvae of Plassey borer (variations). The tubercles and the setae are very prominent. *Prothoracic shield* is very prominent, *meso* and *metathoracic shields* are visible. Variations in the pattern of stripes in relation to the tubercles.

Plate 12.12. Immediately after moulting head capsule remains white except the mandibular area which remains dark coloured. Area of *ocelli* is also dark. Four stripes remain visible during the moulting. Colour of the head capsule changes to orange red to brown after certain growth.

Plate 12.13. *Larvae during hibernation* (A,B &C). Colour of the stripes and tubercles fades away and tubercles appear body coloured. *Larvae before ecdysis* (D), old head capsule is pushed forward exposing the white developing head capsule. Variations in the pattern of stripes on the larval body.

Plate 12.14. Larva (dorsal view). Three *thoracic shields* are present. Large dorsal tubercle (D_1) is roundish square and smaller tubercle (D_2) is relatively larger. Setae are very prominent. Four oil marks on D_1. At the 9th abdominal segment all the four tubercles fuse to form large *sclerotized* area. Sometimes a false median stripe also develops.

Plate 12.15. Dorsal view of larva (close up). Three *thoracic shields* are present. Large dorsal tubercle (D_1) is roundish square and smaller tubercle (D_2) is relatively larger than that of other moth borers. Setae are very prominent. Four oil marks or dark dots present on D_1.

Plate 12.16. Side view of larva. *Tracheal trunk* is visible as white band. *Spiracles* are elongated oval, dark rimmed with white space inside. Tubercle just above a *spiracle* takes a triangular shape. Sometimes in each segment *tiny asetose tubercles* may also develop just near the *spiracle*.

Plate 12.17. Side view of larva. *Tracheal trunk* is visible as white band. *Spiracles* are elongated oval, dark rimmed with white space inside. Tubercle just above a *spiracle* takes a triangular shape. Very *tiny asetose tubercles* may also develop just near the *spiracle*.

Plate 12.18. Anal end of the larva (close up). On the 8th segment each dorsal tubercle pair is fused together. On the 9th segment all the dorsal and subdorsal tubercles fused together and form a large *sclerotized* area. The anal plate is highly *sclerotized*. The *spiracle* pair present on the 8th segment is relatively larger, black rimmed and appears open.

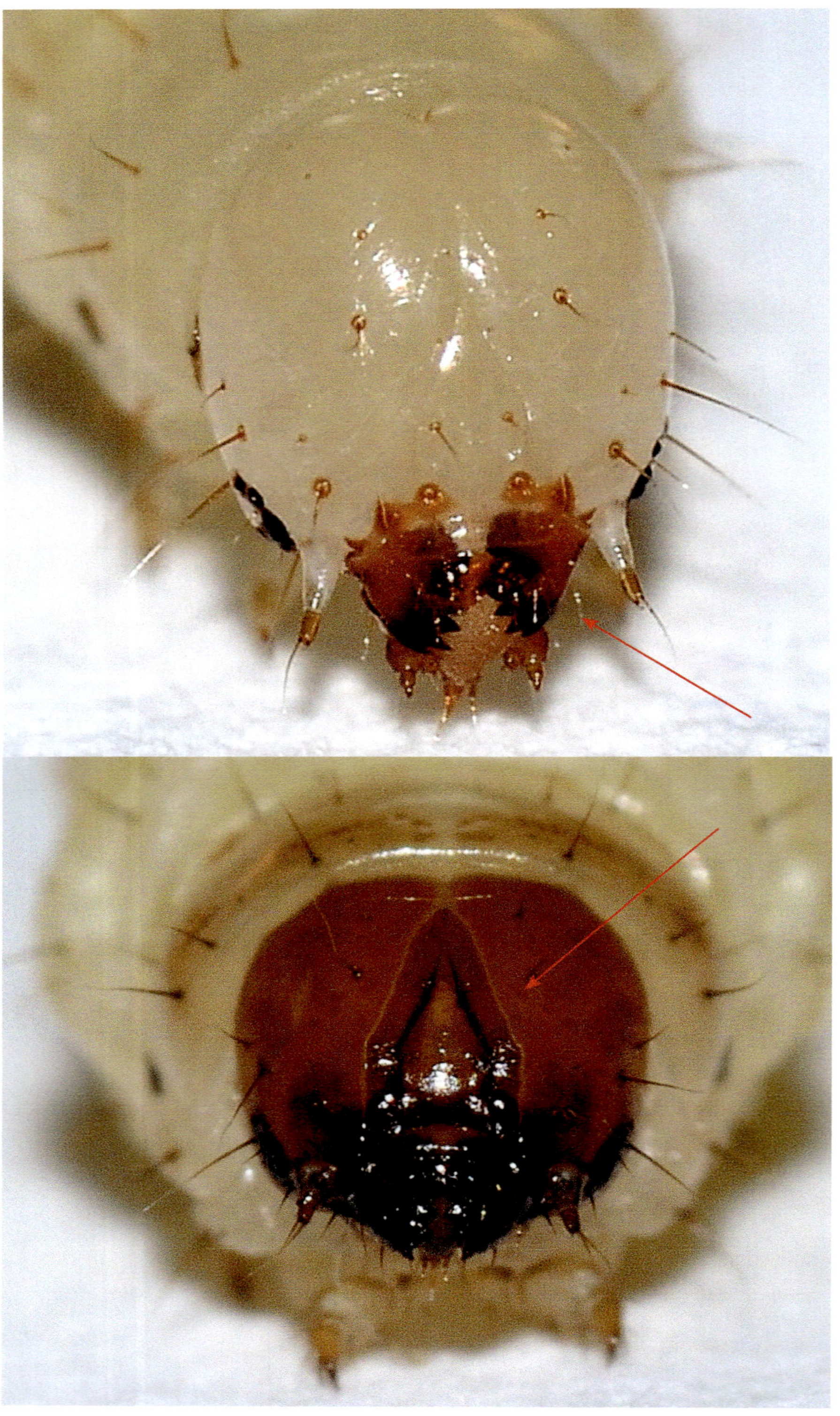

Plate 12.19. Head of the of the larva. Immediately after moulting, head capsule remains colourless (water soaked white) except the mandibular area. *Epicranial suture* is marked with prominent white line. On the face, the white triangle appears temple like.

Plate 12.20. Crochets in the abdominal *prolegs* are *biserial to multiserial* type. Crochets are multiordinal and arranged in a complete circle. Hooks at the lateral side are smaller. Crochets are brown to dark brown in colour.

Plate 12.21. Crochets in the abdominal prolegs are *biserial* to *multiserial* type. Crochets are multiordinal and are arranged in a complete circle. Hooks at the *meson* are larger and at the lateral side are smaller. Crochets are brown to dark brown in colour.

Plate 12.22. Freshly formed pupa of Plassey borer. Pupa is yellowish white and bears four pinkish stripes. Head has two prominent projections and two smaller projections. *Cremaster* remains dark coloured.

Plate 12.23. Male and female pupae of Plassey borer at different phases of maturity. A pupa, immediately after formation remains yellowish white and gradually its colour changes to dark brown. The spiny ornamentations on the body always remain prominent.

Plate 12.24. Size of the pupa on centimetre scale. Female pupa is about 16 mm and male pupa is about 13 mm. But there is a great difference in thickness and volume.

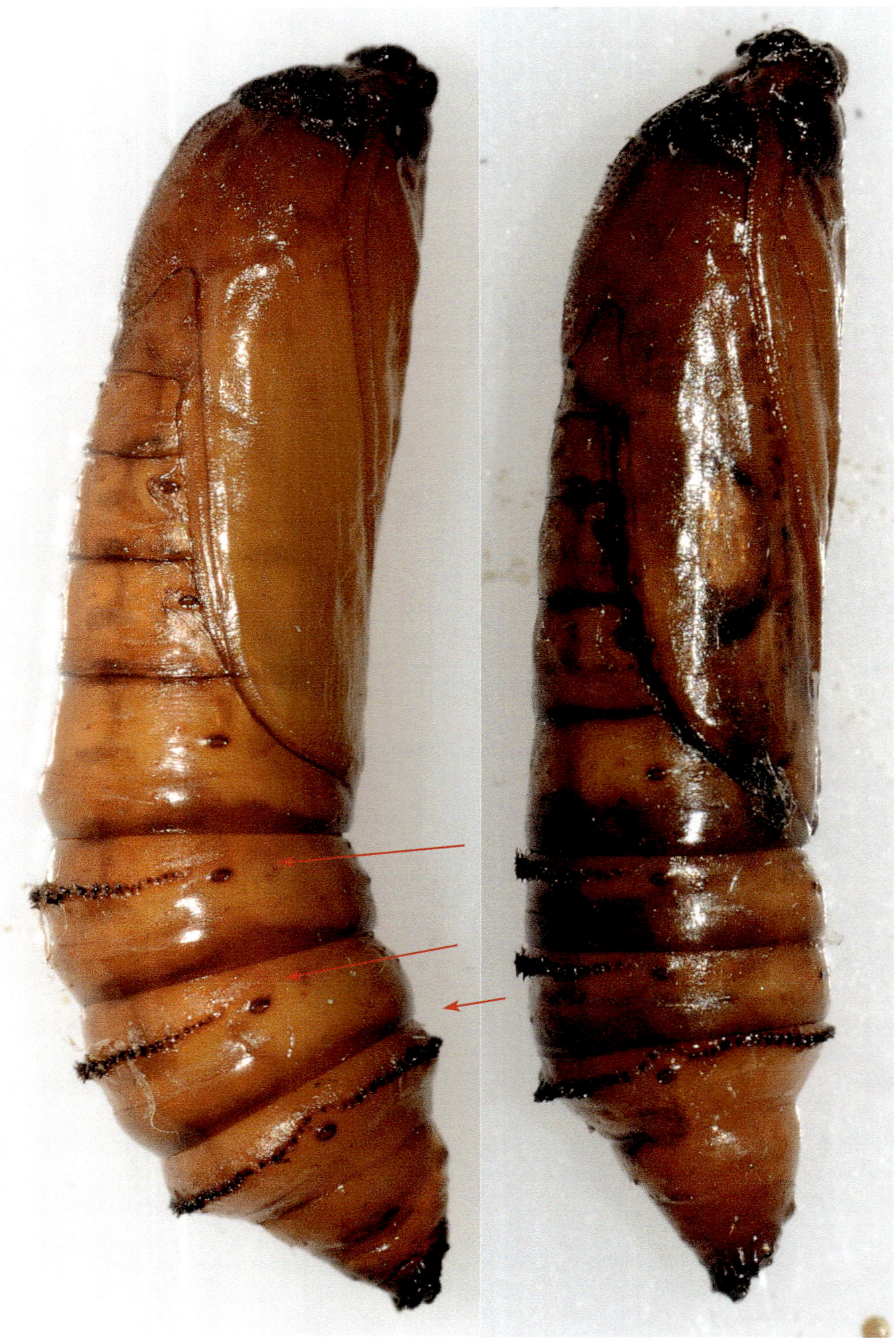

Plate 12.25. Female and male pupae (side view). Prominent spiny ornamentations on the 5th, 6th and 7th abdominal segments. Head is a bit projected forward. At the *cremaster,* dorsal half is clumsy and is bit projected.

Plate 12.26. Male and female pupae (ventral view). Prominent spiny ornamentations encircle the 7th segment. Ventral side is smooth on 5th and 6th segment except the marks of prolegs.

Plate 12.27. Pupae (dorsal and ventral views). Head with four horn like projections. The *meta* and *meso thoracic* legs are visible in between the wing pad area and thus, impart a distinctive appearance.

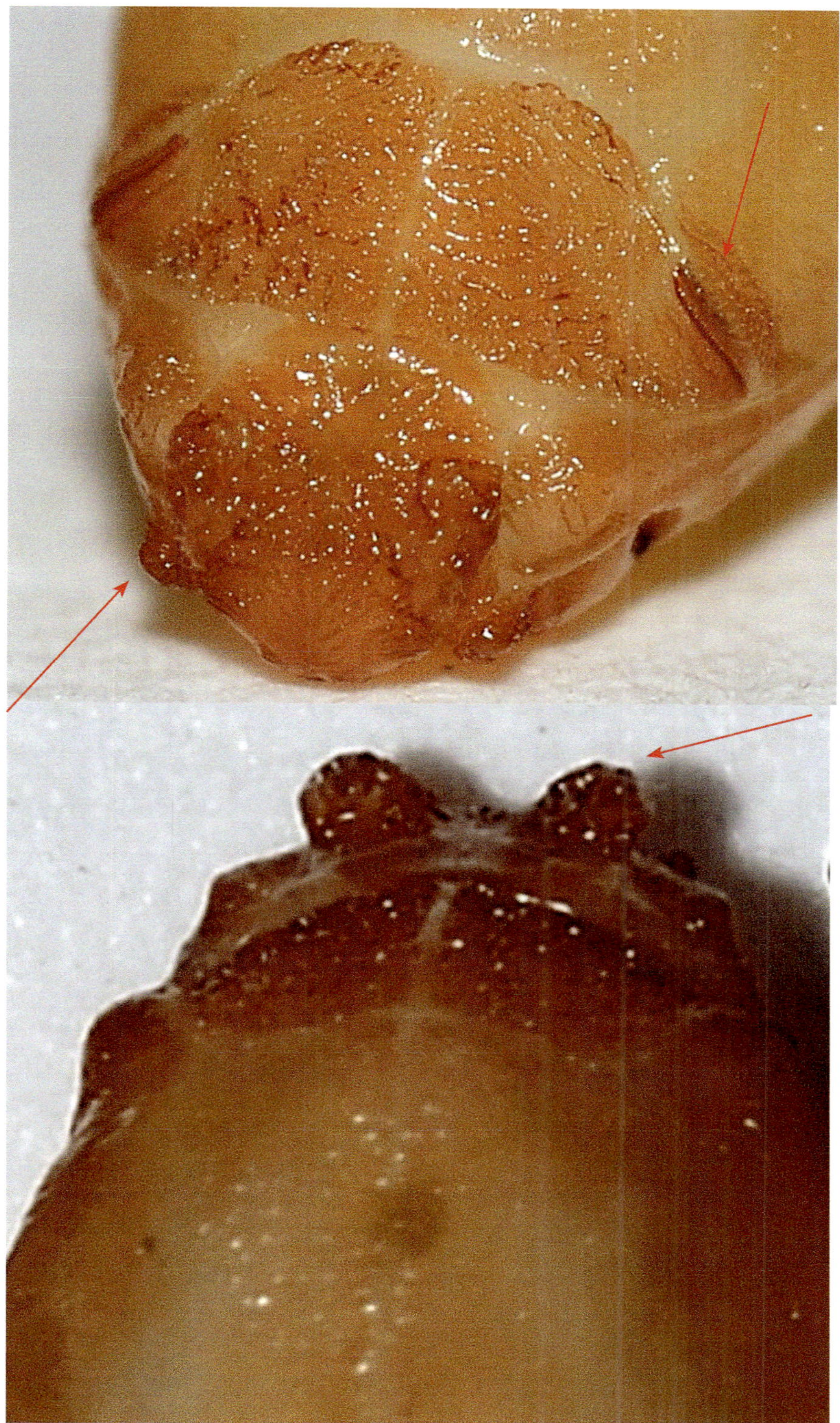

Plate 12.28. Head of an immature pupa. Head with four horn like projections. *Mesothoracic spiracles* are closed type.

Plate 12.29. Head of a mature pupa. Head with four horn like projections. *Mesothoracic spiracles* are with dark terminal portion and are closed type. Prominent median ridge is present on the *mesothoracic plate* and *prothoracic plate* has roughness.

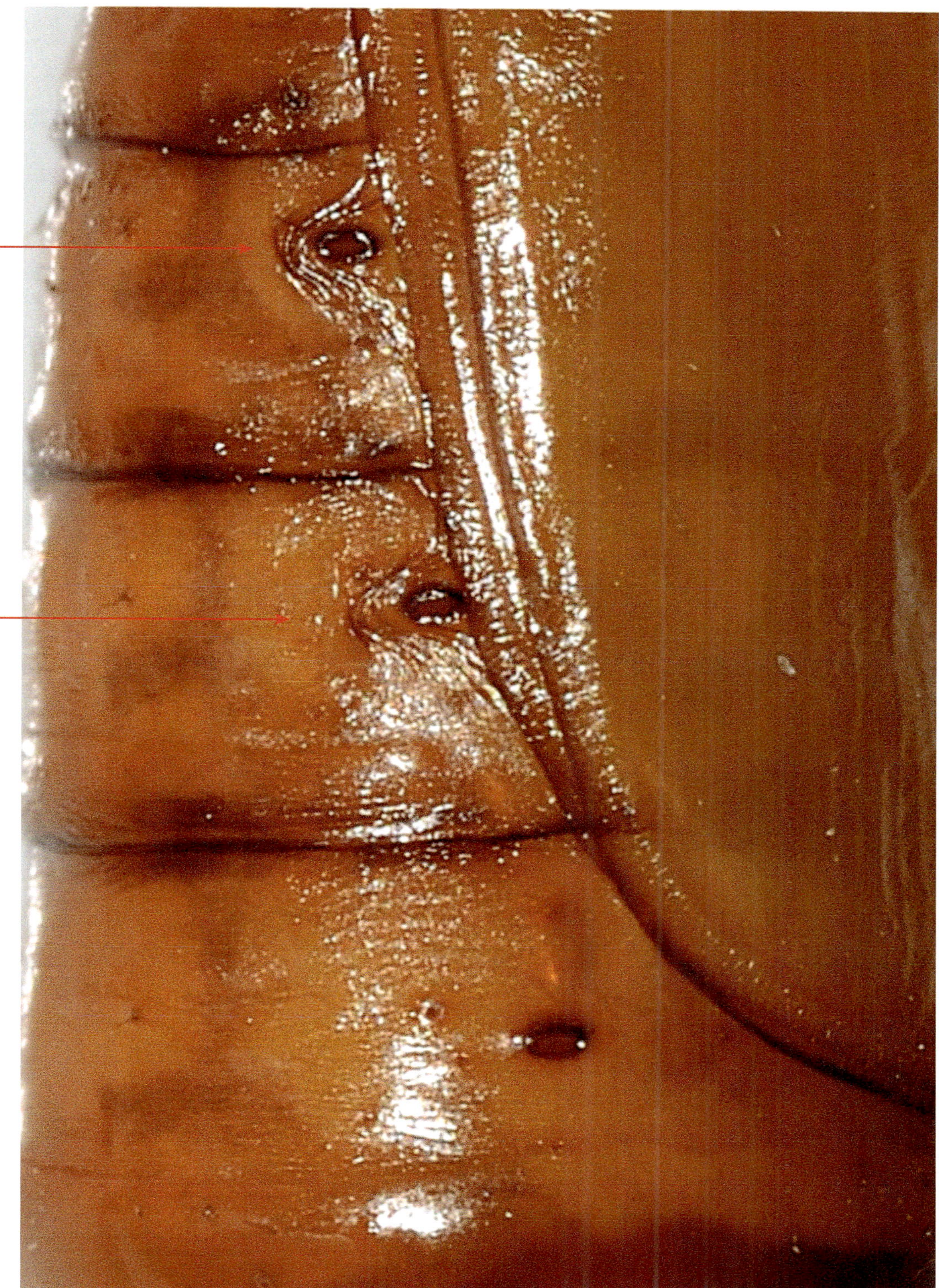

Plate 12.30. Thoracic region of a pupa (close up). Prominent arch formation above the *spiracles* on 2nd and 3rd abdominal segments. *Spiracle* on the 4th segment is elongated oval with raised rim.

Plate 12.31. Pupal abdomen (close up). Prominent spiny ornamentations on 5th 6th and 7th abdominal segments. Usually, 7th abdominal segment bears one row of spines whereas, 5th and 6th segments bear two rows of spines. Spines are sharp and some spines are bi or trifurcate. The row of spine on 7th segment form an arch to avoid the *spiracle.*

Plate 12.32. Pupal abdomen (close up). Prominent spiny ornamentations on 5th 6th and 7th abdominal segments. Usually, 7th abdominal segment bears one row of spines whereas, 5th and 6th segments bear two rows of spines. Spines are sharp and some spines are bi or trifurcate. The row of spine on 7th segment form an arch to avoid the *spiracle*.

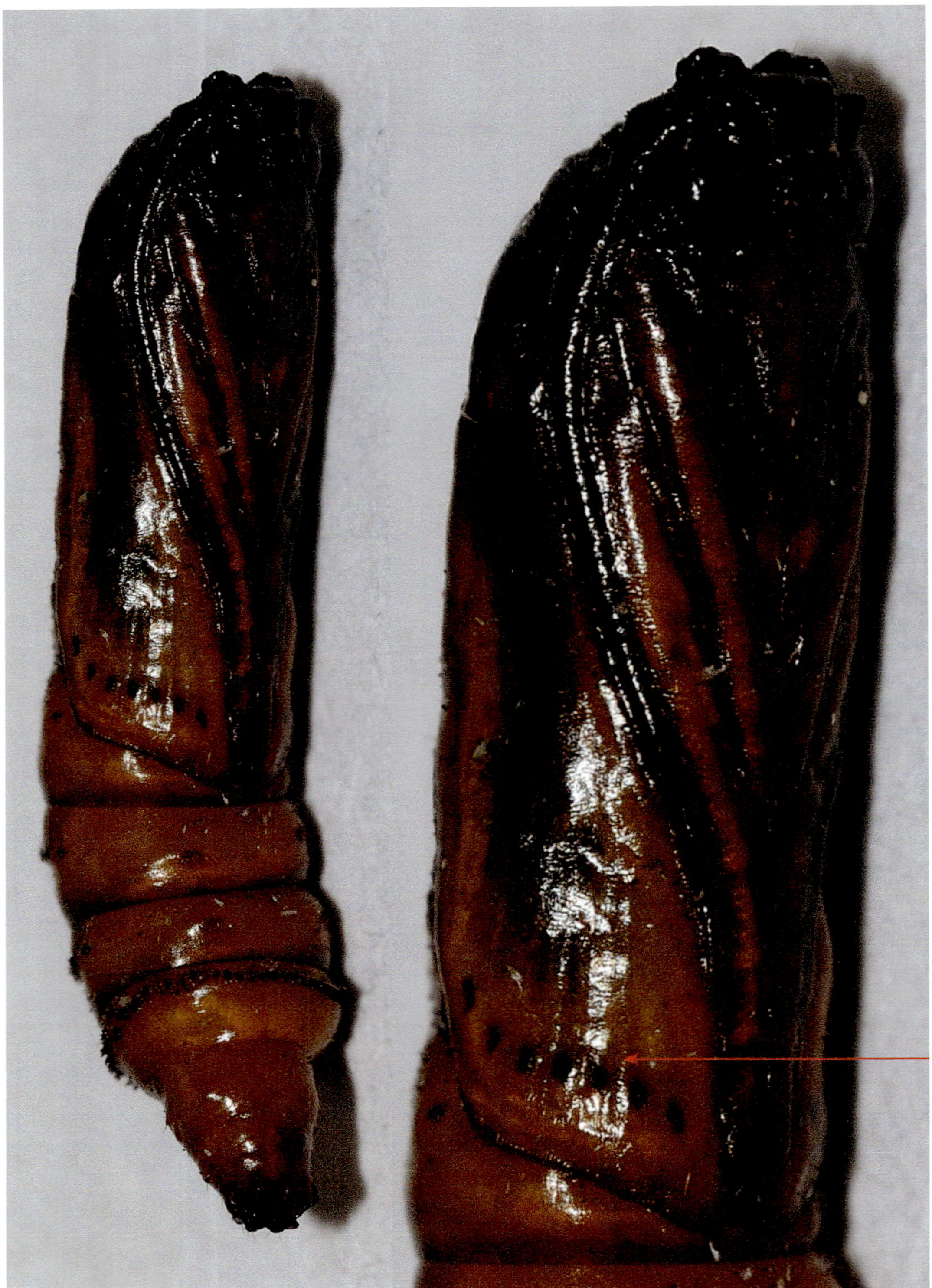

Plate 12.33. Male pupa. Pharate adult showing the formation of terminal black dots on the forewings.

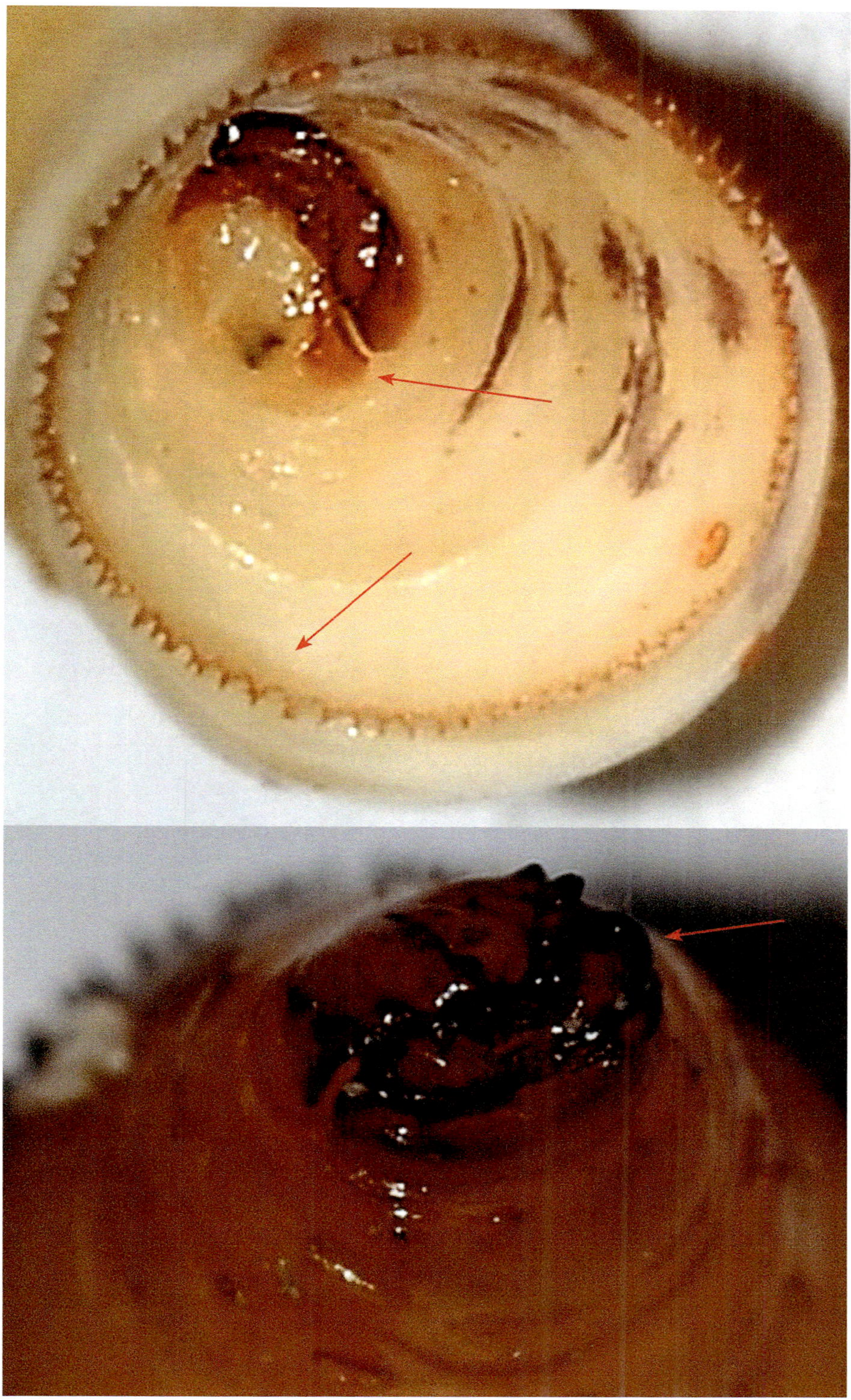

Plate 12.34. *Cremaster* of the pupa is not very distinctive. Dorsal half has two *spoon like* dark rimmed projections. Ventral side has two minor projections in each half.

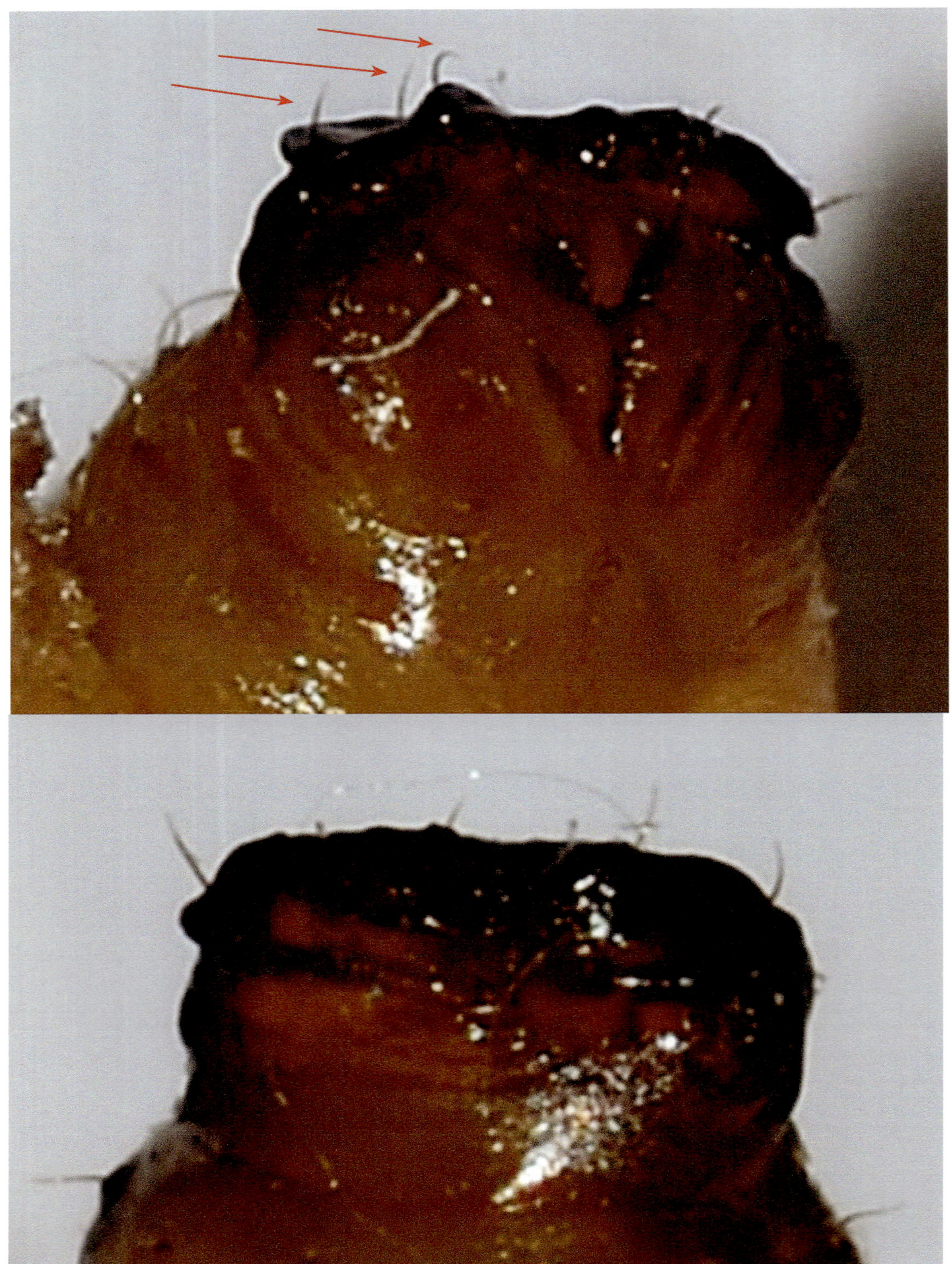

Plate 12.35. *Cremaster* (close up). In each dorsal half bears three fine spines.

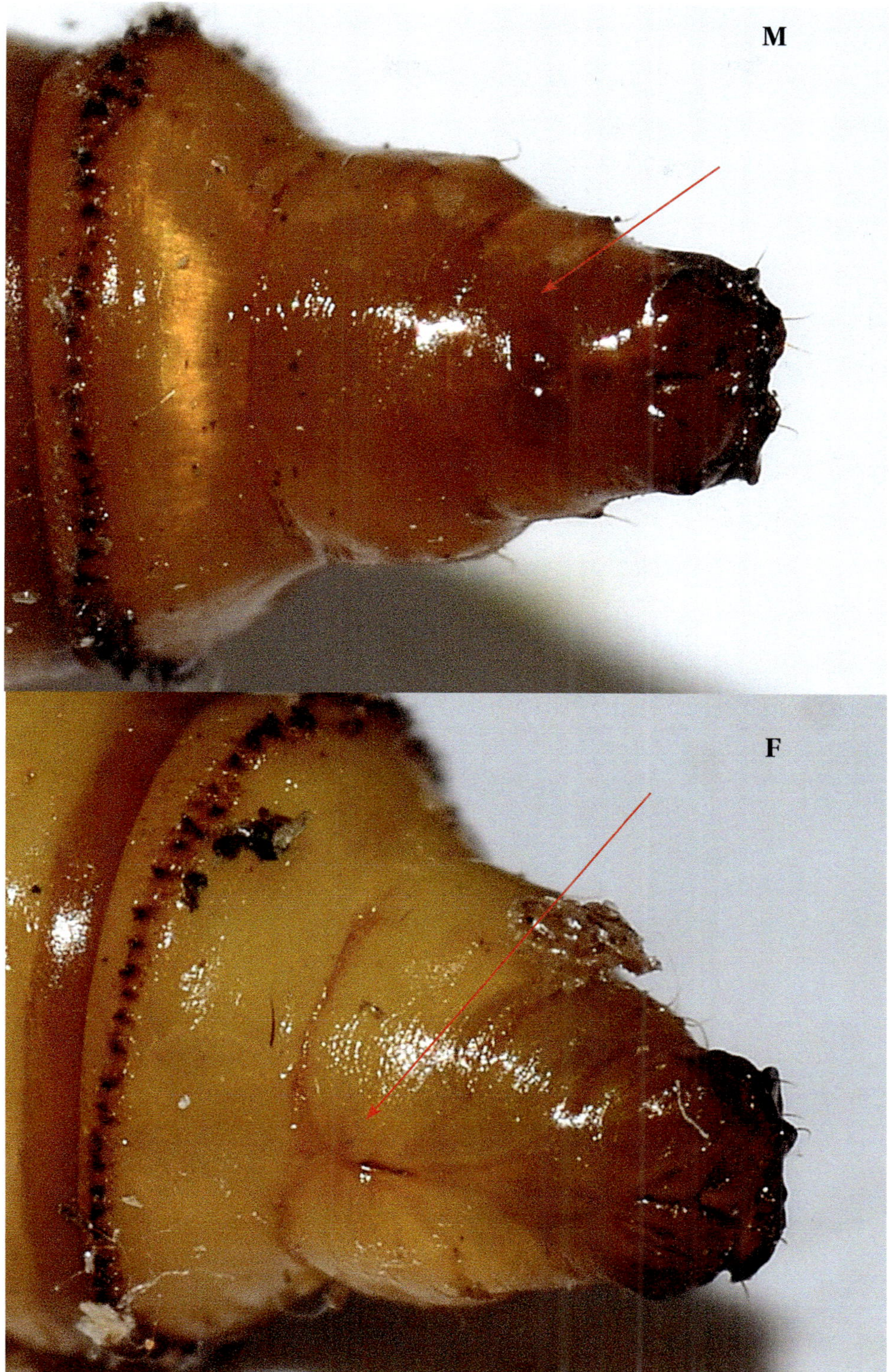

Plate 12.36. Anal end of the pupa imparts a conical appearance. Genital area of male pupa is marked by the presence of two lobes on the 9th abdominal segment and in female pupa, the genital area is marked by a depression in the middle of the 8th segment.

Plate 12.37. Female and male moths of Plassey borer (side view). The colour of the moth as well as the white cut mark on the terminal black dots on the forewings are very characteristic of this species.

Plate 12.38. Female and male moths of Plassey borer (dorsal view). The colour of the moth as well as the white cut mark on the terminal black dots are very characteristic of this species. Antennae in male moth is stout, ribbed with smooth lateral side whereas in female antennae are *filiform* and ribbed.

Plate 12.39. Male moth of Plassey borer (side and ventral views).

Plate 12.40. Female moth of Plassey borer (side and ventral views).

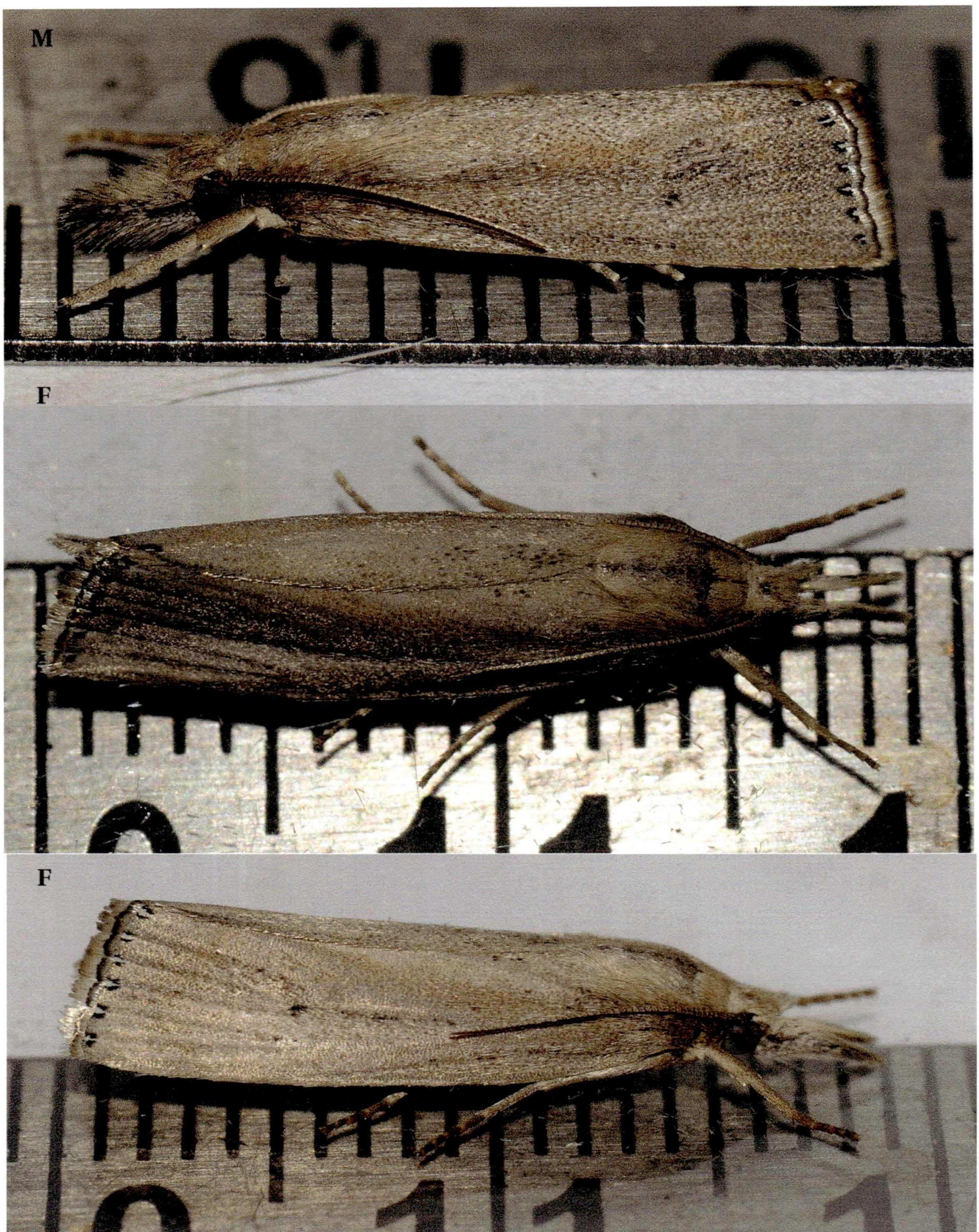

Plate 12.41. Size of the moths on a centimetre scale. (male & female).

Plate 12.42. Terminal portion of the fore wing(close up, side view). The colour of the moth as well as the white cut mark on the terminal black dots are very characteristic of this species.

Plate 12.43. Antennae are dorsally ribbed. In male, antennae are stout, thicker and devoid of scale at the lateral side which appear smooth and light brown in colour. In female, antennae are *filiform. Labial palpi* are porrect and around 3 times (male) to 3.5 times (female) of diameter of its compound eye.

Plate 12.44. Face is mostly rounded with a broad corneous point. *Ocellus* is present but not well developed.

Plate 12.45. The corrugation patter of forewings. Terminal portion of the ventral side of the wings also bear the black spotting but white area remains absent.

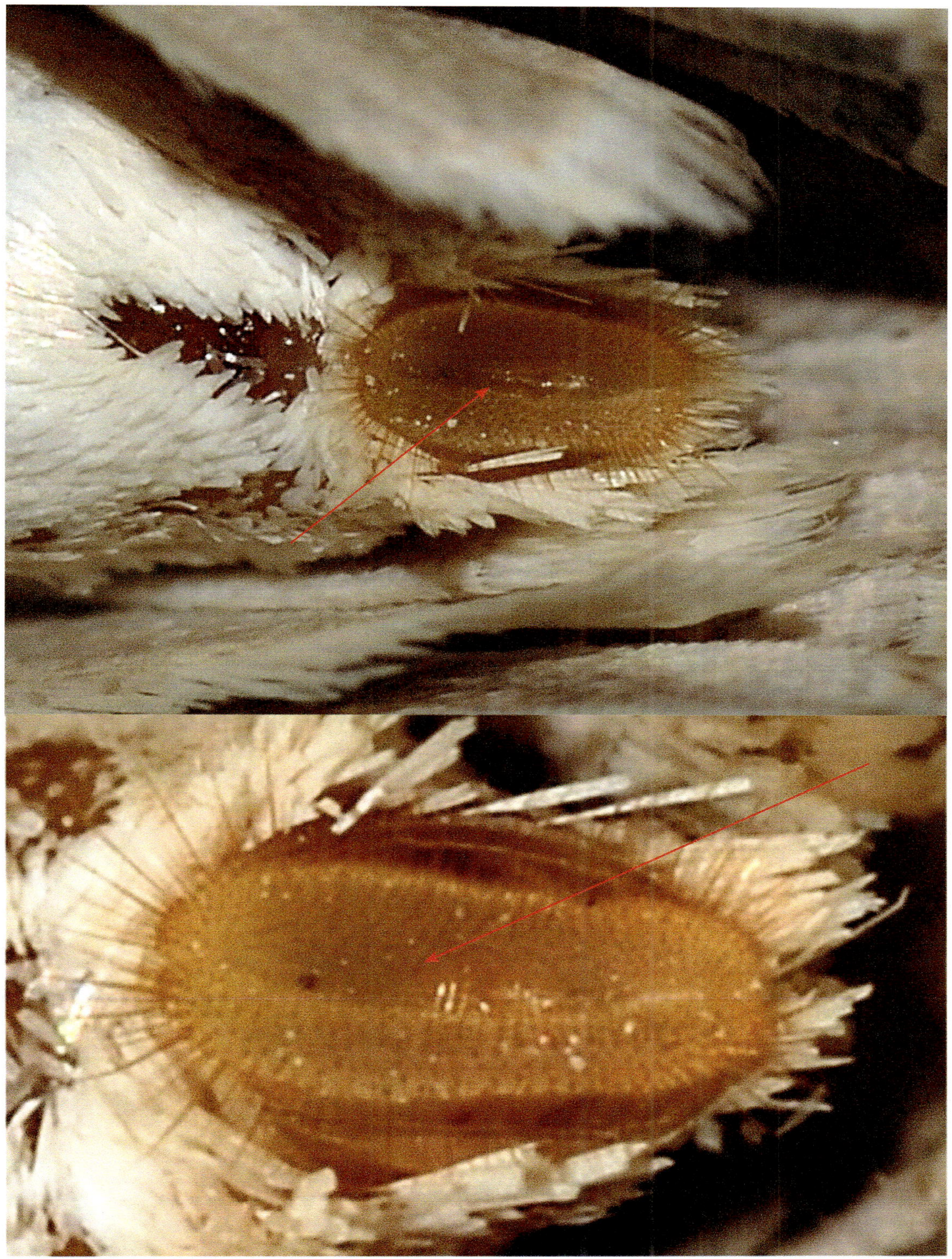

Plate 12.46. *Papillae anales* are elongated, clothed with numerous sensitive setae and prominent median depression in the entire length.

Plate 12.47. Placement of veins on the forewing (stained and unstained). Costal area is prominent. Sc and R_1 run close together. R_2 is nearer the stalk of R_3+R_4. R_3 is short, about 40% of R_4. M_1 arises very close to R_5. M_2 and M_3 are very close at the base but arise independently. A_1 is absent A_2 is prominent and A_3 is rudimentary. *Termen* area is nicely curved with a dip between M_1 and M_2.

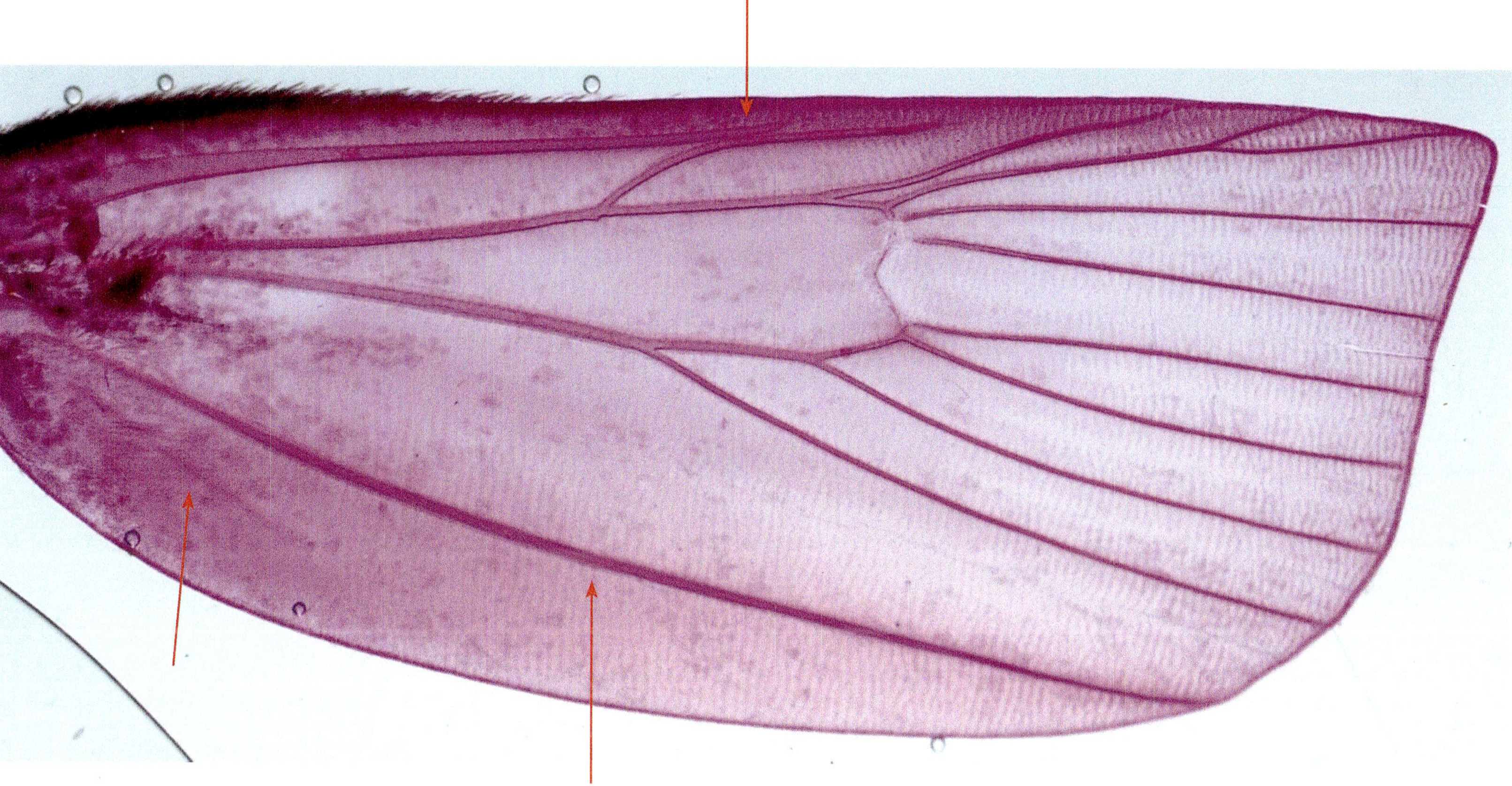

Plate 12.48. Placement of veins on the forewing. Sc and R_1 run close together. R_2 is nearer the stalk of R_3+R_4. R_3 is short, about 40% of R_4. M_1 arises very close to R_5. M_2 and M_3 are very close at the base but arise independently. To discal angles are almost at the same plane. A_1 is absent. A_2 is prominent and A_3 is rudimentary.

Plate 12.49. *Discal region* showing placement of veins. Sc and R_1 run close together M_1 arises very close to R_5, M_2 and M_3 are very close at the base but arises independently. Two discal angles are almost at the same plane. R_2 and corresponding Cu1a arise almost at same level on the *discal cell*.

Plate 12.50. Close up of *discal region* showing placement of veins. Sc and R_1 run close together. M_1 arises very close to R_5, M_2 and M_3 are very close at the base but arise independently. To *discal angles* are almost at the same plane and similarly R_2 and Cu1a are also at the same plane.

Plate 12.51. Close up of veins Sc and R_1. Tangential placement of Sc and R_1 and passing of *tracheal tube* from Sc to R_1.

13

Maize or Sorghum borer, *Chilo partellus* (Swinhoe)

The *Chilo partellus* (Swinhoe), commonly known as Maize or Sorghum borer or '*spotted stem borer*' of sorghum is hardly a pest of sugarcane. In fact, this moth borer does not qualify to be called as a sugarcane moth borer. However, due to its consideration as a sugarcane pest in olden literature, a brief description is provided here to retain the continuity. Moreover, the larvae of this moth borer are often confused with that of Plassey borer and a handy guide will also help in separating and establishing the identity of larvae of both these borers with confidence. Occasionally, it may infest and damage sugarcane stalk like that of Internode borer or Pink borer. The infestation often takes place in a situation where maize or sorghum is grown in the neighbourhood or at the adjoining fields or used as an intercrop with sugarcane (at shoot stage). With the availability of sugarcane in the vicinity, the female moth borer also tries to avail the hospitality of sugarcane and lays eggs on sugarcane leaves. Obviously, the neonate larvae have little choice in the food selection, and try to thrive on the sugarcane plant. However, larvae of this borer do not relish sugarcane and its population always remains very limited in comparison to that of nearby infested fields of sorghum or maize. Laboratory rearing studies using cut canes have also indicated that sugarcane is not a preferred host for this borer. If choice is available, a larva always prefers to eat the soft tissues of sorghum stalk.

This moth borer was first named by Swinhoe (1884) as *Crambus zonellus* from the collection made from Karachi, Pakistan. In the next year, Swinhoe renamed it as *Crambus partellus* Swinhoe (1885) after studying moth collections from Pune, Bombay. Cotes (1889) identified this borer as sorghum borer due to its prevalence in the Central Provinces of India (Please refer **Appendix III**). Fletcher and Ghosh (1919) described this borer as *Chilo simplex* Butler with excellent descriptions and drawings (plates 45-47, page 385; Please refer **Appendix IV**). Fletcher (1928) recognised this borer as *Chilo zonellus* (Swinhoe). Bleszynski and Collins (1962) and Bleszynski (1965) revalidated the currently accepted name *Chilo partellus* (Swinhoe). [Earlier in 1847, a different moth borer was named *Crambus zonellus* by Zeller, which is now known as *Euchromius ramburiellus* (Duponchel, 1836)]

13.1 Currently accepted name

Chilo partellus (Swinhoe)

13.2 Synonyms

= *Crambus zonellus* Swinhoe

= *Crambus partellus* Swinhoe

= *Chilo simplex* (Butler) Hampson

13.3 Symptoms of infestation and brief biology

Symptoms of infestation are quite similar to that of Internode borer, causing *dead heart* and damage like that of Pink borer.

13.4 Eggs and neonate larva

Eggs are creamy white, flattened and scale like. Eggs are laid in an egg mass on leaves in three to five overlapping rows like those of Early shoot borer.

13.5 Larva

A full grown larva of *Chilo partellus* is stout, four striped, and measures around 20-25 mm. Like the larvae of Plassey borer, larvae of maize borer also show variations in the pattern of stripes and colouration of tubercles. Due to the very similar appearance of the larvae with that of Plassey borer, these two borers pose considerable difficulty in distinguishing the larvae with confidence. A mature larva of maize or sorghum borer has reddish brown head and has four stripes on the body (two dorsal and two subdorsal). The dorsal stripes are broader, wavy and run through the larger (D_1) and smaller (D_2) dorsal tubercles. Sometimes these two stripes coalesce and cover the entire dorsal area; the large dorsal tubercles appear as islands. The lateral stripes when prominent, usually cover the *tracheal trunk*; the *spiracles* remain surrounded by a white or body coloured halo area. The *tracheal trunk* is visible as a white tube bearing the prominent dark *spiracles*. *Spiracles* are black rimmed and closed type (no visible space inside). When the colour of the stripes becomes lighter, stripes delicately avoid the tubercles. Position of tubercles on the larval body helps its separation from larva of Plassey borer. The large dorsal tubercles often appear truncated at the central area or *meson*. The following smaller tubercles are not in line; a smaller tubercle always starts horizontally away from the central line and extends to the subdorsal area (the seta of the smaller tubercle is further away laterally from the corresponding seta of dorsal tubercle; trapezoid). The larger dorsal tubercles often bear two to four dark spots. Presence of *asetose* tubercle (tubercle without a seta) on the lateral side of the larval body is the characteristic feature of this species. The *asetose* tubercles (1st to 7th abdominal segments) are in the line with the *spiracles* i.e., part of the *asetose* tubercles are on the *tracheal trunk*. *Acetose* tubercles are also present on both *meso* and *meta thorax*. The tubercle just above a *spiracle* is triangular.

On the 8th abdominal segment, two large dorsal tubercles are conjoined; the two smaller tubercles following the larger tubercles are also completely fused together. The *spiracle* pair on the 8th abdominal segment is black rimmed. However, the size of this *spiracle* pair is relatively smaller in comparison to that of Plassey borer. On the 9th abdominal segment, all the four tubercles join together and form a large *sclerotized* area. Thus, the anal plate and the *sclerotized* area of the 9th segment impart a very distinctive appearance to the anal end. The three *thoracic shields* are well demarcated and easily visible in a mature larva. *Prothoarcic shield* is very prominent and light brown in colour with numerous dark markings. *Epicranial suture* is marked by wavy white line. The median line of the *prothoracic shield* is broad and whitish. Six *ocelli* are present and follow 1 + 4 + 1 arrangement. Crochets in the four abdominal prolegs are sharp and light brown in colour (in Plassey borer are much darker). In an abdominal proleg hooks are arranged in a complete *uniserial* circle. Hooks are *triordinal* towards the interior (towards *meson*) and are *biordinal* to *uniordinal* at the lateral side (exterior).

13.6 Pupa

The pupa of maize borer is very distinctive. It is slender but always appears bent or curved at the abdominal region (from the 5th segment). A mature pupa is reddish yellow to light brown in colour. Pupa also bears the four stripes (two dorsal and two lateral) on the body. On 2nd and 3rd abdominal segments prominent arch formation above the *spiracles* takes place. Ornamentations on the 6th, 7th and 8th abdominal segments are in the form of innumerable (several rows) and intricately arranged tiny spines. These spines reach up to the *spiracle* on these three segments. The ventral side of the pupa including the 7th abdominal segment is smooth (marks of larval prolegs are present on 5th and 6th abdominal segment). All the dorsal ornamentations appear very similar in size. Among these tiny dorsal spines, terminal row of spines are larger and prominent and all these tiny spines collectively appear as a thick spiny dorsal band. At the anal end (*cremaster*), both the dorsal and ventral projections are well separated. There are three pairs of dorsal and one pair of ventral petal like projections at the *cremaster*. Terminal portion of a petal like projection ends in a thick spine. In a male pupa, genital area is marked as a small depression between the two broadly raised lobes on the 9th segment and placed very close to the ventral projection at *cremaster*. In female pupa, the genital area is marked on the 8th segment as a dark depression above the triangular area, almost touching the segmental boundary.

13.7 Moth(Adult)

Moths of maize or sorghum borer are rather ash to pale yellow or mixed-straw coloured. Female moths are larger in size and always a shade lighter in colour than male moths. On the forewings of male moths, many dark scales starting from the *discal* region form a certain patchy area; rest of the portion of the wing remains free, except for the terminal dark patch. It appears that certain light patch (ashy area) has formed between the two dark patchy areas. In female moth, the corresponding patch of lighter area is much wider, often the central portion is interspersed with smaller patches of dark areas. Metallic scales are characteristically absent on the forewings. *Discal* spots are present. The forewing corrugation is more prominent at the terminal end. Terminal black spots are prominent and present at the furrow areas. There is no white scale in or around the black spots. The face of the moth is very characteristic of this species. It is distinctly conical with very prominent corneous point. This face character helps in its quick separation and identification from other sugarcane moth borers. *Ocellus* is present and distinct. Length of *labial palpus* is around 3 to 3.5 times of the diameter of the compound eye.

13.8 Forewing venation

The forewing venation of *Chilo partellus* is quite similar to that of *Chilo tumidicostalis*. In the forewing, *subcostal* vein (Sc) runs straight to *costa*. The vein R_1 takes a curvature towards Sc, both Sc and R_1 run close together and R_1 almost touches the Sc near the terminal end. In most of the cases, the *tracheal tube* of Sc crosses to reach R_1 and goes until the end (This is very similar to that of Plassey borer). R_2 is free and ends at *costal area*. R_3 and R_4 are stalked and both end in *costal area*. R_3 is almost 2/3rd of R_4. R_5 is free and ends in *termen*. Three median veins, M_1, M_2, M_3 are present and well developed. Cu1a and Cu1b are present. A_1 is present and support the development of *retinaculum*. A_2 is well developed and ends at *dorsum*. A_3 is rudimentary. Costal area and the *termen* almost form a right angle at the *apex*.

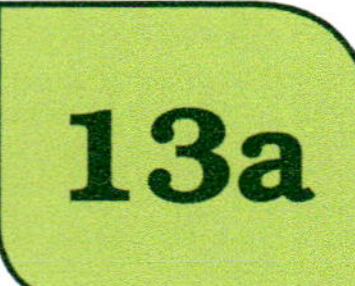

List of plates on Maize or Sorghum borer, *Chilo partellus* (Swinhoe)

Description of plates

Plate 13.1. Larvae of Maize or Sorghum borer (*Chilo partellus*). Larvae are relatively smaller and thicker. Four pinkish stripes, two dorsal and two lateral are present. Dorsal stripes often coalesce and appear as large dorsal band. Colour of the stripes vary widely. Stripes run through the dorsal tubercles (tubercles retain their individuality). In an immature larva the stripes are darker and prominent and gradually the stripes become lighter with maturity.

Plate 13.2. Mature larva of Maize or Sorghum borer (dorsal and side views). Larvae are smaller with relatively more girth. There are four light pinkish stripes. Lateral stripes run through the *tracheal tube*. *Spiracles* are black rimmed and prominent. *Tracheal tube* appear as white tube with faintly dark outline. Head capsule is light brown (reddish yellow). Setae are prominent on the body.

Plate 13.3. Variations in the patterns of stripes, colouration of tubercles, head capsule and thoracic shields of larvae. Thickness appears more in relation to body length of larva.

Plate 13.4. Larva (dorsal view). *Prothoracic shield* is well developed. *Meso* and *metathoracic shields* are prominent. The large dorsal tubercles (D_1) appear almost truncated towards *meson* and small tubercles (D_2) are relatively narrow towards *meson*. The tubercles on the 9th abdominal segment fuse to form a large *sclerotized* area.

Plate 13.5. Larva (side view). *Tracheal trunk* is prominent as white tube. Lateral stripe partially covers the *tracheal tube*. *Spiracles* are elongated, dark rimmed and surrounded by white or colourless area . *Asetose tubercles* are present in line with the *tracheal trunk* (1 to 7th abdominal segment).

Plate 13.6. Side view of the larva (close up). *Spiracles* are elongated and dark rimmed surrounded by colourless or body coloured area. Lateral stripe partially covers the *tracheal tube.* Characteristic *asetose tubercle* is present on 1 to 7th abdominal segment and located on *tracheal trunk* near the end of a segment. Horizontally aligned with small dorsal tubercle (D_2) .

Plate 13.7. Side view of the larva (close up, anal side). *Tracheal trunk* is visible. Wavy lateral stripe partially covers the *tracheal trunk*. *Spiracles* are elongated and dark rimmed surrounded by colourless or body coloured area. *Asetose tubercles* are present near the end of a segment and horizontally aligned with small dorsal tubercle (D_2).

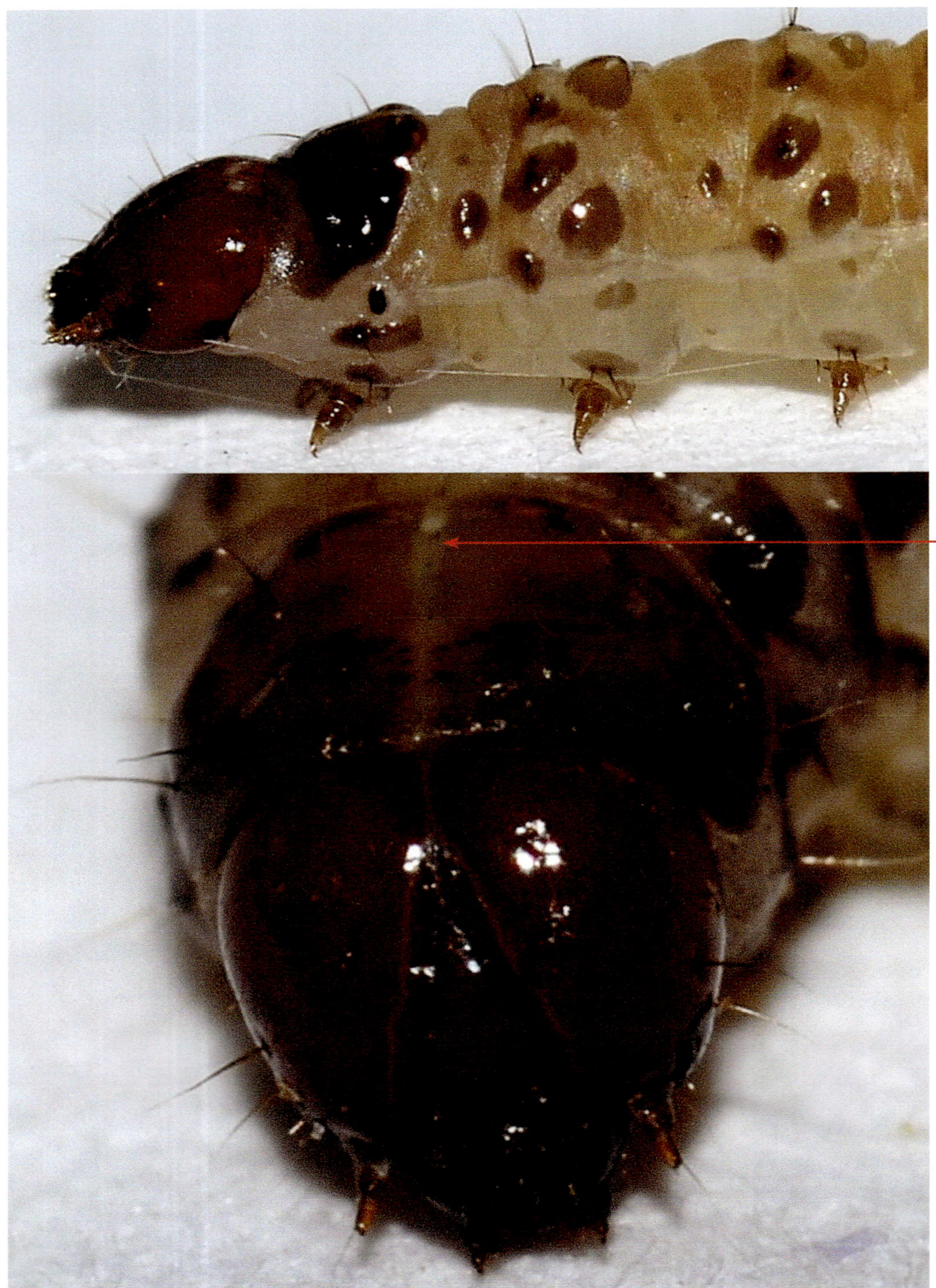

Plate 13.8. *Tracheal trunk* appears as white tube, thoracic *spiracle* is black rimmed and closed type and slight larger than the abdominal *spiracles*. Small *asetose tubercles* are also present on the thoracic segments touching the *tracheal tube*. Head capsule is reddish brown. The *epicranial suture* is marked white creamy white, wavy line. In *prothoracic* shield median area is whitish.

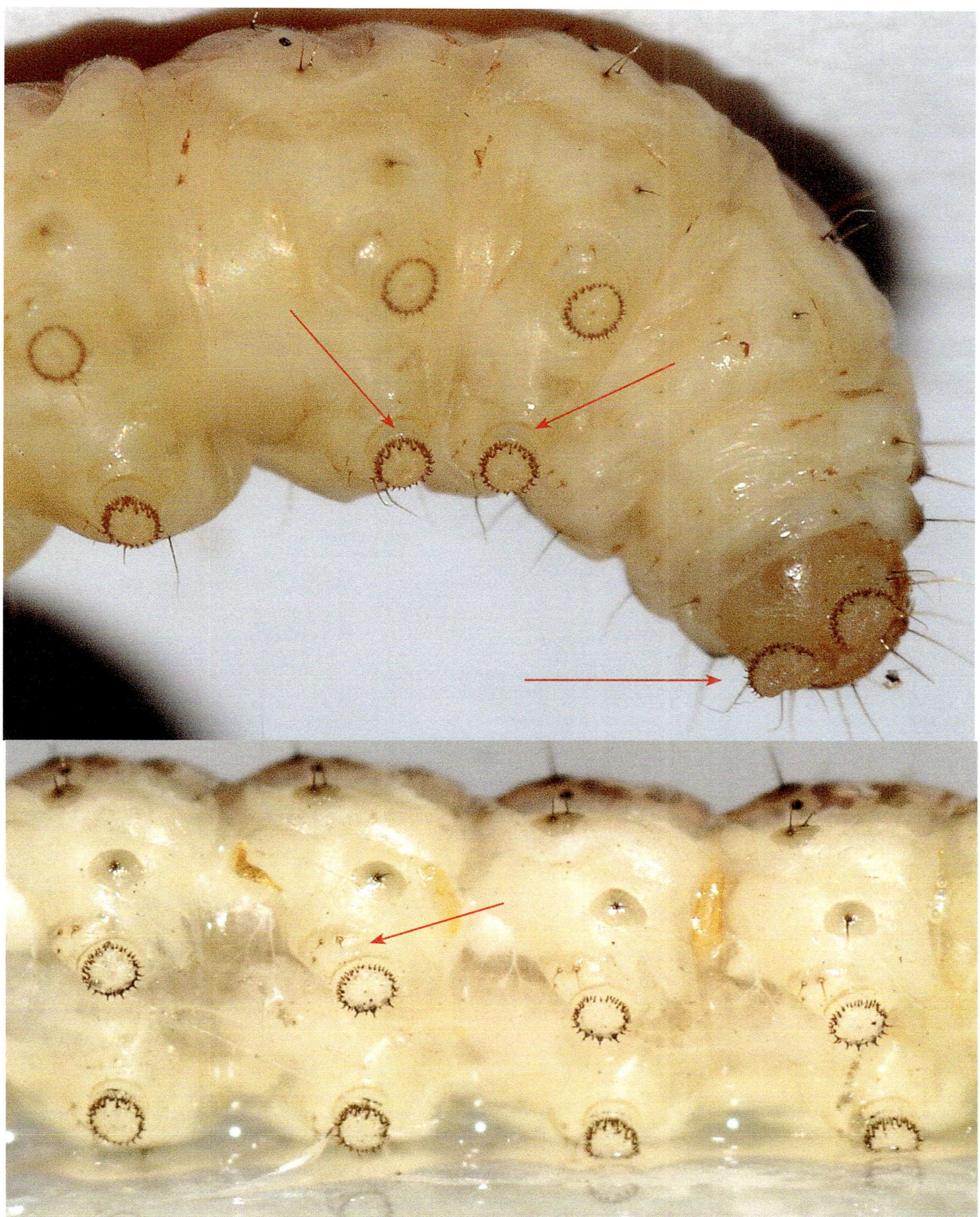

Plate 13.9. Crochets or hooks on the four abdominal prolegs are arranged in a complete circle. Crochets are *triordinal,* larger towards *meson* and smaller towards the lateral side. On the anal prolegs hooks cover more than 50% of the circle.

Plate 13.10. Crochets or hooks (close up) form a complete circle on a proleg. Crochets are *triordinal,* larger towards *meson* and smaller towards the lateral side.

Plate 13.11. A pupa is reddish yellow-brown in colour and bears four stripes. It characteristically forms an inward curvature from the 5th abdominal segment. Dorsal side has small spine like ornamentations on 5th, 6th and 7th abdominal segments. Ventral side is smooth. *Cremaster* with petal like projections.

Plate 13.12. Ventral side is smooth except the marks of prolegs. Ornamentations as small spines. The last row spines are larger and this row always stays above the *spiracle. Cremaster* with petal like projections.

Plate 13.13. Size of male and female pupae on a centimetre scale. Pupa is curved from the 5th abdominal segment.

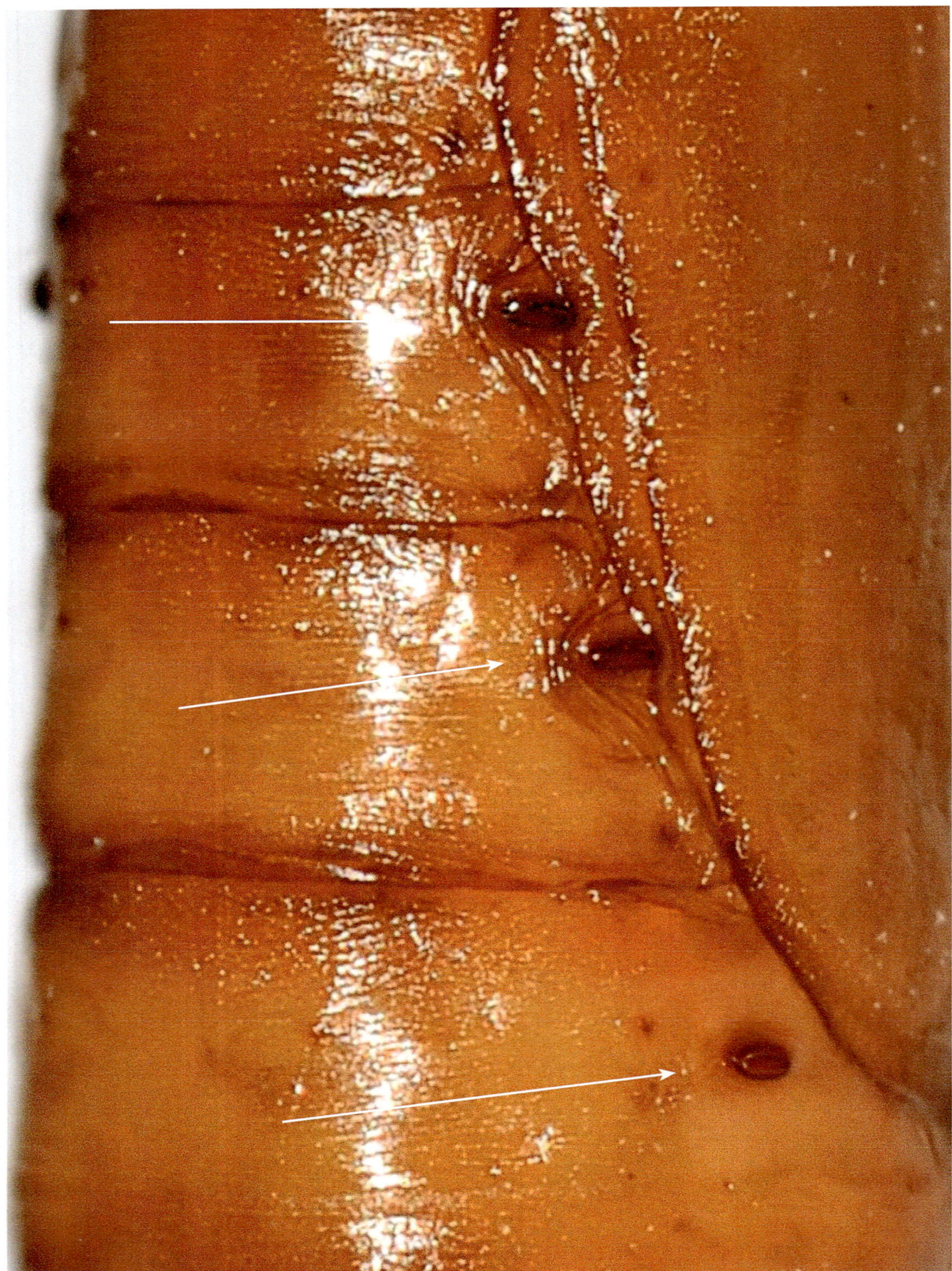

Plate 13.14. Formation of arch above 2nd and 3rd abdominal *spiracles*. *Spiracle* on the 4th segment is elongated with prominent central hollow. Rim of the *spiracle* is distinctly raised from the body.

Plate 13.15. Pupalornamentations(closeup).Onthe(5,6and7th)abdominalsegmentsornamentations are in the form of several rows of mall spines, terminal row of spines in each segment is much longer than the rest spines . These spines appear from the base of the segment and reach up to the *spiracle*.

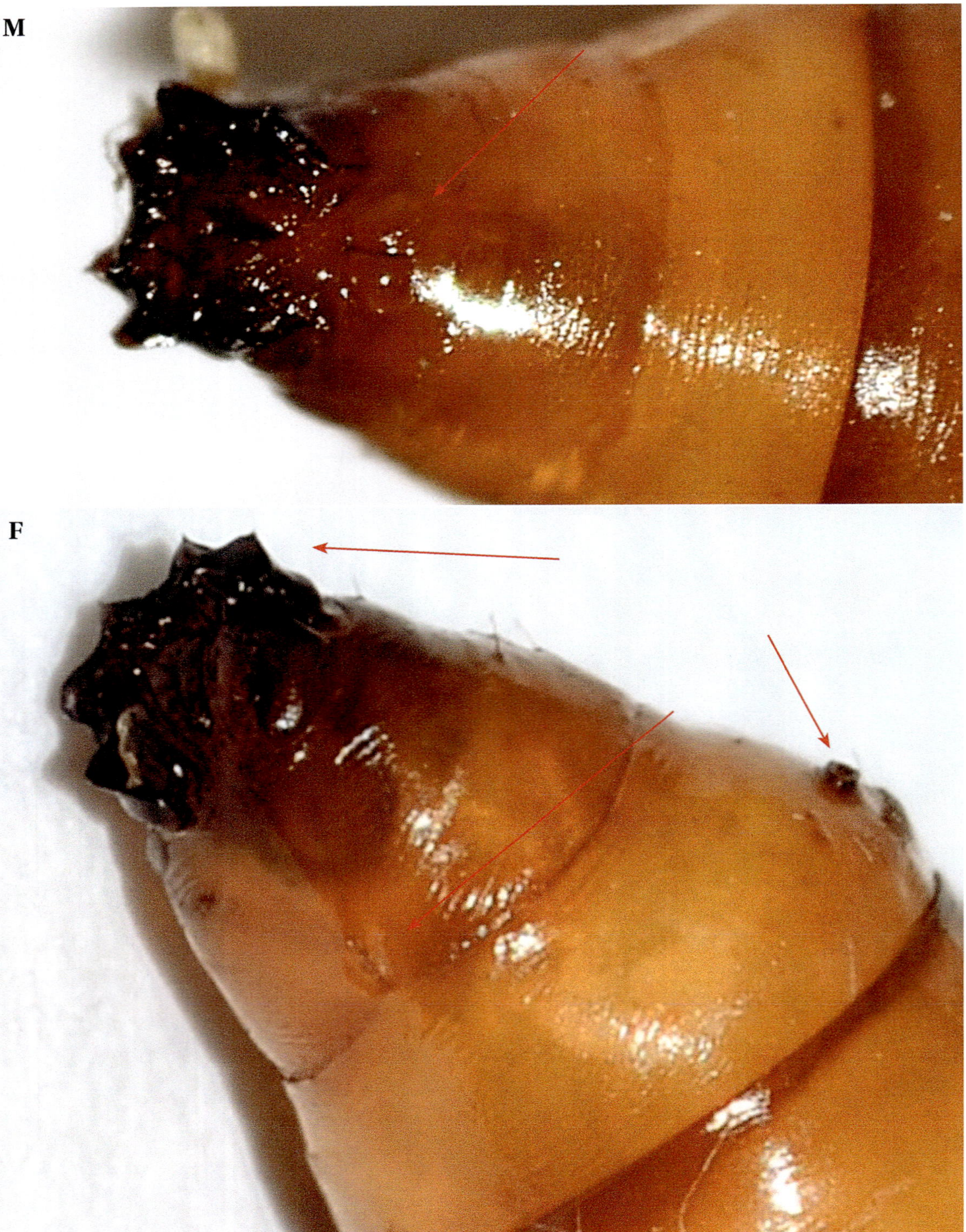

Plate 13.16. In male, genital area is present on the 9th segment as two large lobes with a small depressed line between them . In female, genital area is present on the 8th segment as a depressed area touching lower boundary of 7th segment. Anal marking is Y shaped.

Plate 13.17. Female moth of Maize or Sorghum borer. Terminal black spots are feeble and present in the furrow area. Female moths are a shade lighter than the male moths. A patch of dark scales near *discal cell* and three smaller patches up to the apex are present. Labial palpi are porrect and the antennae are *filiform*.

Plate 13.18. Female moth. Terminal black dots are feeble. Female moths are a shade lighter than the male moths. Variable patterns and colour of scales on the forewings..

Plate 13.19. Male moth of Maize or Sorghum borer. The terminal black spots are more prominent than that of female. Male moths are always one shade darker than the female moths. Like other *Chilo* male moth borers antenna is stout, lateral side is devoid of scales and appears light brown in colour. *Labial plapi* porrect but a bit raised upward.

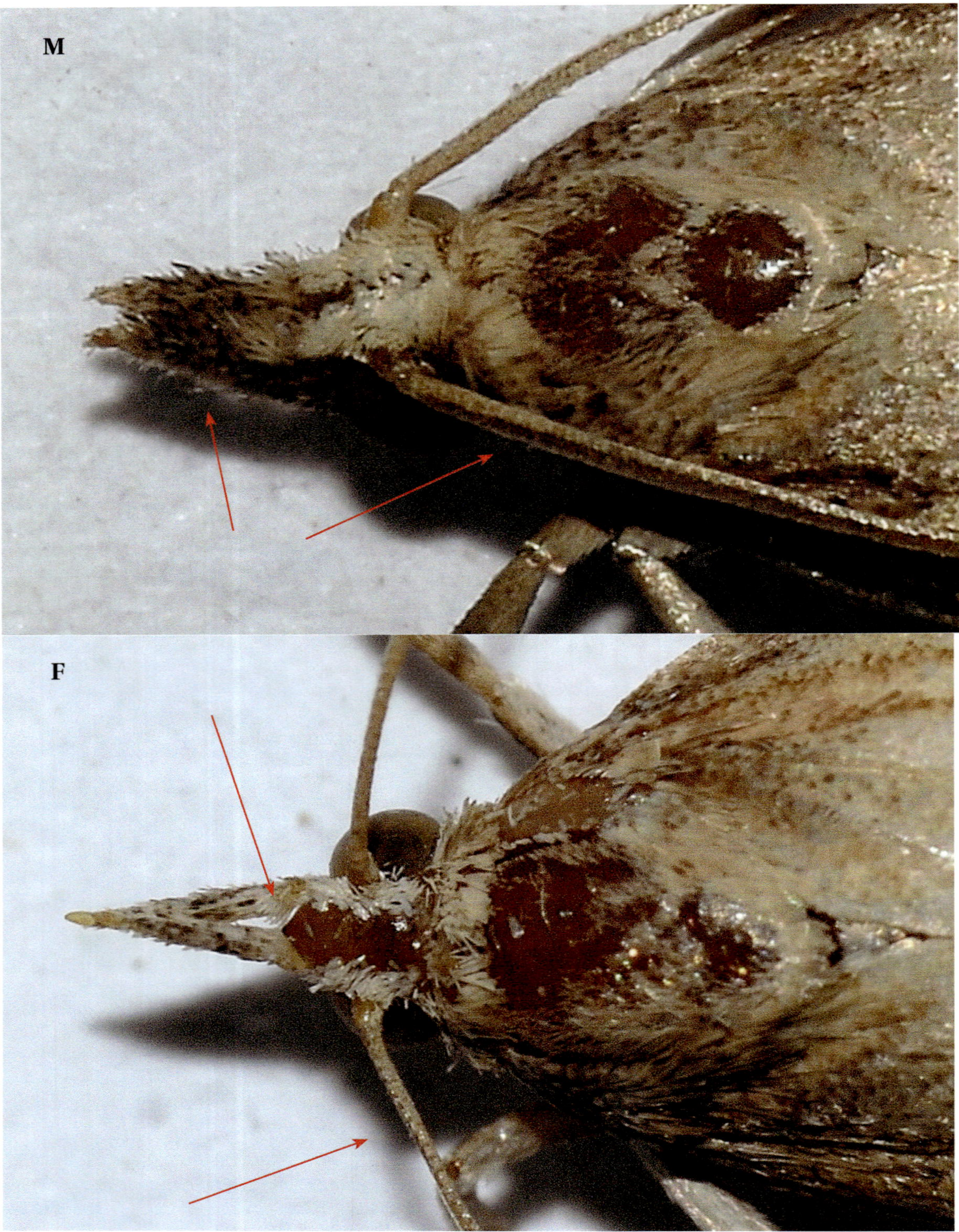

Plate 13.20. Head portion of male and female moths (close up). *Labial palpi* are porrect and conical. In male, *labial palpi* are a bit projected upward. Face is projected forward and forms a sharp corneous point.

Plate 13.21. Male and female moths (ventral view).

Plate 13.22. Head of a female moth showing strong corneous point of face and dark *ocellus* just behind the antenna.

Plate 13.23. Antenna. In female, antenna is *filiform* and clothed with scales. Like other *Chilo* male moth borers, antenna is stout, lateral side is devoid of scales and appears light brown in colour.

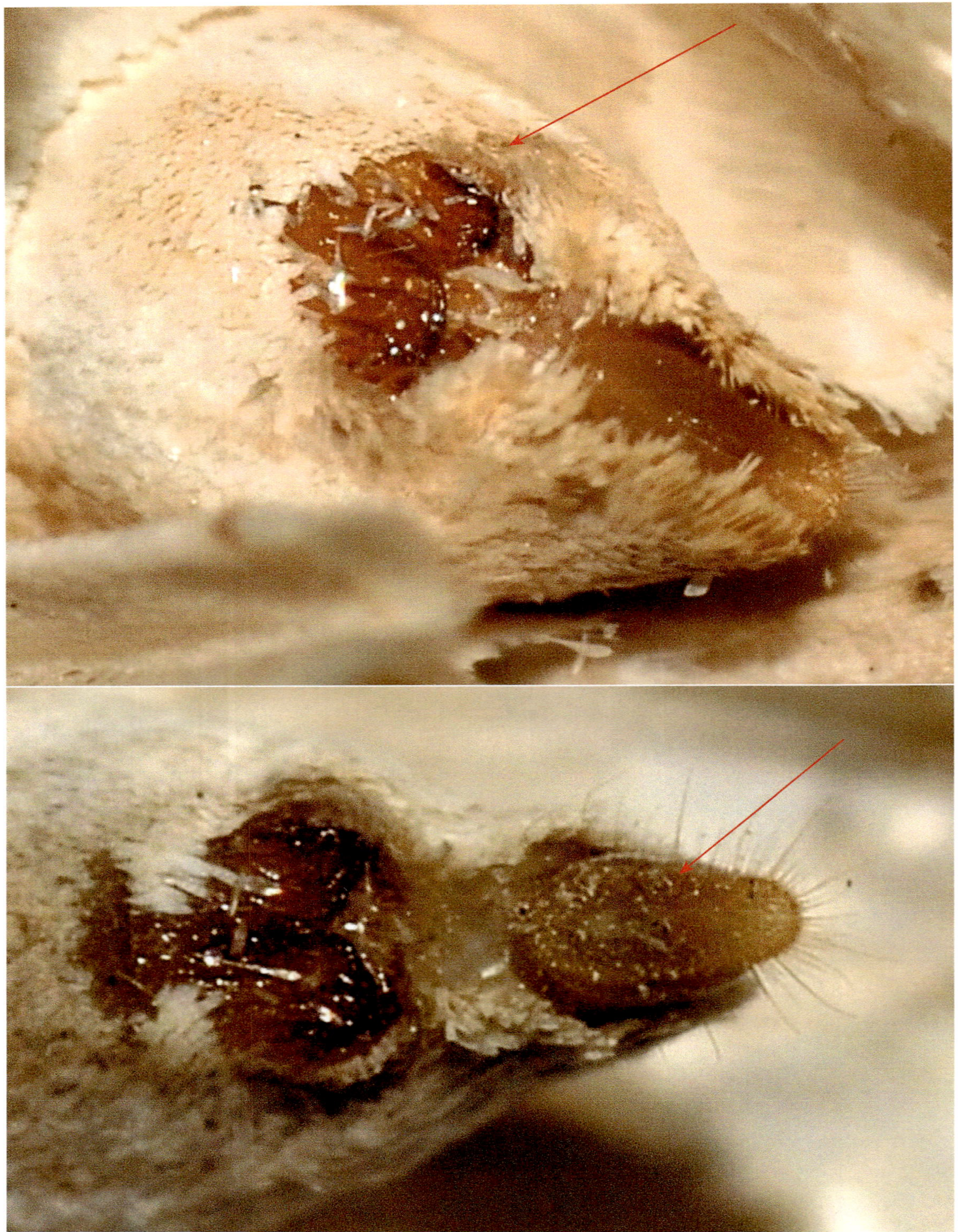

Plate 13.24. *Papillae anales* are elongated and broader at the ventral side. A long central gap along with numerous long sensitive setae are present on the *papillae anales.* Two *sclerotized* plates are prominent.

Plate 13.25. Fore wing venation (female) stained and unstained. Wing venation is similar to that of Plassey borer. Sc and R_1 run close together for a long distance and terminate separately. R_2 is free. R_3 and R_4 are stalked. R_3 is about 3/4th of R_4. R_1, R_2 and Cu1b and Cu1a almost corresponds, respectively at their places of origin on the boundary of *discal cell.* A_3 is visible and support the *retinaculum.* A_2 is very prominent and A_3 is feeble.

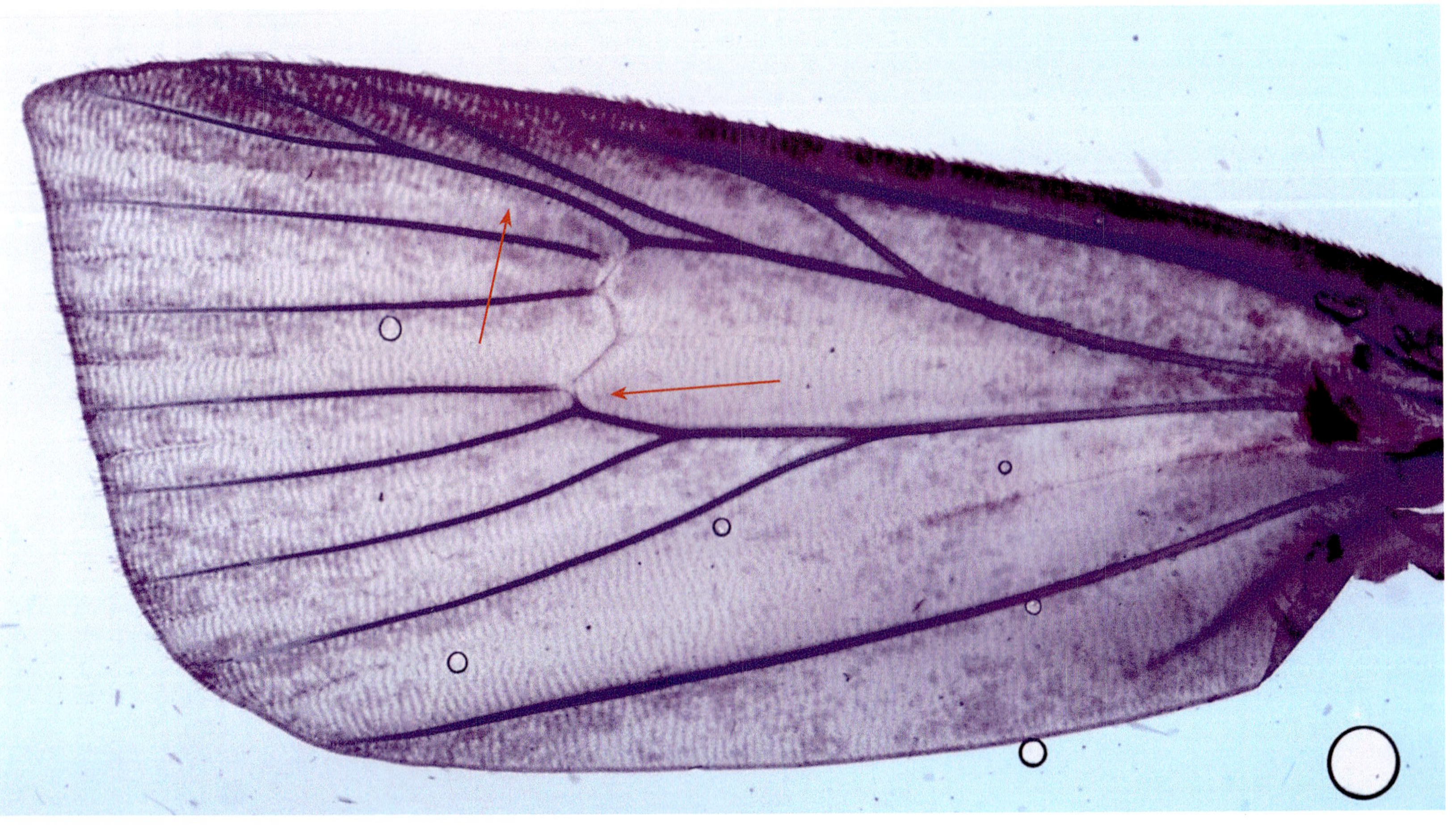

Plate 13.26. Fore wing venation (female). Wing venation is similar to that of Plassey borer. Sc and R_1 run close together and terminate separately. R_2 is free. R_3 and R_4 are stalked. R_3 is about 3/4th of R_4. R_1, R_2 and Cu1b and Cu1a almost corresponds, respectively at their places of origin on the boundary of *discal cell*. A_3 is visible and support the *retinaculum*. A_2 is very prominent and A_3 is feeble.

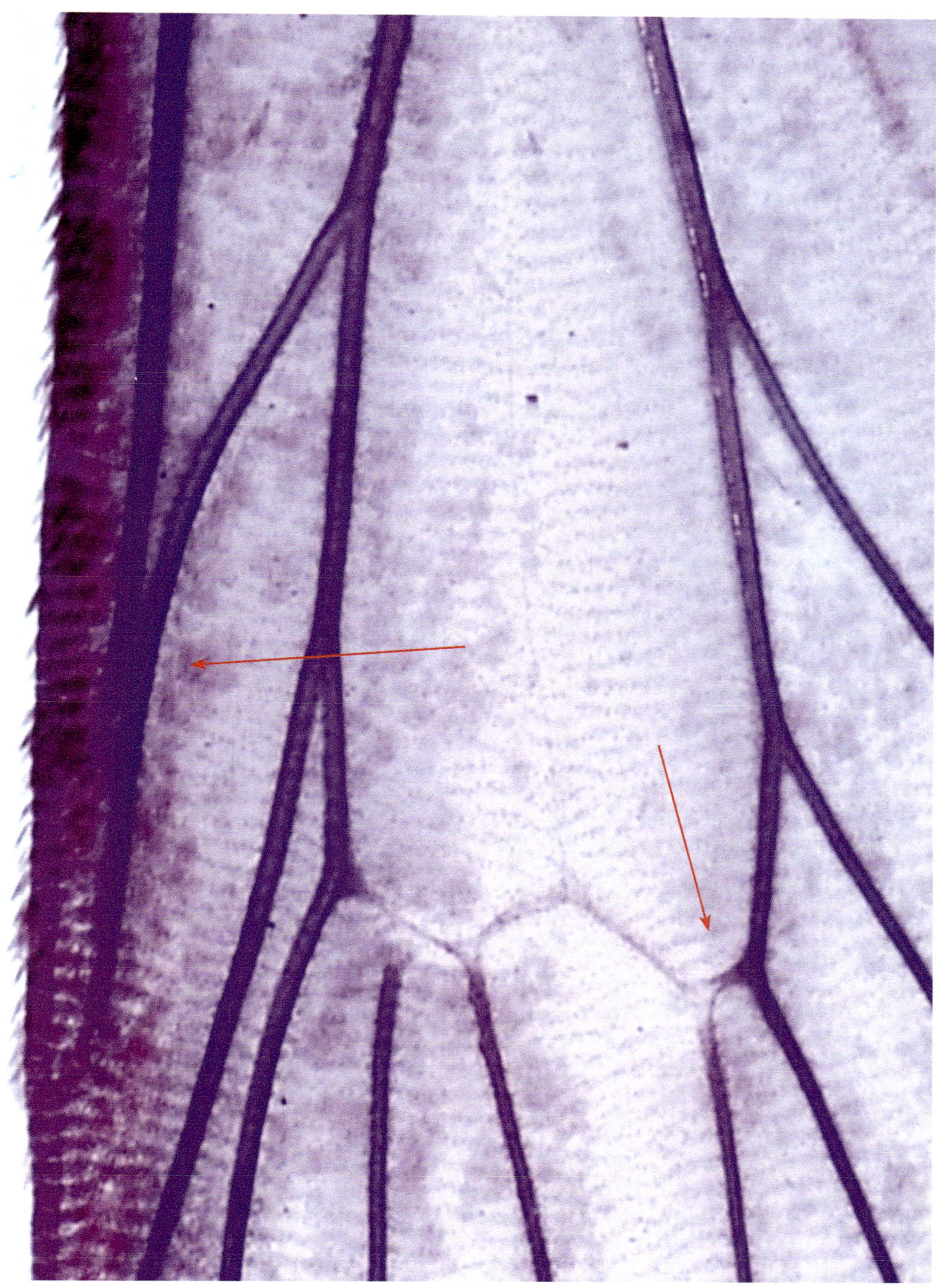

Plate 13.27. Wing venation (close up). Sc and R_1 remain separate. R_2 maintain good distance from the stalk of R_3+R_4. M_1 arises from the cell angle. Similarly M_2 also arise from cubital peak and M_3 arises just below .

14

Keys to the identification of sugarcane moth borers (Based on larval and pupal characters)

In a field situation, very similar appearance of the larvae of moth borers of sugarcane renders their identification relatively difficult and this is more so for the field entomologists, especially when many borer species share the same space and time. An easy and accurate identification at field level may be facilitated by careful observation of the symptoms of infestation, nature of damage in combination with both larval and pupal characters. Apparently, to a sugarcane farmer, larvae of different borers, especially the striped borers appear very similar. However, for a trained person, the subtle differences in the larval characters are sufficient to establish the identity of the moth borer species involved in causing the damage. In case of very similar appeance of larvae, pupal characters always confirm the actual identity. Due to this reason, emphasis is given here to identify the moth borer species with the help of larval characters and fine-tune the identification further with the help of pupal characters (if required). As a larva passes through several instars in the process of its growth and development, the morphological characters also change accordingly. The characters of a mature larva (usually having a length more than 20-25 mm) are taken into consideration for the identification of a species of moth borer. Moreover, larvae of most of the moth borers also pass considerable time in hibernation and in this period, many larvae exhibit fading or non-appearance of the colour stripes on the body. Many a times, in a prepupal larva, coloured stripes also fade away. Besides the morphological differences, the habits of the larvae and their feeding behaviour along with the symptoms of crop damage further help in establishing the identity of a moth borer species in the field. For example, if mid-rib tunnelling symptom is present, one can identify the moth borer species as the Top borer even without actually observing the presence of a larva within the affected cane plant.

For easy identification, the larvae or caterpillars may be broadly divided into two broad categories depending on the colour of the body. The larvae may be white (milky or creamy) or may have some form of colour on the body; if the larvae are coloured then they are further subdivided into two subcategories, i.e., fully coloured (no stripe) or partially coloured (presence of coloured stripes). If striped, then larvae are again separated on the basis of number and colour of the stripes on the body (presence or absence of the median stripe quickly indicates the number).

Colour of the tubercles on the larval body (dark patchy raised area usually bearing setae) also becomes handy in the quick field identification.

Shape, size and colour of the tubercles, along with the presence of *asetose* tubercles (tubercle without setae) become important. Similarly, shape and colour of the thoracic shields also helps in quick separation of the larvae of different borers. Shape of the spiracle and colour of its rim; spiracle of the anal end in relation to the body stripes, etc. are also important. Shape, size and arrangements of crochets in the four abdominal *prolegs* form very important identifying characteristics.

1. ***Colour of the Body***:	White or creamy (W) or coloured (C)
2. ***Body coloured***:	Fully coloured (FC) or partially coloured or striped (S)
3. ***Body partially coloured (striped)***:	Four striped(S4) or Five striped (S5)
4. ***Crochets*** :	Complete (O) or incomplete (U)

W. Larval body is white to creamy in colour (Top borer and Root borer)

W1. Top borer

In an infested cane, larva damages only the crown portion of the plant and remains restricted at the crown portion in all the broods.

A Larva passes several instars, and colour of the larva varies from white to creamy yellow in colour depending on the instar a larva is passing through. Larval body is cylindrical, smooth (without any perceptible hairs on the body except the head). The body abruptly ends at both ends. Head is relatively smaller with respect to rest of the body. In a mature larva *tracheal trunk* and *spiracles* are inconspicuous. Crochets or hooks on the *prolegs* are very small and are arranged in a complete elongated circle (like just two first small brackets). On an anal proleg around 22 hooks are present. A third instar larva, which remains above the growing point of the crown, is whitish in colour, smooth and very fragile. The larva is sluggish and on extrication from the tissue, it moves very slowly.

A Pupa is typically cylindrical and colour varies from yellowish to whitish depending on the stage of the development of the pupa. The 5th, 6th and 7th abdominal segments are smooth i.e., without any specialized ornamentations. In a female pupa, the anal segments are larger and have distinctive whiteness. With maturity whiteness changes to orange colour due to the formation of anal hairs. At maturity, body of a female pupa becomes white whereas, a male pupa turns light straw in colour with sharp conical anal end. The tip of the hind legs of a female pupa always remains free from the body surface. In male pupa, hind legs are quite long and reach the 7th segment and terminal part remains free. Pupation takes places in a silken pupal chamber within the feeding tunnel and remains protected behind several silken discs.

W2. Root borer

In an infested cane shoot or in a mature cane a larva of Root borer restricts itself to the bottom portion of the cane and feeds in the stalk portion of the plant at or below the soil surface. Due to its habit of staying in the stalk at the basal portion of the cane, which mostly remains below the soil surface, a prepepual larva prepares a long silken tunnel up to the ground surface for the safe exit of the moth in the aerial environment.

A Larva of Root borer is shiny or glistening milky white to pale yellowish white in colour (depending on the instar it is passing through). A larva always remains shiny white on extrication from the cane tissue, the larva appears relatively more active than the larva of Top borer. On provocation, it responds aggressively and occasionally it tries to bite. A larva bears perceptible hairs on the body and has a habit of taking a quick and sharp bend, if disturbed. Larval body is broader at the thoracic region and gradually tapers towards the anal end. Thus, in dorsal view a larva imparts a conical appearance. *Prothoracic shield* is relatively larger and light yellow in colour. *Spiracles* are roundish with reddish orange rim and appear prominent on the white body. The crochets or hooks on the prolegs are small and arranged in a complete obovate circle/pear shaped circle (circle is narrower towards *meson*/inner side). In the anal proleg only 7-8 hooks are present.

A Pupa is also quite broad at the thoracic region and gradually tapers towards the anal end. A freshly formed pupa is yellowish in the wing pad region but whitish at the abdominal region. With maturity, a pupa turns yellowish brown. A pupa has numerous characteristic depressions or pits on the body. On the dorsal side of the anal end it has a characteristic dark coloured hump like protuberance. The 5th, 6th and 7th abdominal segments are smooth i.e., without any specialized ornamentations except the depressions. Pupation takes places in a silken pupal chamber.

C. Larval body is coloured (with or without stripes)

The coloured larvae are divided again into two broad categories, i.e.,

FC. Larval body is coloured but without any stripe; a larva may have variations in colour patterns (Pink borer, Green borer).

FC1. Pink borer

Larva damages the sugarcane crop mostly at shoot stage and produces easily pullable *dead heart* that emits foul smell (less intense). After cane formation, larva may act as a cane borer (very low intensity).

A Larva is flesh coloured. Dorsal side is pink and ventral side is creamy yellowish-white. Larva is very agile and on extrication quickly moves away in search of hiding place. Larval body suddenly tapers at the anal end. The *spiracles* are elongated oval, black rimmed and are very prominent on the pink body. The four pairs of abdominal *prolegs* are stout and *duck foot* type. Crochets are stout and present in an incomplete circle. Hooks cover only one third of the imaginary circle, only around 15-16 hooks are present in a *proleg*.

A Pupa is stout and light to dark brown in colour and has marked depressions or pits on the pupal body. The 5th, 6th and 7th segments are smooth (except the pits). Four triangular spiny black projections are present at the anal end. Eye area of the pupa is bulged and the head area is marked with a bluish tinge. Occasionally, head and thoracic portion of the pupal body is covered with ashy white substances. No specialized pupal chamber is formed.

FC2. Green borer

Larva damages the sugarcane crop only at shoot stage and its infestation is mostly restricted to the early part of the season (overlaps with Early shoot borer). Larva passes the monsoon as well as winter months in hibernation at the base of the clump in soil.

A Larva is characteristically very active; in fact, it is even more active than the larva of Early shoot borer. Larva is relatively smaller in size. The second instar larva has greenish thorax with light pinkish abdomen. The dorsal side of a mature larva is dark copperish (as if copper is mixed with green) and the ventral abdominal side is green. *Prothoracic shield* is well developed and greenish in colour and this imparts a distinctive appearance to the larva. *Spiracles* are dark rimmed. Crochets on the abdominal prolegs are arranged in a complete circle.

A Pupa is greenish in colour when freshly formed and with maturity changes to greenish brown. Pupa is much broader at the thoracic area and gradually tapers towards the anal end. Abdominal segments are lighter in colour and have numerous oval pits. No specialized ornamentations are present on the 5th, 6th and 7th abdominal segments. Anal end is made up of a lightly chitinized dorsal half with two lobes pointing posteriorly and carrying usually six (six to ten) long characteristic brownish *circinate* hairs.

S. Larval body is partially coloured due to presence of prominent coloured stripes on the body.

Depending on the number of *stripes present on the body* the larva of striped borers may be further divided into two sub categories i.e.,

Larval body bearing four coloured stripes (S4) and larval body bearing five coloured stripes (S5).

S4. Larva with four prominent stripes on the body; the dorsal median stripe is absent (Internode borer, Maize borer, Plassey borer and Gurdaspur borer);

Depending on the arrangement of crochets on the abdominal prolegs these larvae may be divided again i.e., crochets form a complete circle and crochets form an incomplete circle on the prolegs.

Crochets are arranged in an incomplete circle (U) i.e., part of a proleg remains without crochets.

S4U 1. Gurdaspur borer

A Larva is slender and agile and on extrication moves away quickly in search of a hiding place. Larval body is basically white in colour and appears dorsally flat. The presence of broad and dark subdorsal stripes imparts a squarely appearance when viewed from the dorsal side. The two lateral stripes are much wider and fully cover the *tracheal trunks*. Characteristically these stripes are very dark brown and placed a bit lower in relation to the *spiracle* i.e. placed nearer towards the prolegs. Upper side of a lateral stripe covers the *tracheal trunk* completely and thus, renders the *spiracles* obscure to the naked eye. *Spiracles* are oval with dark rim. Crochets or hooks are stouter and are arranged in an incomplete circle (hooks just cover $2/3^{rd}$ of the imaginary circle of the *proleg*).

A Pupa is slender and does not possess any specialized ornamentation on 5th, 6th and 7th abdominal segments. However, microscopic spines/processs are present on these segments. The anal segment is short and on each half of the dorsal side, one upwardly curved prominent projection is present. Pupation takes places in a silken pupal chamber.

Crochets or hooks in the prolegs are arranged in a complete circle (O) (Internode, Plassey and Maize borer)

S4O 2. Internode borer

A Larva is slender and very active. The two pink dorsal stripes are straight and run through both the dorsal tubercles pairs (D_1 and D_2). The stripes mostly have a straight outline (not wavy). Dorsal tubercles are jet black and are very prominent during rainy season. Generally, the colour of the tubercles fades away during hibernation and before pupation. During winter months the larva appears whitish, tubercles take the body colour, and pink stripes fade away. At this period, the dorsal stripes are visible even on the 9th abdominal segment. In each body segment, a characteristically horizontal *white cut mark/* faded area is present on the pink dorsal stripes. The cut mark appears just in front of the large dorsal tubercle (D_1) near the upper segmental boundary. This horizontal cut mark helps in differentiating a larva during hibernation. Lateral stripes run through the *tracheal trunk*. *Spiracles* are elongated oval and black rimmed. Crochets or hooks on the *prolegs* are finer and are arranged in a complete circle. Occasionally, a few hooks towards the outer/lateral side remain absent and thus the arrangement of crochets appears incomplete (especially during hibernation in winter).

A Pupa is slender and light to dark reddish brown in colour. Ornamentations on the 5th, 6th and 7th abdominal segment are in the form of velvety areas, slightly darker than the body colour (formed due to presence of irregular rows of velvety growth). The velvety patch is restricted to the dorsal side; even on the 7th segment this patch does not reach the *spiracles*. The ventral side is smooth. The *cremaster* has six dorsal spines; four are prominent and two are small. There are two ventral projections. Pupation takes places in a loosely formed silken pupal chamber.

S40 .3. Plassey borer

A Larva of Plassey borer shows variations in colour of the stripes. Four distinct larval types may be observed. Larvae with lighter stripes and distinct tubercles are common. Due to these variations, larvae of Plassey borer are quite similar to those of the larvae of *Chilo partellus*. General colour of the larval body is pale water-washed yellowish white. *Prothoracic shield* is large and dark brown. In a typical larva, large dorsal tubercles (D_1) are prominent and impart a squarish appearance (like fur markings on a cheetah). Larval body bears prominent setae. The smaller dorsal tubercles (D_2) always remain aligned with larger tubercles at the median area (*meson*) but remain significantly extended towards the subdorsal area. On the 9th segment, all the four dorsal tubercles are fused together to form a very large *sclerotized* area (four setae present appear in a straight line on the fused tubercles). *Tracheal trunk* is visible. The lateral stripes always remain above the *spiracles*. *Spiracles* are black rimmed, elongated oval and appear open inside. The crochets on the abdominal *prolegs* are stout, dark brown in colour and are arranged in a complete circle.

A Pupa when freshly formed appears yellowish white with prominent spiny ornamentations on the three abdominal segments. Head of the pupa is a bit protruded forward. A mature pupa appears stouter (imparts a robust appearance with prominent spines) and deep brown in colour. The distinct spiny ornamentations on the 5th, 6th and 7th abdominal segments are very characteristic of the pupa. The ornamentations are in the form one to two rows of thick and well-developed spines, which can be felt by the touch of a finger. Occasionally, tip of some of the spines gets bifurcated or branched. On the 5th and 6th abdominal segments, spines only extend up to the *spiracle*, and thus, the ventral area in these two segments remains smooth. On the 7th abdominal segment, the spines completely girdle the segment and circumvent the *spiracle* by forming an arch over it. The *cremaster* is not very distinctive. No specialized pupal chamber is formed for pupation.

S40.4. Maize borer

A Larva appears very similar to that of Plassey borer. The two dorsal stripes are relatively wider and often cover the entire dorsal area. When the dorsal stripes become lighter, they characteristically leave the tubercles untouched. Presence of *asetose* lateral tubercles on the *tracheal trunk* (from 1st to 7th abdominal segments) is the characteristic feature of this species. These *asetose* tubercles remain in line with the respective *spiracles* on the *tracheal trunk*. The tubercle just above a *spiracle* is triangular.

A Pupa is slender and light to deep brown in colour. Ornamentations are present on the 5th, 6th and 7th abdominal segments and are in the form of several irregular rows of tiny spines (at the terminal row, spines are larger and darker), slightly darker than the body colour. These ornamentations appear more or less uniform on the three segments. Ventral side is smooth. The rows of spines on the 7th abdominal segment just cross the *spiracle*. Projections in *cremaster* appear like a petal of lotus when viewed from the ventral side. No specialized pupal chamber is formed for pupation.

S5. Larva with median or five prominent stripes on the body (Early shoot borer and Stalk borer).

S5U .1. Early shoot borer

A Larva of Early shoot borer is slender and very active and appears a shade darker than that of Stalk borer. Characteristically the crochets or hooks in the abdominal *prolegs* are arranged in an incomplete circle. Hooks cover only 75%of the inner circle and the outer 25% area remain without any hook. The visibility of the *tracheal tube* is variable, sometimes feeble, and sometimes prominent. The body stripes are always a shade darker than those of Stalk borer. The stripes are relatively dark-pinkish, more irregular or wavier at the margins than those of Stalk borer. Large dorsal tubercles (D_1) are relatively smaller in comparison to that of Stalk borer. The smaller dorsal tubercles (D_2) are more elongated (narrow). The horizontal distance between two large dorsal tubercles in an abdominal segment is relatively wider. Due to this gap at the *meson* area, the median stripe does not touch the large dorsal tubercles. The three dorsal stripes converge at the two conjoined tubercles of the 8th abdominal segment and the median stripe crosses the tubercle pair and reaches the 9th abdominal segment. The lateral tubercles present at the thoracic area are smaller than that of Stalk borer.

A Pupa is slender and yellowish to light brown in colour. Pupa also bears the corresponding five stripes as observed in larva. The abdominal ornamentations are in the form of characteristic discontinuous projected ridges on the 5th, 6th and the 7th abdominal segments. Terminal area of these ridges are dark. On the 5th and 6th segments the discontinuous ridges just cross the *spiracle* of that segment; the rest of the ventral surface remains smooth. In the 7th abdominal segment, the discontinuous ridges completely encircle the segment. The dorsal side of the *cremaster* has four raised areas with four spines. The ventral side bears only two projections.

S5O. 2. Stalk borer

A Larva is slender and active but relatively fragile in comparison to that of Early shoot borer. Crochets or hooks in the *prolegs* are arranged in a complete circle. The three dorsal stripes do not converge at the 8th abdominal segment. The median stripe just reaches there but may or may not touch the conjoined tubercles of the 8th segment (median stripe never crosses the tubercles). The two subdorsal stripes are wider and deep in colour. The lateral stripes are relatively narrower and light in colour. Dorsal tubercles (D_1) are relatively larger, roundish-square and just a shade darker than the body colour. Distance between the two large dorsal tubercles (D_1 and D_1) in an abdominal segment (*meson* area) is relatively narrower. Due to the close proximity of the tubercles, the median stripe often partially runs through the dorsal tubercles (D_1). The lateral tubercles present at the thoracic area are distinctively larger. The *tracheal tube* is very prominent like a white tube (more prominent than Early shoot borer). Anal end of a larva appears more setose than that of the Early shoot borer.

A Pupa of Stalk borer is slender and possesses two characteristic horn-like projections on the head. The mesothoracic *spiracle* is very prominent and distinctively raised. Five stripes, corresponding to the larval stripes, are also present on the pupa. The veins on the wing pads of the *pharate adult* are very prominent. Pupal ornamentations on the 5th, 6th and 7th abdominal

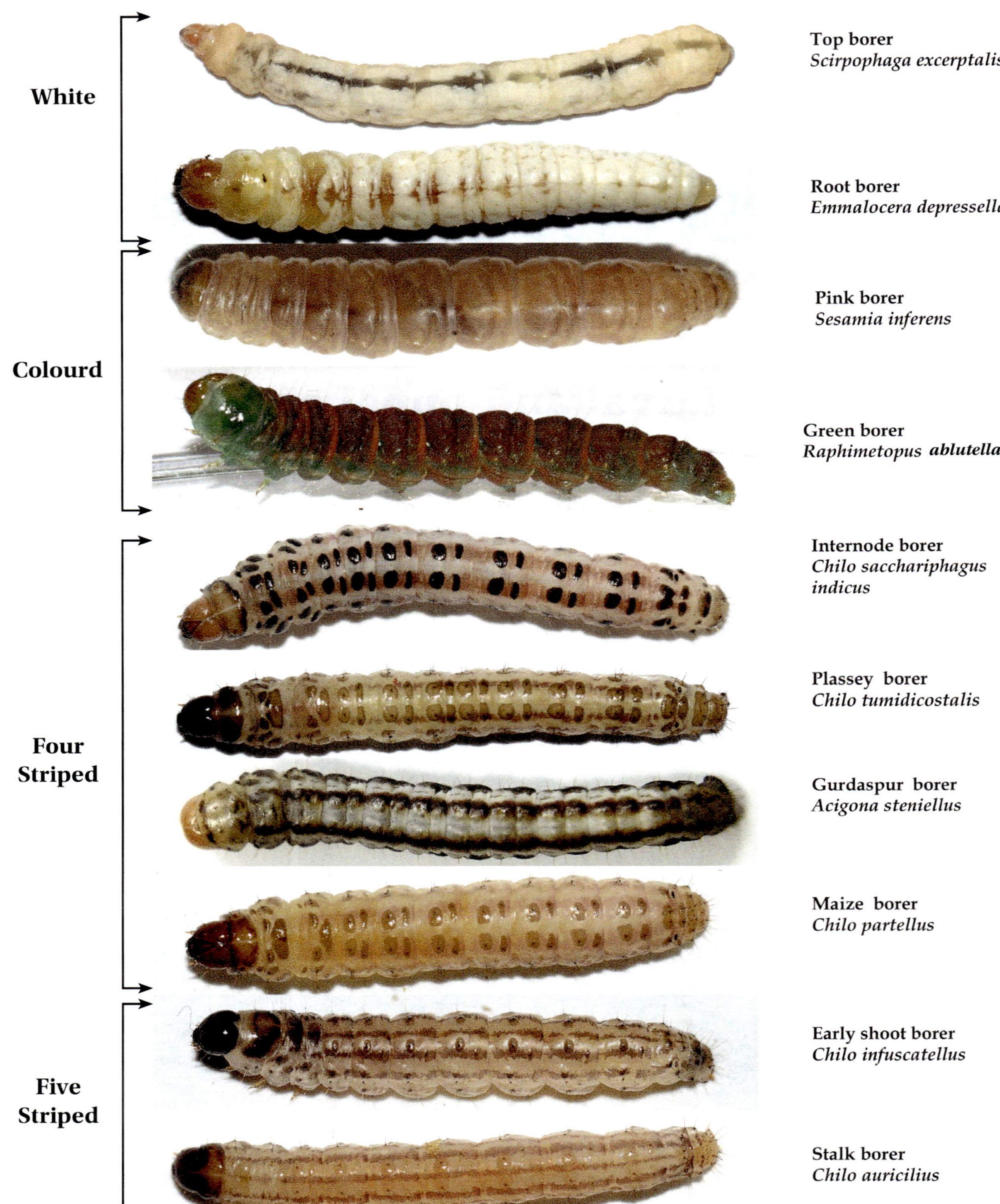

Plate 14.1. Larvae of different moth borers (not according to size)

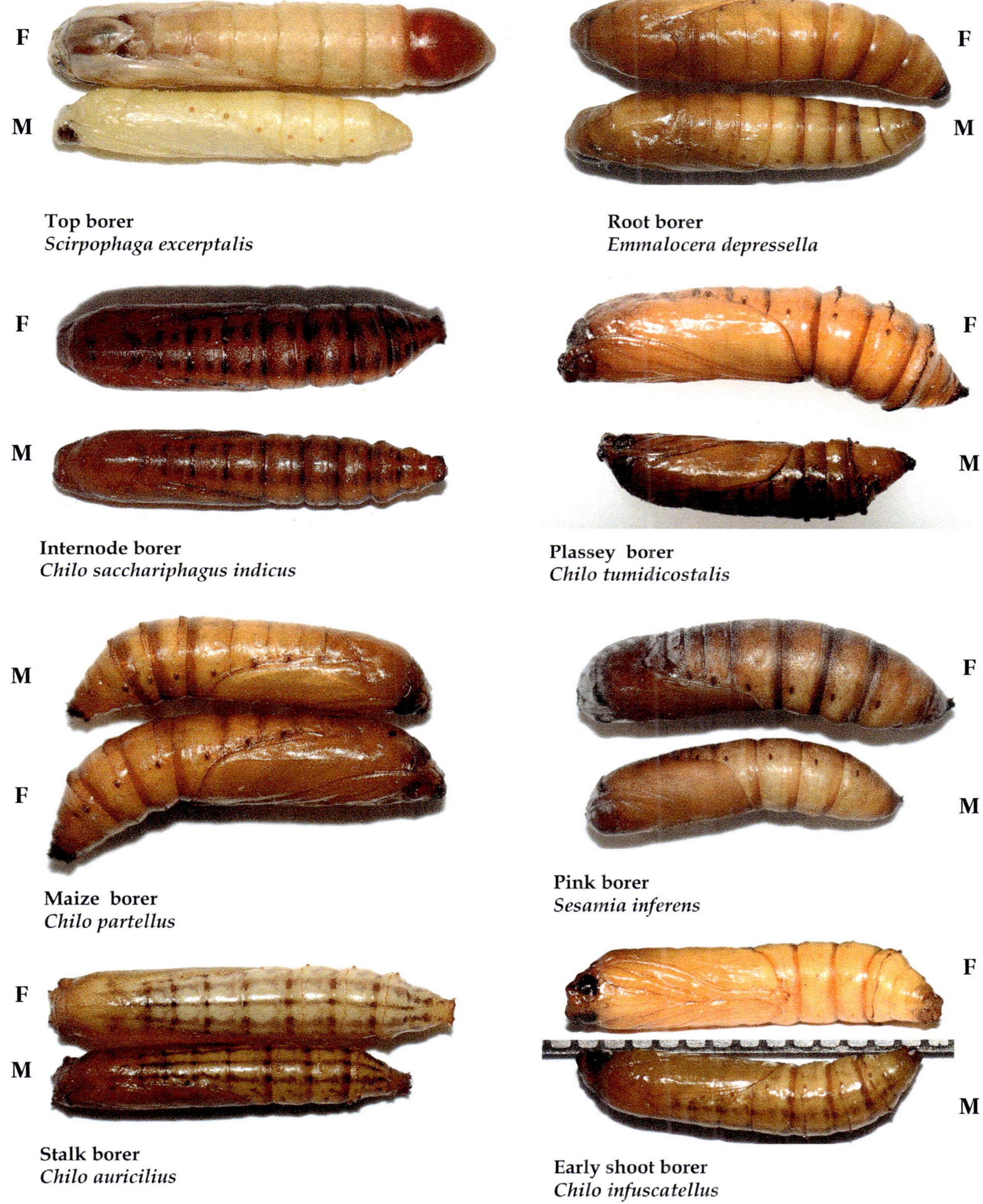

Plate 13.25. Pupae of different moth borers

15

Important bio-agents of sugarcane moth borers

The biological world of animals is largely governed by the *trophic* dependence i.e. dependence on food - a relationship of host-predators and parasites. In the insect world, many organisms live at the cost of the other organisms i.e., one group of organism specifically utilizes (consumes) another group of organism(s) for its growth and development. The former group of organisms that subsists on the latter is categorised as predators, parasites or commensals. A *predator* catches a living organism for food and tries to consume it immediately. Whereas, a *parasite* draws nutrition from the host organism in a protracted manner and gradually debilitates the host organism (may or may not kill the organism).

The specialized term '*parasitoid*' within the fold of parasite was coined to explain the behaviour of *Hymenopterous* parasitic insects. Adults of these insects are free living and the female lay eggs outside or inside the body of host insect. The eggs on hatching become maggots and feed on the host. Usually, the parasitoids kill their hosts. Parasitoids may be external feeders i.e., draw the nutrition by staying outside the host body (*ectoparasitoid*) or may be internal feeders i.e., draw the nutrition by staying inside the host body (*endoparasitoid*).

Most of the important bio-agents of insect pests of crops belong to the *Hymenoptera* i.e., insects with membranous wings. In common terms, these bio-agents are known as *wasps*. The parasitic wasps mainly belong to the three families, viz., *Chalcidoidea, Proctotrupoidea* and *Ichneumonoidea*. These wasps are generally grouped depending on the stage of the host they utilize as their preferred food i.e., those that attack eggs are grouped as *egg parasitoids*, those that utilize larva are called *larval parasites* and those that target pupa are termed as *pupal parasites*. Some parasitoids can utilize more than one life stages of the host; in that situation, it is grouped on the basis of the predominant stage of the host it parasitizes. However, due to short life span of an adult moth borer, no parasitoids have ventured to utilize it as the food base; only the predators and scavengers utilize the adults as their food.

Eggs of *Chilo* sp. are small, scale like and usually less than 1 mm in size. Obviously, the egg parasitoids of *Chilo* borers are tiny wasps (*Trichogramma, Telenomus*) as because these parasitoids have to sustain the total development of an adult by utilizing the nutrition available in a tiny egg of the *Chilo* moth borers. Moreover, the size of the adult egg *parasitoid* is again limited by the boundary of the eggshell. However, there is no such restriction on the size in case of *larval parasites*. Thus, two types of larval parasites (medium and small) are present. In case of small sized parasite, one host larva is adequate to support the development of many adults (*Rhaconotus, Cotesia*). In large sized wasps, one host larva supports the development of only one adult (*Isotima, Stenobracon*). In case of pupa, only tiny parasitoids are present. In case of the *Eulophid* endoparasitoid *Tetrastichus* that mostly uses pupa as the preferred food, a single host pupa supports the development of many adults (>250 adults develop within a female pupa of Top borer).

Among the important bio-agents utilized in sugarcane agro-ecosystem for the containment of moth borers, *Trichogramma* is the smallest one (around 0.6 mm), followed by the *Tetrastichus* and *Telenomus*, which are around 1.5 to 2 times (around 1 to 1.2 mm) of the size of *Trichogramma*. These tiny egg parasitoids are followed by the intermediate sized gregarious larval parasitoids like *Rhaconotus, Cotesia* (2 times of *Telenomus* or *Tetrastichus*). Larger solitary wasps like *Isotima* and *Stenobracon* (body length 10 mm or more) fall in this category. In these two relatively large *ectoparasites*, one host larva is consumed to produce one adult wasp.

In the laboratory mass rearing, good success has been achieved with the gregarious parasitoids like *Trichogramma, Tetrastichus, Cotesia,* etc. and this success has facilitated their effective use in the biological control of different insect pests of crops. However, a severe limitation exists in the laboratory mass rearing and faster multiplication of large sized solitary parasitoids and this hinders their effective utilization in bio-control. Conservation of these parasitoids in the sugarcane field is usually practiced.

The tiny trio viz., *Trichogramma, Telenomus* and *Tetrastichus* are extremely difficult to identify in the field. These parasitoids need at least the aid of a hand lens for their identification. However, one can very well identify the wasps of intermediate size (*Cotesia, Rhaconotus*) in the field. The large sized wasps (*Isotima, Stenobracon*) are most easily identifiable in the field and often farmers confuse them as poisonous wasps.

Despite a large number of parasitoids reported to be present in the sugarcane agro-ecosystem, only a few could be exploited commercially for the containment of moth borers of sugarcane. It is hoped that with more intensive research in this direction, new parasitoids might be found or new technique of mass multiplication for utilising the existing ones would be developed for the containment of the borer pests effectively and economically sans the pesticides.

Important bio-agents

E. Egg parasitoids

The egg parasitoids are *ployphagus* and use eggs of different moth species. These wasps, by destroying the eggs of insect pests, reduce the initial population of neonate larvae. Obviously, egg parasitoids get a preference over the larval or pupal parasitoids as they check the pest before it reaches the larval stage, which actually cause the damage to the crop. Theoretically, egg parasitoids are most suited as the bio-control agent, but practically it is not so. Egg parasitoids have their own limitations. Probably nature has maintained these parasitoids in that manner; otherwise, these parasitoids would have completely wiped out the insect pests from the face of the earth.

E1. *Trichogramma chilonis* Ishii

Trichogramma species are minute (0.35 to 0.6 mm in length) polyphagus wasps parasitizing the eggs of different Lepidopteron pests. The first description of a *Trichogramma* species was given by Riley in 1871 from North America. He found this tiny wasp emerging from the eggs of a butterfly and identified the new wasp as *Trichogramma minutum* Riley. Identification of a species of *Trichogramma* with reasonable confidence is difficult due to its small size and very similar morphological characters. As the *Trichogramma* females are very similar in appearance, taxonomists mainly rely on the morphology of the males for the identification of a species. One has to rely on the specialists for its accurate identification.

Trichogramma chilonis is a small, light yellowish bodied insect with orange compound eyes. Antenna is also club shaped.

In India, deployment of *Trichogramma* for the control of sugarcane moth borers (*Chilo infuscatellus*) on field scale was started in Mandya, Karnataka during 1930's using the eggs of rice grain moth *Cocyra cephalonica* Stainton.

The Mysore sugar Factory was started at Mandya in 1933 to utilise the sugarcane grown in the area commanded by the newly opened canal irrigation system. In 1934 planting (July - November) the sugarcane crop suffered a serious setback due to the borer attack. To contain the borer problem with bio-agent *Trichogramma,* a laboratory was established at Mandya for the large-scale production of *Trichogramma* and the laboratory started functioning from 1935.

However, during all these years since 1935, little progress has been made in the natural establishment of *Trichogramma* in the sugarcane agro-ecosystem (as recovery of *Trichogramma* from the sugarcane field always remained a conscientious issue to be resolved) and thus, *repeated inundative releases* have become the norm for its use to check the moth borers of sugarcane. Host searching efficiency of *Trichogramma* is also very poor and it mostly utilizes the naked egg mass of moth borers. Despite poor efficiency of parasitisation and requirement of repeated inundative releases, *Trichogramma* still holds the most important position in the bio-control of the Lepidopterous pests of various crops due to its amenability to the large scale multiplication on the eggs of factitious hosts, their storability and easy deployment.

E2. *Telenomus beneficiens* Zehntner

Telenomus are very dark brown to black tiny wasps (almost 1.5 to 2 times of *Trichogramma*) which can utilize the hair-covered egg mass of Top borer in addition to the naked egg masses of other borers. In the sugarcane agro-ecosystem, the presence of *Telenomus* may be easily ascertained on the basis of antennae and their colour. In the first two broods of Top borer, *Telenomus* predominates in parasitizing the egg masses. In this period, a random collection of around 100 egg masses of Top borer as well as their incubation to maturity, in most cases, results in the emergence of the wasp *Telenomus* from a few egg masses.

Sexual dimorphism is very prominent in the antenna. Antenna is 12 segmented. Five top segments of the female antenna are larger and impart a club shaped appearance. Male antenna is longer with uniform *moniliform* segments throughout. Abdomen of *Telenomus* is definitely conical.

Telenomus is an efficient naturally occurring egg parasitoid in the sugarcane agro-ecosystem that needs to be exploited for the bio-control of moth borers. An efficient and fast laboratory multiplication technique is still awaited for this species to realize its full potential, especially in the management of Top borer in the first two broods.

L. Larval parasite

The arena of larval parasites are dominated by the Braconids. The Braconids are most frequently encountered in the borer complexes of any crop, including sugarcane. Braconids are in the habit of swaying their antennae frequently.

L1. *Cotesia flavipes* Cameron

Cotesia flavipes (formerly known as *Apanteles flavipes*) is a Braconoid wasp with distinctly yellowish legs. *Cotesia* is a gregarious larval *endoparasitoid* (many grubs feed staying within the larval body) of many *Chilo* borers. These wasps locate the host larva and initiate egg laying by pushing the ovipositor inside the body of the host larva. Once the feeding period of the grubs is complete, mature grubs come out of the host and gregariously spin loose silken white *cocoons*. Despite the insidious feeding habit of borer larvae in the cane, an adult of *Cotesia* waits for its chance to get an access to the larva and to paralyze it. In the field, identification *Cotesia* becomes easy when it produces cocoons in the feeding tunnel of a larva. Presence of distinctive milky white, fluffy and loosely attached cocoons outside the host skin is very characteristic of *Cotesia*. The cocoons remain independent of each other.

The adult female wasp has reddish to yellowish brown body colour, around 2.5 mm long body with a small ovipositor. Head and thorax are deep brown to black. The dorsal side of the abdomen are two shades lighter in colour and bears yellowish white bands. First few abdominal segments of the ventral side (sternite) appear yellowish white. The hind leg is darker (reddish brown) in colour than the fore and mid leg (yellowish); *tarsus* is always a shade darker than the *tibia* or *femur*. Antennae are characteristically *moniliform*, with 18 relatively thicker segments. Sexual dimorphism is evident in the antennae. In male, segments of flagellum are larger, elongated like a glass tumbler, and have a darker band at the centre. In the female the segments are like tea cups. On the death of a female wasp, these antennae usually bend backwards imparting a horn-like appearance (horn of a ewe) on the head.

L2. *Rhaconotus scirpophagae* Wilkinson

Rhaconotus is a gregarious larval *endoparasitoid* (many grubs feed staying within the larval body) of larva of Top borer. It is slightly larger than *Cotesia*. The long *filiform* antennae of *Rhaconotus* immediately distinguish it. Prepupal larvae of *Rhaconotus* come out of the host body and gregariously spin *translucent silken cocoons* and the *cocoons* remain firmly attached together. Body size of the female is around 3 to 4 mm. *Filiform* antennae are characteristically longer than the body (around 7 mm). Ovipositor is short, stout and measures around 1.5 mm (one-third of the body length).

In the field, presence of *Rhaconotus* may be judged by looking for the distinctive closely attached *translucent cocoons* in the feeding tunnel of the Top borer larva. The silks of the cocoons are so closely woven that the cocoon walls appear as fine sheets of tracing paper, smooth and translucent. The *pharate adult* is white with black eyes; it gradually turns brown with maturity. After completion of the pupation an adult wasp, by cutting holes in the silken disks (prepared by the Top borer larva) emerges through the exit hole. For the direct escape, the wasps may also cut nice round holes directly on the stalk tissue or leaf sheath tissues, as the case may be, and come out of the stalk.

In the female wasp, the *ovipositor* appears as single black stout structure initially (due to protective cover) but with maturity, the sheath cover of the *ovipositor* separates out and the female wasp become ready for laying eggs. For laying eggs, a female wasp thrusts its *ovipositor* to the selected larval body and deposit a large number of eggs.

L3. *Isotima javensis* Rohwer

The larval parasite *Isotima* is 5 to 6 times larger than the *Cotesia* or *Rhaconotus*. Only one parasite adult emerges consuming one host larva. Antennae are dark brown/black and *filiform*. Characteristically, at dorsal side four median segments of an antenna appear white or ashy grey in colour. Abdominal segments (sternites) have definite white bands or markings. The anal segment is fully white. Usually at the dorsal side, a female wasp bears one white abdominal band and a male wasp bears two to three dorsal abdominal bands in addition to white bands of the anal segments, which are uniform in both sexes. The ovipositor of *Isotima* is stout, spear shaped and finely serrated. Immediately after emergence, the ovipositor remains sheathed and with maturity sheath cover separate out exposing the *spear shaped serrated ovipositor*.

The availability of favourable temperature regime in the tropical sugarcane belt provides an advantage to the biological control. Successful biological control of Top borer was achieved through the deployment of *Isotima*. Mr. B. D. Gupta from IISR, Lucknow utilized this curious fact as early as 1956; through the introduction of *Isotima javensis* from subtropical north India to the tropical south India. He took *Isotima* from Muzaffarnagar, western Uttar Pradesh and introduced it in the cane fields of Pugalur area of Trichy district of Tamil Nadu. In a short period of time, *Isotima* not only successfully established itself in the new area but also effectively reduced the Top borer incidence. Later, a similar success in the management of Top borer was replicated in tropical areas wherever *Isotima* was introduced (other parts of Tamil Nadu, Karnataka and Madhya Pradesh). In fact, in the Indian context, it remains the first success story of biological control in sugarcane through introduction of a parasitoid from one location to another.

L4. *Stenobracon nicevillei* Bingham

Stenobracon is a large *braconid* wasp and is an important larval parasite of the Top borer of sugarcane. It can also utilize the larva of other important borers. Its presence in the field may be easily observed during summer months. It continues to be present in the sugarcane field during the rainy season. Adult wasp has a yellowish body. Antennae are dark/black and *filiform*. Part of the head is dark/black. A large black area is present at the dorsal segment (*tergite*) of the anal end. On the forewings, three rows of dark/black areas are present. The dark/black area meeting the *termen* is much lighter. *Stenobracon* has a very characteristic *long ovipositor*. The ovipositor is dark coloured, flexible and longer than its body length and measures around 15-16 mm. This long and flexible ovipositor helps the wasp to push the eggs on or near the larval body breaching through the protective discs. On hatching it attaches itself on the host body and draw the nourishment staying outside. One larva of Top borer supports the development of only one parasite.

C. Pupal parasite

C1. *Tetrastichus howardi* (Olliff)

Tetrastichus howardi (Olliff) is a gregarious Eulophid pupal *endoparasitoid*, which predominantly utilizes pupa of a large number of moth borers. The female pushes its ovipositor inside the pupa and lays several eggs. The grubs feed inside the pupal shell and completes its full development there. Finally, adult wasps come out by cutting a small hole in the wall of the pupal shell. A female wasp has an uncanny ability to distinguish between a parasitized and non-parasitized pupa and the number of eggs to be pushed within a pupa , if sufficient food is available.

These wasps are also black coloured and are quite similar to *Telenomus* in physical appearance. However, the shape of abdomen and the nature of antennae immediately differentiate these two species. In *Tetrastichus*, antenna is only seven segmented (in *Telenomus* the antenna is 12 segmented). Out of these seven segments, five segments forming the terminal end are black in female. In male, only the terminal segment of the antenna is black. Terminal segment of the antenna is conical in both sexes. The abdomen of a female *Tetrastichus* appears wider, flat (square type) in most of its length and abruptly ends up in a sharp pointed structure at the anal end. In *Telenomus*, the abdomen is distinctly conical and gradually tapers to the anal end.

List of plates on important bio-agents of sugarcane moth borers

Description of plates

Plate 15.1. **A, B & C** . *Trichogramma adult.* **D** . Emerging *Trichogramma* adult from a *Corcyra* egg. **E.** White *Corcyra* eggs. **F.** Black parasitized *Corcyra* eggs

Plate 15.2. *Telenomus* adult **A**. Male **B**. Female. **C**. Parasitized eggs of Internode borer. **D**. Exit hole of the wasps on egg shell of Internode borer & **E**. Visible emergence hole in the midst of coverings of anal hairs (Top borer egg mass).

Plate 15.3. A, B & C. *Telenomus* female prasitizing top borer egg mass by pushing its abdomen through the layers of anal hair covering. **D &E** Emergence of *Telenomus* adult (male) from an egg mass of Top borer.

Plate 15.4. *Rhaconotus* larvae. **A & B**. Gregarious feeding behaviour. Coming out after completion of feeding in the larva of top borer. **C &D**. Close-up of larvae

Plate 15.5. Formation of closely attached cocoons of *Rhaconotus*. **A**. It got access to the host larva through the damaged epidermis. **B**. Got access to ther host larva through the exit hole.

Plate 15.6. A.&B. Formation of closely attached cocoons of *Rhaconotus*. Developing adults are white with distinct black eye.

Plate 15.7. Emergence of *Rhaconotus*. Adult wasp can cut nice round hole on the cane and comes outside. Adults are light to dark brown with long antennae.

Plate 15.8. *Rhaconotus* female. Light coloured legs, stout ovipositor and long *filiform* antennae (1.2 times the body length).

Plate 15.9. *Cotesia flavipes* with distinctly yellow legs and *moniliform* antennae, black head and thoracic segments.

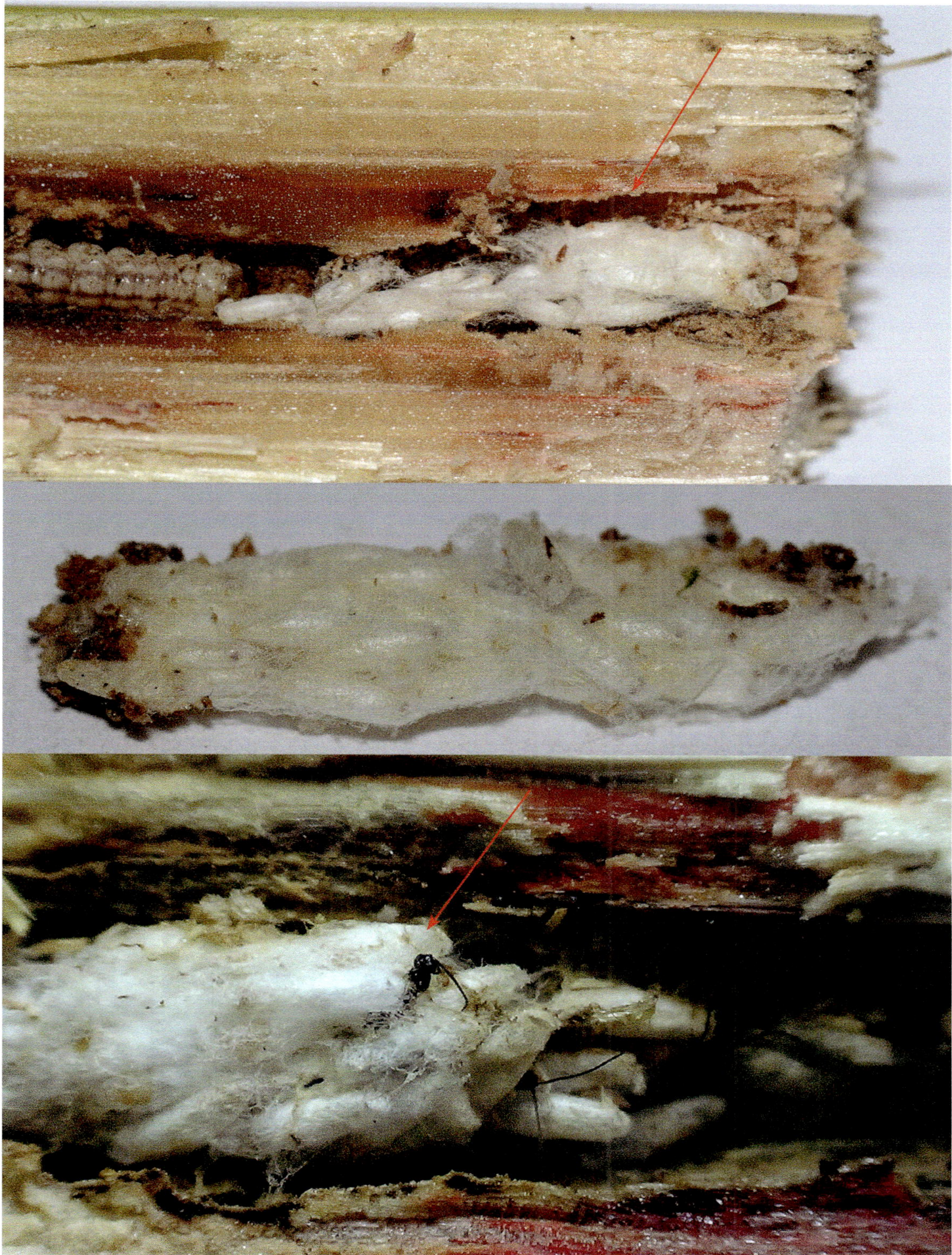

Plate 15.10. Loosely attached cottony cocoons of *Cotesia* within the feeding tunnel of Stalk borer larva. Masses of cocoon taken out. Adults emerging from the cocoons.

Plate 15.11. *Cotesia flavipes* with distinctly yellow legs and *moniliform* antennae. Forewings with dark area (*pterostigma*). Abdomen with white bands.

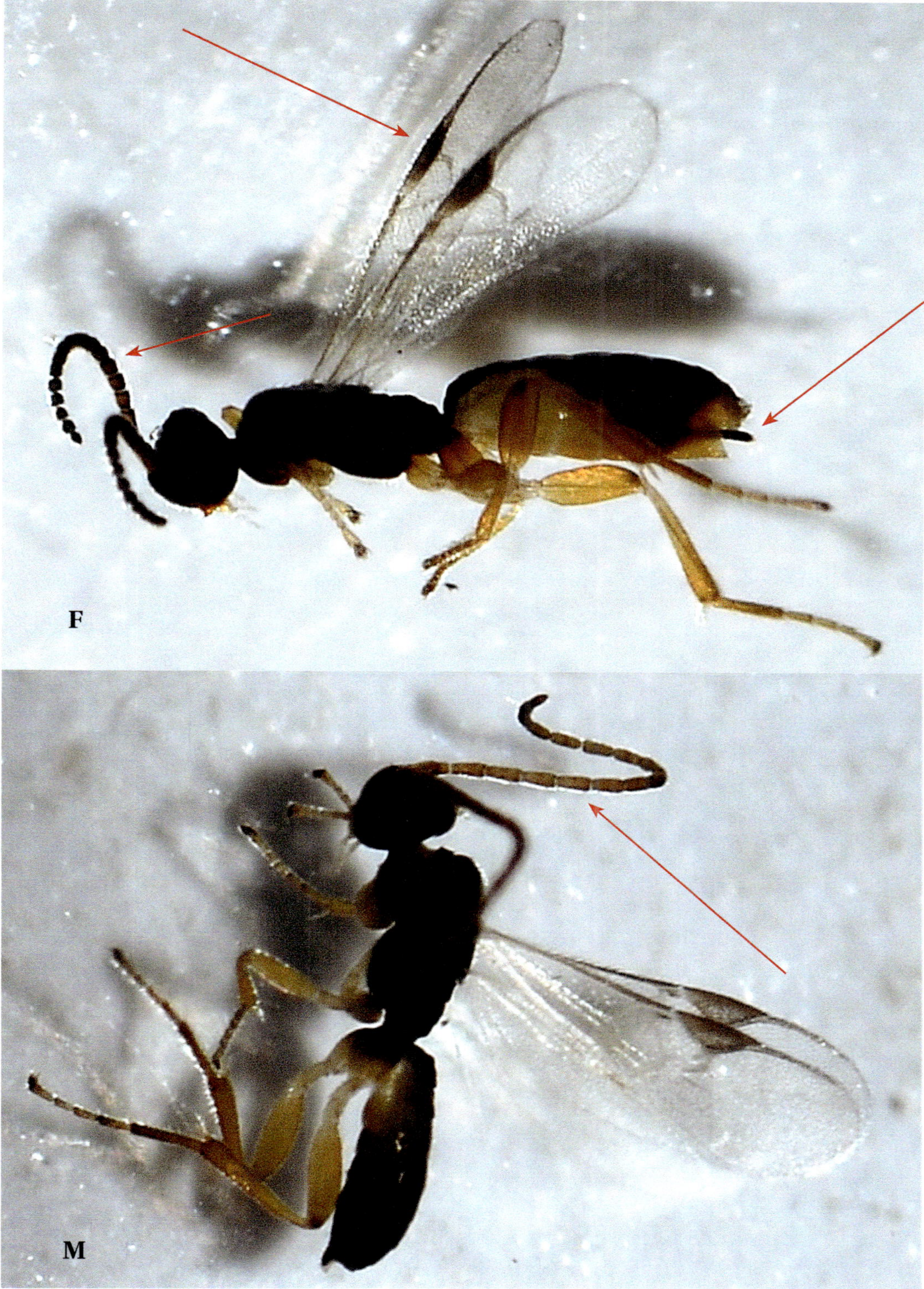

Plate 15.12. *Cotesia flavipes* with distinctly yellow legs and *moniliform* antennae. Forewings with dark area (*pterostigma*). Abdomen with white bands. In female, flagellar segments are shorter, tea cup type whereas, in male these seggments are larger, small tumbler type.

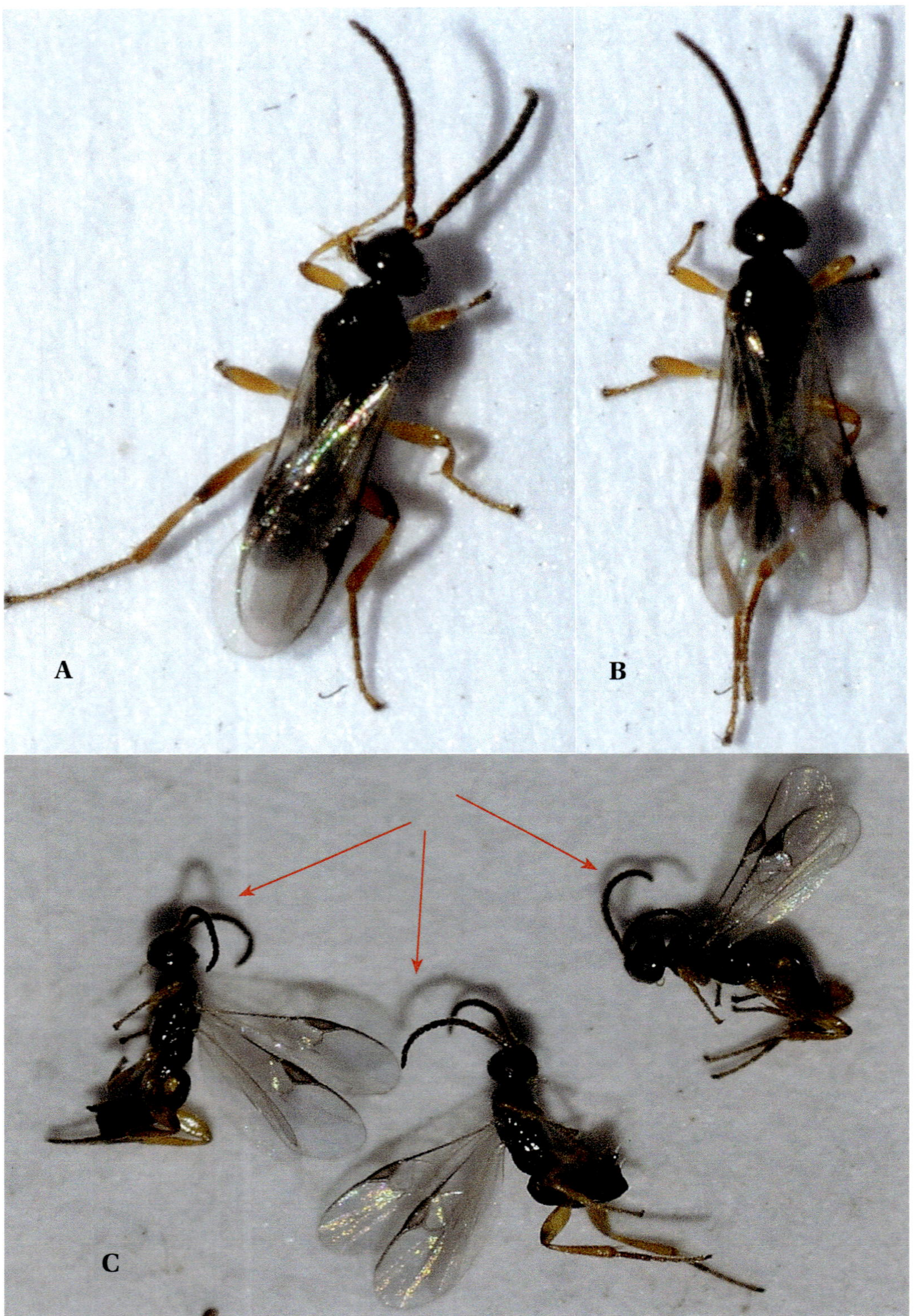

Plate 15.13. *Cotesia flavipes* with distinctly yellow legs and *moniliform* antennae. Forewings with dark area (*pterostigma*). Horn like bending of antennae on death.

Plate 15.14. *Isotima* female (A & B) and male (C) sitting on sugarcane leaf (in field).

Plate 15.15. *Isotima* female (A & B) and male (C). Distinct white bands are present in the abdomen; *tibia* and *tarsus* of the hind leg are dark , *femur* is light reddish brown. Anal segment also bears white colouration.

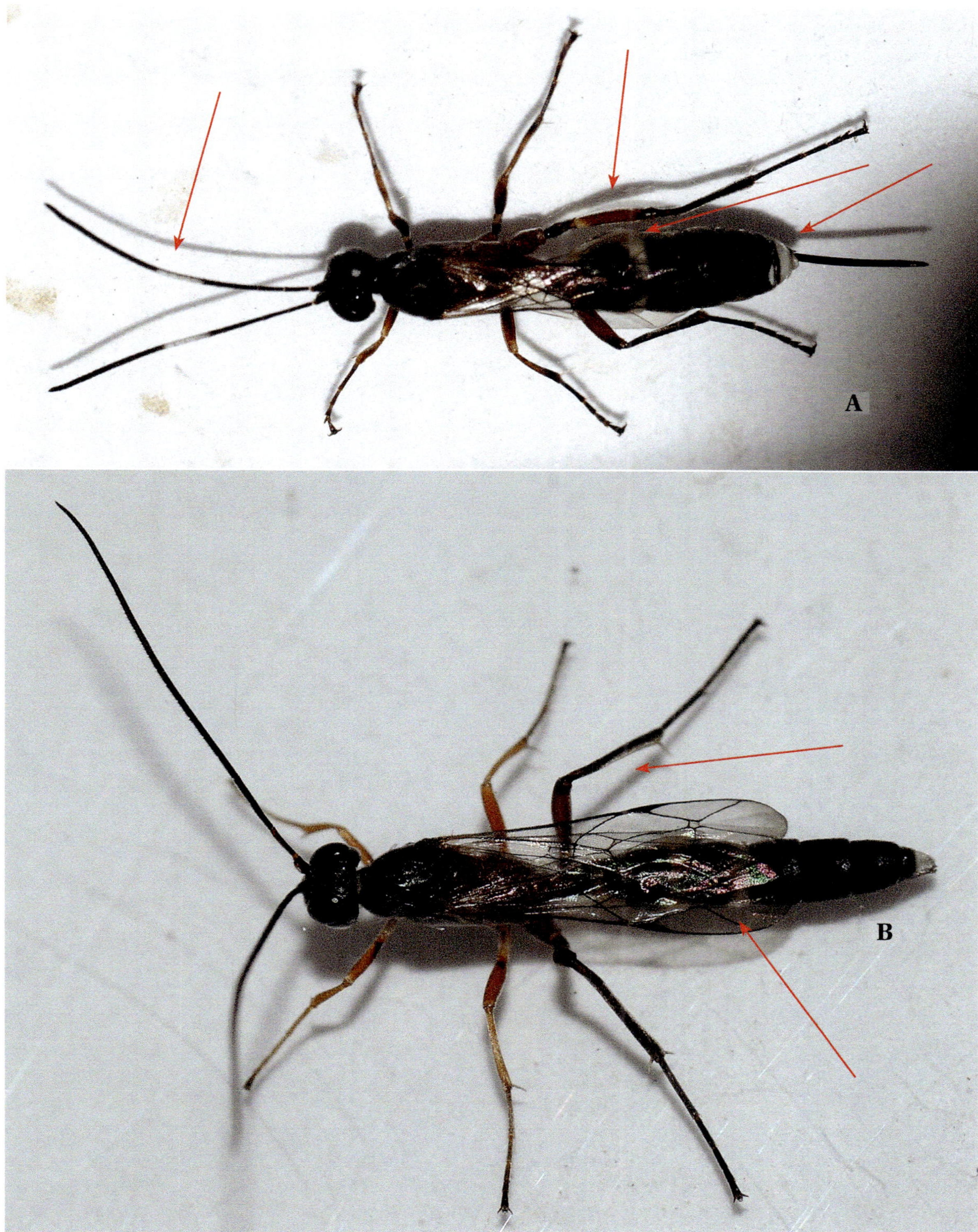

Plate 15.16. Dorsal view of *Isotima* female (A) and male (B). Wings partially cover the abdomen. Colour variations of the legs. *Tibia* and *tarsus* of the hind leg are dark, *femur* is light reddish brown. Antennae are *filiform*.

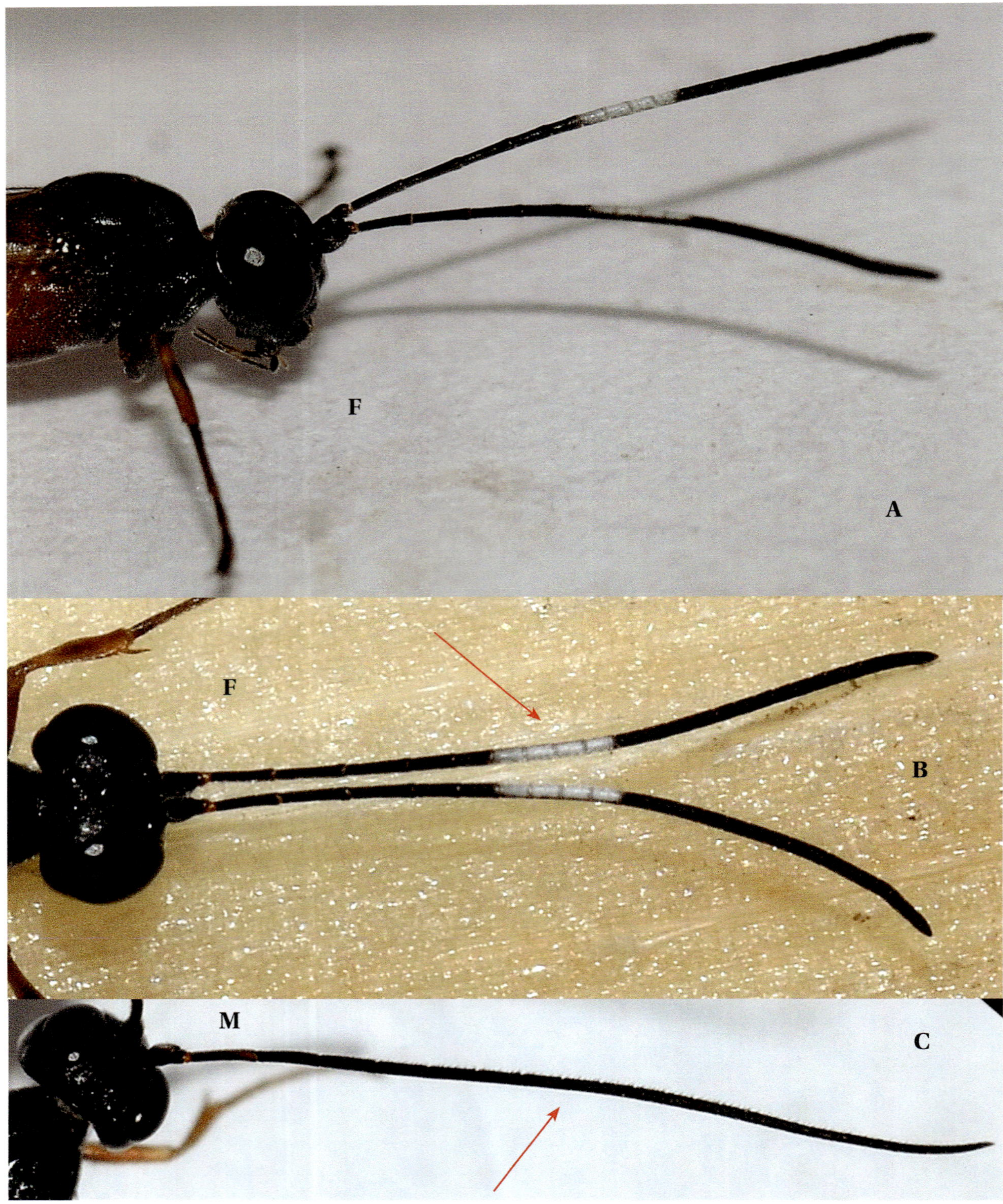

Plate 15.17. *Isotima* female (A & B) showing four central white segments of the antennae and (C) male antennae without any white segment.

Plate 15.18. *Isotima* female. Ovipositor (A) after emergence remains sheathed. (B) Ovipositor is spear like, sharp and serrated (C) Position of ovipositor and sheaths at the anal end.

Plate 15.19. *Stenobracon* male. Body is yellowish, head and anal end with dark area. Antennae are long and *filiform*. Head is black except the face area. Wings with three black or dark patches.

Plate 15.20. *Stenobracon* female. Body is yellowish, head is balck except the face area. Anal end with dark area (*tergite*). Ovipositor is very long (more than body length), flexible and black.

Plate 15.21. *Stenobracon* female. Body yellowish, head black except the face area. Forewing with three black patches and hind wing with two black patches. Ovispositor is very long (more than body length), flexible black/dark. Antennae *filiform*.

Plate 15.22. *Tetrastichus* (A) immediately after emergence from a pupa. (B) In male, only the conical terminal segment of antenna is black . (C &D) In female, five terminal segments are dark brown to black. E. Male mounting on a female.

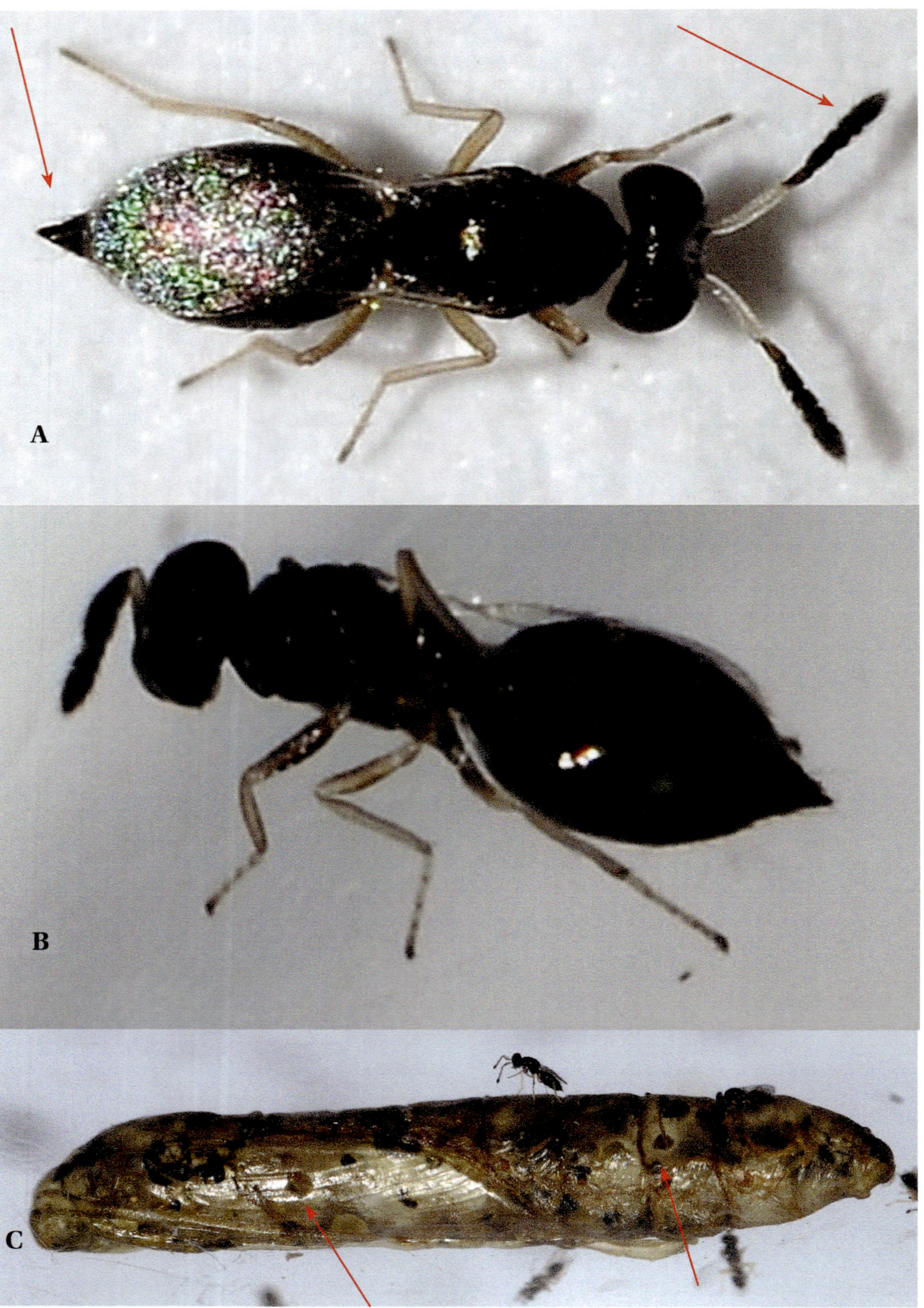

Plate 15.23. *Tetrastichus* female. (A) Dorsal view; shape of abodomen. (B) Side view (C) Emergence holes on the exhausted pupa of Top borer.

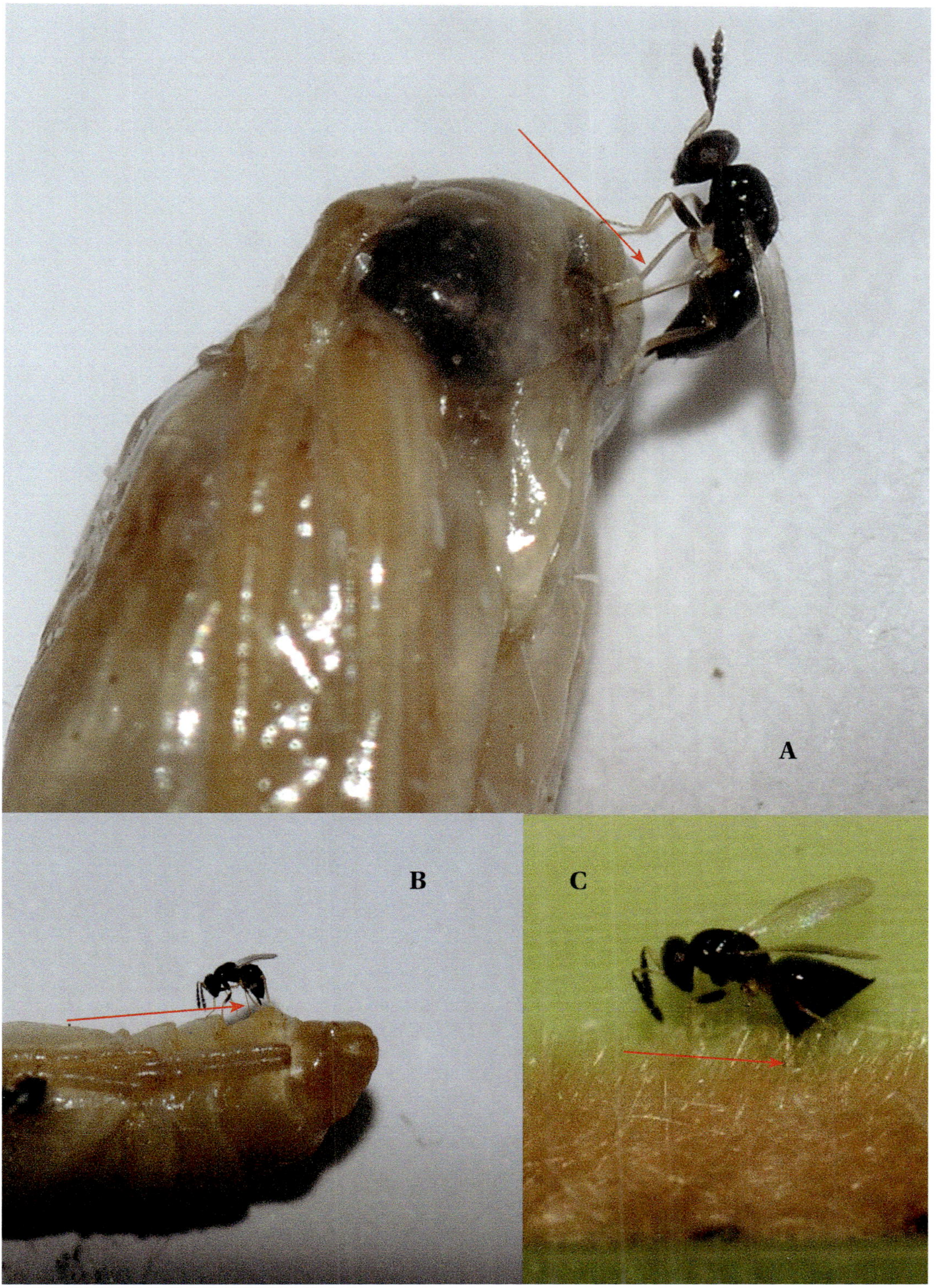

Plate 15.24. *Tetrastichus* female. (A & B) Laying eggs in the body of a pupa of Top borer. (C) Laying eggs in the egg mass of Top borer.

16

Selected References

Agarwal, R. A. 1964. Biology of sugarcane internode borer (*Proceras indicus* K.), *Indian Sugar*, **14:** 145-156.

Agarwal, R. A. 1975. Moth borers and cane damage. Identifying characters of the larvae, adults and the damage caused by moth borers to sugarcane in India. *Intern. Sugar Journal.*, **77:** 232-234.

Agarwal, R. A. and Siddiqi, Z. A. 1964. Sugarcane pests. In, *Entomology in India* (1938-63). New Delhi. The Entomological Society of India, pp. 149-186. (Silver Jubilee Number of *Indian J. Entomology*, 1964, p. 529).

Agarwal, R. A. and Tiwari, C. B. 1969. Identity of Indian species of sugarcane moth borers through their genitalic characters. *J. Research*, **6:** 372-382.

Arora, G. S. 2000. *Studies on some Indian Pyralid species, Rec. Zool. Surv. India, Occ. Paper* No. **181**, i-vii, p. 169. (ZSI, Calcutta).

Avasthy, P. N. 1972. Biological control of the insect pests of sugarcane: A review. *Indian Sugar*, **13:** 345-368.

Avasthy, P. N. and Chaudhary, J. P. 1962. External morphology of the larva of sugarcane moth borer *Raphimetopus ablutellus* Zell. *Indian J. Sugarcane Res & Development*, **7:**30-35.

Bisset, G. A. 1939. A new species in the genus *Chilo* [*tryptes*]sp. n. bred from sugarcane in Punjab (Lepidoptera, Pyralidae). *Proc. Roy. Ento. Soc.*, London. (B), **8:**47-48.

Bleszynski, S. 1962. Studies on the *Crambidae* (Lepidoptera) part 36. On some species of Chilo Zincken. Acta Zool. Cracov. **7:** 107-133.

Bleszynski, S. 1965. *Crambinae*. In, *Microlepidoptera Palaearctica* 1. (eds.) Amsel, H. G., Reisser, H. and Gregor, F. Vienna, G. Formme, p 533.

Bleszynski, S. 1969. The taxonomy of the *Crambinae* moth borers of sugar cane (Studies on the Crambinae (Lepidoptera) Part 45). In, *Pests of sugarcane*. (eds.) Williams, J. R., Metcalfe, J. R., Mungomery, R. W. and Mathes, R. Elsevier, Amsterdam. pp. 11-59.

Bleszynski, S. 1970. A revision of the world species of *Chilo* Zincken (Lepidoptera, Pyralidae). *Bull. Brit. Museum* (*Natural History*), **25:** 99-195.

Bojer, W. 1856. Report of the select committee appointed to examine the extent of damage done by the cane borer in Mauritius. p. 46.

Box, H. E. 1953. *List of sugarcane insects*. C.I.E., London. p. 101.

Butani, D. K. 1956. A key for the identification of sugarcane borers. *Indian J. Entomology*, 18:303-304.

Butani, D. K. 1967. Systemic, morphology, bionomics and control of sugarcane root borer, *Emmalocera depressella* Swinhoe. *Labdev J. Sci. & Technology*, **5:** 182-191.

Butani, D. K. 1972. Parasites and predators recorded on insect pests of sugarcane in India. I*ndian Sugar*, **22:** 17-32.

Cheema, P. S. 1947. Biology of sugarcane root-borer, *Emmalocera depressella* (Swinhoe) in the Punjab. *Indian J. Entomology*, **9:** 167-175.

Cheema, P. S. 1951. A note on the morphology of the sugarcane root-borer, *Emmalocera depressella* (Swinhoe). *Indian J. Entomology*, **13:** 79-85.

Cheema, P. S. 1953. Nature and extent of damage caused by the sugarcane root-borer, *Emmalocera depressella* (Swinhoe). *Indian J. Entomology*, **15:** 139-145.

Cheema, P. S. 1948. Differentiation between the sugarcane root borer and stem borer attack in the early stages of crop growth as well as harvest. *Indian J. Entomology*, **10:**130-131.

Cherian, M. C. and Subramanian, C. K. 1940. Studies on *Diatraea venosata*, a pyralid pest of sugarcane in India. *Madras Agric. Journal*, **30:**44-49.

Common, I. F. B. 1990. *Moths of Australia*. Melbourne Univ. Press. p. 544.

Cotes, E. C. 1889. Notes on Indian insect pests. Further notes II. The sugarcane borer moth. *Indian Museum notes*, Vol 1. pp. 22-27.

David, H. and Kalra, A. N. 1965. A study of the sugarcane internode borer, *Proceras indicus* Kapur. *Proc. ISSCT*, **12:**1444-1453.

David, H., Easwaramoorthy, S. and Jayanthi, R. 1986. *Sugarcane Entomology in India*. (Eds.). Sugarcane Breeding Institute, Coimbatore. p. 564.

Dudgeon, G. C. 1905. Description of new species of moths from India and Burma. *J. Bombay Natural History Society*, **16:** 399-405.

Duttamajumder, S. K. 2018. *Top borer of sugarcane*. Published by Online Gatha, Lucknow. p. 281.

Fletcher, T. B. 1914. *Some south Indian insects and other animals of importance*. Published by the Superintendent, Madras. p. 565.

Fletcher, T. B. and Ghosh, C. C. 1919. Borers in sugarcane, rice etc., In, *Proc. 3rd Entomological Meetings, Pusa*. In Vol. **1:** 354-417.

Gupta, B. D. 1940. The anatomy, life and seasonal histories of striped moth-borers of sugarcane in North Bihar and west United Provinces. *Indian J. agric. Science*, **10:** 787-817.

Gupta, B. D. 1954. Observation on the life history and habits of the stem borer, *Bissetia steniellus* Hampson. *Proc. Second Bienn. Conf. Sug. Res. & Dev. Workers*. **2:** 198-204.

Gupta, B. D. 1958. Identification of the striped moth borers of sugarcane in India. *Tech Bull. No.* 1, IISR, Lucknow. p. 10 +10 plates..

Gupta, B. D. 1959. Insect pests of sugarcane in India. I. Sugarcane root borer, *Emmalocera depressella* Swinhoe. *Indian Sugar,* **8:** 775-782.

Gupta, B. D. 1960. An indigenous parasite succeeds in controlling the top borer. *Proc. Annual. Conv. Sug. Tech. Assoc., India.* **28:** 1-24.

Gupta, B. D. and Avasthy, P. N. 1952. Biology of sugarcane borer, *Emmalocera depressella* in Uttar Pradesh, *J. Agri. and Animal husbandry,* U. P., **12:**19-25.

Gupta, B. D. and Avasthy, P. N. 1959. Biology and control of the stem borer, *Chilo tumidicostalis* Hmpsn., *Proc. ISSCT,* **10:** 886- 901.

Gupta, B. D. and Avasthy, P. N. 1960. *Insect pests of sugarcane and their control.* Indian Central Sugarcane Committee, New Delhi. p. 34.

Gupta, B. D. and Garg, R. L. 1943. Observations on the distribution and Biology of *Chilo tryptes* Bisset in the Dehra Dun valley of the U. P. *Indian J. Entomology,* **5:** 173-176.

Gupta, B. D. and Singh, K. G. 1951. Biology of sugarcane stem borer, *Diatraea auricilia* Ddgn in the Uttar Pradesh. *J. Agri. and Animal husbandry,* U.P. **11:**12-20.

Hampson, G. F. 1895. On the classification of the *Schoenobiinae* and *Crambinae,* two subfamilies of moths of the family *Pyralidae. Proc. Zool. Soc.* London for the year 1895. pp. 897-974.

Hattori, I. 1971. Stemborers of Gramineous crops in Southeast Asia. (Chap. 6), In, *Proc. Symp. Rice Insects, Kitaku, Tokyo, Japan.* pp. 145-153.

Isaac, P. V. and Rao, K. V. 1941. A key for the identification of the larvae of the known Lepidopterous borers of sugarcane in India based on morphological characters. *Indian J. agric Sci.,* **11:** 795-803.

Isaac, P. V. and Venkatraman, T. V. 1941. A key for the identification of the pupae of the known Lepidopterous borers of sugarcane in India based on morphological characters. *Indian J. agric Sci.,* **11:** 804-815.

Jepson, W. F. 1954. *A critical review of the world literature on the Lepidopterous stalk borers of tropical Graminaceous Crops.* Commonwealth Institute of Entomology, London, p. 127.

Kalra, A. N. 1979. *Sugarcane pests and their control.* ICAR, New Delhi. p. 53.

Kalra, A. N. and Sidhu, A. S. 1965. Biology of sugarcane top borer, *Scirpophaga nivella* F., in the Punjab. *Indian Sugar,* **15:** 37-43.

Kalra, A. N. and Srivastava, R. C. 1964. Biology of *Apanteles flavipes* Cam.,(Braconidae : Hymenoptera) a parasite of moth borers of sugarcane , under micro-ecological conditions. *Proc. All India Conf. Sugarcane Res. & Dev. Workers.,* **7:** 89-95.

Kapoor, M. S. 1957. Studies on the bionomics and control of *Bissetia steniellus* Hampson in the Punjab. *Indian J. Entomology,* **19:** 132-143.

Kapur, A. P. 1950. The identity of some Crambinae associated with sugar cane in India and of certain species related to them (Lepidoptera: Pyralidae). *Trans. Roy. Ento. Soc.* London, **101:** 389-434.

Khanna, K. L. 1956. On tiller mortality and compensation due to borers in sugarcane. *Proc. ISSCT.*, **9:** 936-971.

Khanna, K. L., Nigam, L. N. and Puri, V. D. 1958. *Chilo tumidicostalis* Hampson - A serious stem borer pest of sugarcane in Bihar. *Proc. Indian Acad. Sci.* (B)., **46:** 75-95.

Krisnamurti, B. and Usman, S. 1952. The *ragi* stem borer, *Sesamia inferens* Walker. *Bull. Ento. Ser. Dept Agric. Mysore.* No. 15. p. 70.

Kumar, S. 1975. Behaviour of Gurdaspur borer, *Bissetia steniellus* Hampson (Lepidoptera, Pyralidae) in Rajasthan. *Indian Sugar,* **25:** 1-8.

Lefroy, H. M and Howlett, F. M. 1909. *Indian Insect Life. A manual of the insects of the plants.* Thacker, Spink & Co., Calcutta. p. 786.

Lefroy, H. M. 1906. *Indian Insect Pests.* The Superintendent of Government Printing, India, Calcutta. p. 318.

Lefroy, H. M. 1906. Mothborer in sugarcane, maize and sorghum in western India. *Agric. J. India,* **1:** 97-114.

Lefroy, H. M. 1910. The more important insects injurious to Indian agriculture. *Memoirs Dept. Agriculture, Entomology Series,* **1:** 113-252.

Lewvanich, A. 1981. A revision of the Old World species of *Scirpophaga* (Lepidoptera: Pyralidae). *Bulletin of the British Museum Natural History* (*Entomology*), **42:** 185–298.

Mookerjee, B. J. 1857. Remarks on a disease affecting the Bombay or Red sugarcane in certain districts of Bengal. *Journal Agric. & Hort. Soc., India.* **Vol. 9 (part 3):** 355-358.

Nixon, G. E. J. 1941. The Indian and African species of *Rhaconotus* Ruthe (Hym., Braconidae). *Annales and Magazine of Natural History,* **7:** 473–503.

Pemberton, C. E. 1963. Insect pests affecting sugar cane plantations within the Pacific. *Proc. ISSCT.*, 11: 678-679.

Puri, V. D. 1957. Studies on the anatomy of the sugarcane moth borers. I. Comparative morphology of the antennae. *Proc. Indian Acad. Science,* (B)., **45:** 32-42.

Puri, V. D. 1958 a. Studies on the anatomy of the sugarcane moth borers. II. External anatomy of *Chilo tumidicostalis,* the stem borer of sugarcane. *Proc. Indian Acad. Science,* (B)., **45:** 197-215.

Puri, V. D. 1958 b. Studies on the anatomy of the sugarcane moth borers. III. Comparative study of the head and its appendages. *Proc. Indian Acad. Science,* (B)., **46:** 96-108.

Puri, V. D. 1958 c. Studies on the anatomy of the sugarcane moth borers. IV. Comparative study of the wings. *Proc. Indian Acad. Science,* (B)., **47:** 316-330.

Puri, V. D. 1958 d. Studies on the anatomy of the sugarcane moth borers. V. Comparative study of the adult thorax. *Proc. Indian Acad. Science,* (B)., **48:** 65-78.

Puri, V. D. 1958 e. Studies on the anatomy of the sugarcane moth borers. VI. Comparative study of the female genitalia. *Proc. Indian Acad. Science,* (B)., **48:** 79-88.

Rahman, Khan A. 1940. Parasites of the insect pests of sugarcane in the Punjab. *Indian J. agric. Sci.,* **11:** 119-128.

Rahman, Khan A. 1941. Migration of *Chilo tryptes* Bisset from the top portion of sugarcane to its roots for hibernation. *Proc. 28th Ind. Science Congress.* Section **8:** 193.

Rahman, Khan A. and Singh, D. 1942. Studies on 'dead hearts' caused by different species of sugarcane borers in the Punjab. *Indian J. Entomology,* **4:** 77-85.

Rahman, Khan A. and Tandon, D. N. 1940. *Chilo tryptes* Bisset. (Pyralidae) - A new pest of sugarcane from the Punjab. *Indian J. agric. Sci.,* **10:** 818-823.

Rajarao, S. A. 1961. Introduction of indigenous parasites for the control of the top borer in sugarcane at Pugalur (Madras State). *Proc. Bienn. Conf. Sugarcane Res & Dev. Workers, India,* **4:** 504-513.

Rao, V. P. and Nagraja, H. 1965. A comparative study of the four species of paddy stem-borers belonging to the Genera *Chilotraea* and *Chilo* in Asia (Lepidoptera: Pyralidae: Crambinae). *Proc. Indian Acad. Sci.* (B) **63:** 175-217.

Rohwer, S. A. 1918. Descriptions and notes on some Ichneumon-flies from Java. *Proc. United States Natural Museum,* **54:** 563–570.

Siddiqi, Z. A. and Singh, O. P. 1959. Biology of *Raphimetopus ablutellus* Zeller (Lepidoptera: Pyralidae), the Green borer of sugarcane, in Uttar Pradesh. *Indian J. Entomology,* **21:** 32-37.

Singh, M., Chillar, B. S. and Madan, Y. P. Biology of sugarcane root borer, *Emmalocera depressella* Swinhoe. *Indian Sugar,* **51:** 599-602.

Srivastava, R. C. and P. N. Avasthy. 1969. The morphology of *Apanteles flavipes* Cam. (Hymenoptera: Braconidae). *Labdev. J. Sci. Technol.,* **7:** 202-204.

Stebbing, E. P. 1903. Insect pests of the sugarcane in India. No. 3. *Indian Museum notes.* Vol. **5**: 64-182.

Swinhoe, C. 1884. On Lepidoptera collected at Kurrachee. *Proc. Zool. Soc.,* London, for the year 1884. pp. 503-529.

Swinhoe, C. 1885. On the Lepidoptera of Bombay and Deccan. Proc. *Zool. Soc.,* London, for the year 1885. pp. 825-886.

Tams, W. H. T. 1942. Note on the name of the sugarcane cane borer of Mauritius (Lepidoptera, Pyralidae). *Bull. Ento. Research.* **33:** 67-68.

van Deventer, W. 1906. *De Dierlijke Vijanden van het Suikerriet en Hunne Parasieten. Volume 2. Handboek ten Dienste van de Suikerriet-cultuur en de Rietsuiker Fabricage op Java. Vereenigde Proefstations voor Suikerriet in West-en Oost Java,* Amsterdam, p. 273.

Venkatraman, T. 1961. Varietal behaviour on yield and quality of top borer affected canes. *Indian J. Sugarcane Res. & Dev.* 6: 43-46.

Watt, G. 1893. *A dictionary of economic products of India. Vol IV. part II. (Sabadilla to Silica).* Reprinted by Cosmo Publications, Delhi-6. p. 687

Williams, J. R. 1953. The larvae and pupae of some important Lepidoptera. *Bull. Ento. Research,* **43:** 691-701.

Williams, J. R., Metcalfe, J. R., Mungomery, R. W. and Mathes, R. 1969. *Pests of sugar cane.* Elsevier Pub. Co., Amsterdam. p. 568.

Appendices

These publications are more than a century old and now in Public domain

Appendix I

Mookerjee (1857)

355

Remarks on a disease affecting the Bombay or Red Sugar Cane, in certain districts of Bengal: communicated by BABOO JOYKISSEN MOOKERJEE.

I have the honor to submit, for the information of the Agri-Horticultural Society, the following circumstances connected with the cultivation, growth, and finally entire destruction of the red sugar canes, commonly called Bombay canes, in the districts of Rungpore, Hooghly, and in a portion of Burdwan. I have not been able to ascertain whether the red cane was first brought from Bombay, or from any other place: it is different from the large white canes generally believed to have been imported from the Otaheite islands.

The two species of sugar canes known in this country by the names of Kajlee and Pooree canes have from time immemorial been cultivated in the districts of Rungpore and Hooghly in common with other districts, but since the increase of the demand of sugar for exportation, the culture of these canes has increased considerably. The crops of these canes had almost always given good returns, except at bad seasons, when it failed considerably, but no account of the total failure of these canes, as has been the case with the Bombay red ones, can be gathered from the oldest inhabitant.

About twenty-five years ago, Mr. McDowall, of the Kissoreegunge Indigo concern, introduced the red canes in the district of Rungpore, hence the natives of that place call these canes "Shahiban Koosar". On comparison, the experimental cultivators were convinced that the new canes had more saccharine matter in them than the country ones, and that they grew larger and yielded more juice than the latter, so much so, that the pecuniary gain to the ryot was more than twice the product of the other. Thus in a very few years the neighbouring fields of Kissoreegunge were covered with these canes. In about eight or ten years these canes were introduced in

3 A

most of the Northern parts of the district, and from thence gradually spread over throughout the Southern parts too.

The mode of preparing the lands is the same as that of the country cane. After ploughing, the lands are manured with cowdung and refuse of Indigo vats. The fields being well prepared, the cuttings are planted in lines; when these take their roots, and begin to shoot out, the earth about them is loosened and cleared of common herbs and grass. After a few weeks, when the canes grow a foot and more high, they are weeded, and manured with mustard oil cakes, and as the canes grow higher, four or more of their heads or tufts are taken together, and their leaves wound round them, and four or more of such wound-up bundles of the cane tufts are again tied together in the same manner. In this state they are allowed to grow till ripe for reaping. The canes when ripe are reaped and carried to the mills, where they are cut in small pieces for being pressed, and the fields cleared of grass, &c. A few days after, new shoots begin to make their appearance out of the roots, and the fields are then taken care of, weeded, and the earth loosened and manured, and the heads tied together as in the first instance. In a similar manner a third crop is also reaped from the same field. In the first and third years the produce of the Shahiban canes were moderate, but in the second year they yielded a far superior crop. In the fourth year some of these fields are ploughed and manured, and some other crops are cultivated, but in some instances the lands are left uncultivated for renewing the fertility of the soil. For some years the Shahiban canes were very luxuriant in this district, and the cultivation of the country canes decreased in the proportion the other was introduced.

About four years ago the disease (the natives call it *dhosah*) which proved fatal, and totally destroyed the culture of one of the much esteemed and principal harvests of the district, began to make its appearance. A few cane bushes out of a field of one *don* (thirteen cottahs) appeared withered in the first year. The next year more than two-thirds of each of the cane fields shared the same fate. In the first year very little attention was paid to this subject, but the cultivators were alarmed in the next, and commenced enquiring into the matter. On dissection, the piths of the canes were found of a reddish color, and eaten up by worms of reddish white color with a black mark or spot on their mouth. In the third or last year, whole fields were destroyed. From the stinking smell of the canes thus destroyed, it became hardly possible for men to pass by these fields, much less their being fit for consumption by human beings, or even by cattle. After some experiments, such as putting less or more manure, and shifting the cultivation from one to another field, the disappointed cultivators gave up the culture of the red cane in despair, and re-introduced that of the country canes (Kajlee and Pooree), which seems to be promising.

It is upwards of twenty years since the red or Bombay cane was introduced in the district of Hooghly from Nungy Bunglow, a place a few miles below Calcutta. At first they were cultivated at Bally, Rugoonathpore, Ooterparah, and its neighbourhood; from thence it gradually spread over to the banks of the Damoodah in a few years, and latterly on the western bank of that river close to pergunnah Chunderkonah.

The mode of cultivation, &c., nearly corresponds with that of Rungpore. Instead of mustard oil cake exclusively and refuse of Indigo vats, the cane fields of this district are manured sometimes with mustard oil, and at other times with castor oil cakes. The ryots of Hooghly generally reap from richer alluvial soils three, and in inland soils two crops consecutively from the same cuttings. Afterwards, for a year sometimes two, *dhuncheea* reed, indigo, or other crops, not only of a less exhausting kind, but the juice and leaves of which go a great way in restoring strength to the land, are cultivated on the same field. In this rotation red canes and indigo and other easy crops followed each other. In those places, where the red canes were planted earliest, *i. e.*, about twenty years, the disease appeared slightly about two years ago. Last year the decay increased, and this year total destruction has taken place. Where this cane has been introduced only lately or ten or fifteen years ago, there the crops, though they have somewhat suffered this year from excess of rains, yet they are free from the disease. In the lands of the Burdwan district bordering Hooghly a similar result has taken place. The new cane is only being gradually introduced in the interior of Burdwan. The process of destruction and disease in this district is similar to that of Rungpore. I have not been able to find out so accurately the result of this cane cultivation in other districts, but from what I have heard generally, I fear it is equally distressing.

The people of the affected districts are unable to assign any other cause for this catastrophe, than that of the red cane being of a very exhaustive nature.

The country cane is known to have been cultivated every second year alternately with indigo and other easy crops in the same field from time immemorial. It is yet a moot question whether after a lapse of ten or fifteen years the same fields will again bear the cultivation of red cane from healthy cuttings or not. The subject is worthy the attention of the Society as it affects, and will ultimately affect in a larger proportion, the production of sugar in this country, which is my apology for occupying the time of the Society.

Report on Cotton raised in the Society's Garden from North American stock.

To the Members of the Cotton Committee.

Gentlemen,—I beg to circulate, for the favour of your opinion, three musters of Cotton (as per list annexed) the produce of the Society's garden. I also circulate, for the sake of comparison, three other musters, as detailed.

Appendix II

Swinhoe (1885)

GALLERIIDÆ.

175. MELISSOBLAPTES DEPRESSELLUS, n. sp. (Plate LVII. fig. 5.)

Poona, October.

Antennæ and palpi greyish yellow; thorax grey, with a brown band in front; abdomen whitish. Fore wings greyish yellow, irrorated with grey, with the irrorations intensified into pale streaks on the costa, in the centre and on the hinder margin, also some grey streaks on the veins, on the marginal border. Hind wings pure white, unmarked.

Expanse of wings $1\frac{2}{10}$ inch.

Genus CHILO.

Chilo, *Zinck. Germar's Mag. Ent.* ii, p. 33 (1817).

Type, *C. phragmitellus*, Honrath, from Europe.
Range. Universally distributed.

Proboscis absent. Palpi porrect, clothed with rough hair, and extending from two and a half to three times length of head; maxillary palpi dilated with scales at extremity; frons with a conical projection; antennæ minutely serrate and ciliated; tibiæ with the outer spurs about two-thirds length of inner. Fore wing with the apex rectangular in male, typically acute and produced in female; vein 3 from before angle of cell; 4, 5 well separated at origin; 7 straight and well separated from 8, 9; 10 free; 11 curved and approximated to 12. Hind wing with vein 3 from near angle of cell; 4, 5 from angle; 6, 7 from upper angle.

4203. **Chilo simplex**, *Butl. P. Z. S.* 1880, p. 690.

Crambus zonellus, *Swinh. P. Z. S.* 1884, p. 528, pl. 48, fig. 16; *C. & S.* no. 4703.
Crambus partellus, *Swinh. P. Z. S.* 1885, p. 879; *C. & S.* no. 4700.

♂. Yellowish brown suffused with fuscous. Fore wing with the costal area rather darkest; traces of dark specks below middle of cell and at lower angle; the veins of outer area slightly streaked with fuscous; a marginal series of black specks. Hind wing whitish with slight fuscous tinge.

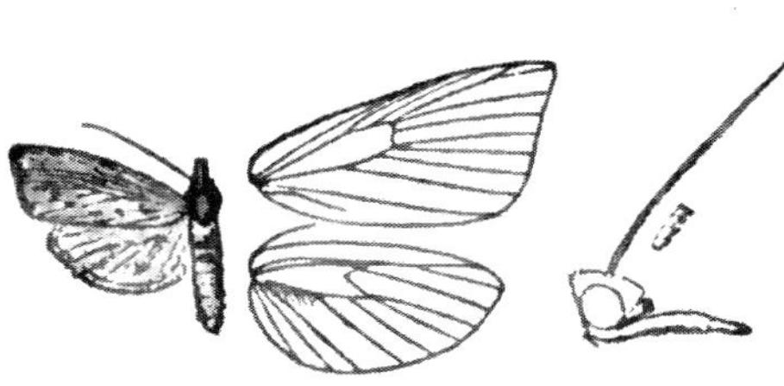

Fig. 17.—*Chilo simplex*, ♂. ½.

♀ paler; the hind wing white.

The form *partellus* has on the fore wing of male a highly curved antemedial series of short fuscous streaks, a slight yellowish patch in end of cell, an oblique series of diffused fuscous streaks from apex to middle of inner margin, and a submarginal series of specks; female with some diffused fuscous from apex round lower angle of cell, or sometimes nearly evenly suffused with fuscous, with a dark fuscous patch beyond cell.

Hab. Japan; Chusan; Formosa; Meean Meer, Punjab; Karáchi. *Exp.* 22 millim.

4204. **Chilo ceylonicus**, *Hmpsn. P. Z. S.* 1895, p. 957.

Ochreous brown irrorated with fuscous. Fore wing with the fuscous irroration forming streaks in the interspaces except on inner margin; an indistinct fulvous medial line curved below costa and with silvery scales on its edges; a submarginal silver line bent inwards below costa; a marginal series of black spots with white

Hampson (1896)

B. *Frons with a long flattened corneous projection more or less buried in scales.*

4285. **Anerastia spinifrontella,** *Rag. Nouv. Gen.* p. 48; *id. Mon. Phyc.* pl. xl, fig. 2; *C. & S.* no. 4636.

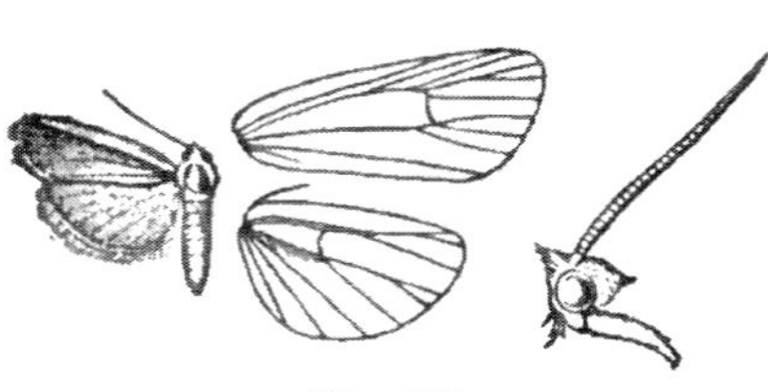

Fig. 38.
Anerastia spinifrontella, ♂. 1/1.

Head whitish, the palpi tinged with brown; throax brown; abdomen ochreous. Fore wing ochreous suffused with reddish brown, especially below the broad costal fascia, which is whitish and narrows to apex. Hind wing white with a slight fuscous tinge.

Hab. Karáchi. *Exp.*, ♂ 18–22, ♀ 24 millim.

4286. **Anerastia ablutella,** *Zell. Isis,* 1839, p. 177.
Anerastia stigmatella, *Rag. Nouv. Gen.* p. 49; *id. Mon. Phyc.* pl. xl, fig. 4; *C. & S.* no. 4637.

Ochreous. Fore wing slightly irrorated with fuscous; a dark speck at lower angle of cell. Hind wing white.

The variety *stigmatella* has a reddish tinge and large discocellular spot.

Hab. Mediterranean subregion; Punjab. *Exp.* 13–18 millim.

4287. **Anerastia pulverulella,** n. sp.

♂. Head white and brown; thorax pale brown; abdomen ochreous. Fore wing ochreous, with broad white costal fascia irrorated with fuscous, tapering to apex, its lower edge defined by brown; a prominent dark speck at lower angle of cell; the veins beyond the cell, vein 1, and inner margin white irrorated with fuscous. Hind wing white.

Hab. Puttalam, Ceylon (*Pole*). *Exp.* 18 millim. Type in B. M.

*4288. **Anerastia brunneovittella,** *Rag. Nouv. Gen.* p. 49; *id. Mon. Phyc.* pl. xl, fig. 6; *C. & S.* no. 4633.

♀. Pale ochreous grey. Fore wing with the costa narrowly ochreous white, and with a subcostal rufous streak. Hind wing dull semihyaline white.

Hab. Bombay. *Exp.* 18 millim.

*4289. **Anerastia bimaculella,** *Rag. Nouv. Gen.* p. 49; *id. Mon. Phyc.* pl. xl, fig. 5; *C. & S.* no. 4632.

♂. Ochreous. Fore wing short, widening towards outer margin; the costal area tinged with rufous; the veins irrorated with a few black scales; a prominent black spot at lower angle of cell; a subbasal black speck on vein 1. Hind wing yellowish grey.

Hab. "S. India." *Exp.* 12 millim.

Hampson (1899)

Habitat.—Khásis. *Exp.* 28 mm.

4156*a*. CRAMBUS CORTICELLUS, n. sp.

♂. Head and thorax dark and golden-brown; abdomen fuscous. Forewing coppery golden-brown, the veins and interno-median fold irrorated with white scales; the costal area suffused with blackish; a black point in end of cell; a marginal series of black points surrounded by white scales; cilia paler. Hindwing fuscous-grey. ♀. With the ground-colour uniform dull brown; hindwing grey-brown.

Habitat.—Khásis; Calcutta. *Exp.* ♂ 24, ♀ 40 mm. *Type*—♂ in British Museum; ♀ in coll. Rothschild.

4159. CRAMBUS PARVELLUS=4178, *C. ochristrigellus.*

4202*a*. CHILO STENIELLUS, n. sp.

♂. Head and thorax pale grey-brown; abdomen whitish, dorsally ochreous towards base. Forewing long and narrow; pale grey-brown irrorated with fuscous and suffused with fuscous in the interspaces, especially the cell and discal fold, leaving the veins slightly paler; traces of two very oblique lines from below apex to middle of inner margin; a marginal series of black points. Hindwing pure white.

Habitat.—Tezpore, Assam. *Exp.* 30 mm. *Type*—In coll. Rothschild.

Genus DIATRÆA.

Diatræa, Lands. Guild. Trans. Soc. Encom. Arts, xlvi, 143 (1832).

Type.—*D. saccharalis*, Fabr.

Range.—Nearctic and Neotropical regions; Mauritius; Ceylon; Borneo; Java.

Palpi extending about three times the length of head and thickly clothed with hair; maxillary palpi triangularly dilated with hair; proboscis absent; frons with a tuft of hair; antennæ of male minutely serrate and fasciculate; tibiæ somewhat hairy, the spurs well developed. Forewing with the apex somewhat acute; vein 3 from before angle of cell; 4, 5 from angle; 6, 7 from near upper angle; 8, 9 stalked; 10 free; 11 anastomosing with 12. Hindwing with vein 3 from before angle of cell; 4, 5 from angle; 6, 7 from upper angle.

4202*a*. DIATRÆA SACCHARALIS, Fabr., Ent. Syst., iii, 2, 238.

Chilo obliteratellus, Zell., Mon. Chil. and Cramb., p. 8.

Crambus leucaniellus, Wlk., Cat., xxvii, p. 161.

lineosellus, Wlk., Cat., xxvii, p. 162.

with pale red-brown and slightly irrorated with darker brown, the veins and costal edge white; a black discoidal point; a terminal series of black points; cilia white with a brown line near base and the tips tinged with brown. Hind wing almost pure silvery white. Underside of fore wing except the costal edge and the costal area of hind wing suffused with pale red-brown.

Hab. TRANSVAAL, Tweefontein (*Janse*), 1 ♂, Van der Merwe's Farm (*Janse*), 1 ♂, 3 ♀ type, Pretoria (*Distant, Janse*), 2 ♂. *Exp.* 20–22 mm.

(23 *e*) *Argyria arœalis*, sp. n.

Head, thorax, and abdomen white tinged with ochreous, the last long and slender; antennæ fuscous at sides; palpi long; fore femora and tibiæ fuscous on inner side. Fore wing long and narrow, the apex produced, ochreous white irrorated with reddish brown; a black discoidal point defined by white; a terminal series of black points; cilia tinged with brown and with a blackish line through them. Hind wing silvery white. Underside of fore wing and costal area of hind wing tinged with red-brown.

Hab. MASHONALAND (*Dobbie*), 3 ♂, 3 ♀ type. *Exp.* 16–20 mm.

(24 *a*) *Argyria albivenalis*, sp. n.

♀. Head, thorax, and abdomen white faintly tinged with rufous, the last with some fulvous yellow at base of dorsum. Fore wing white suffused with rufous leaving the veins white, the interspaces of terminal half slightly irrorated with brown; a slight brown terminal line. Hind wing silvery white. Underside of fore wing and costal area of hind wing tinged with rufous.

Hab. PARAGUAY, Santa Cruz (*S. Moor*), 1 ♀ type. *Exp.* 22 mm.

(26 *g*) *Argyria atrisparsalis*, sp. n.

Diatræa neuricella, Druce, Biol. Centr.-Am., Het. ii. p. 295 (nec Zell).

Head, thorax, and abdomen white tinged with rufous. Fore wing white tinged with rufous and sparsely irrorated with black-brown, the submedian fold pure white; a minute antemedial black-brown spot on inner margin and black discoidal point; a terminal series of prominent black points slightly defined on inner side by white. Hind wing white faintly tinged with red-brown. Underside white tinged with red-brown.

Hab. MEXICO, Vera Cruz, Atoyac (*H. H. Smith*), 1 ♂ type, Tobasco, Teapa (*H. H. Smith*), 1 ♀, Godman-Salvin Coll. *Exp.*, ♂ 26, ♀ 30 mm.

(26 *i*) *Argyria tumidicostalis*, sp. n.

Hind wing of male with glandular swelling at base of costa.

♂. Head and thorax whitish suffused with red-brown, the palpi at sides and patagia irrorated with black-brown; abdomen white tinged with red-brown; pectus, legs, and ventral surface of abdomen white tinged with rufous. Fore wing whitish suffused with red-brown; a rather diffused dark red-brown fascia below median nervure to just beyond end of cell and a few black-brown scales beyond the cell; a terminal series of minute black spots with silvery white points in centres; cilia silvery white with a red-brown line at base. Hind wing white, the costal area tinged with red-brown; a red-brown terminal line from apex to vein 2. Underside of fore wing and costal area of hind wing suffused with red-brown.

Hab. BENGAL, Patna, 2 ♂ type. *Exp.* 24 mm.

Larva in sugar-cane.

(26 *j*). *Argyria sticticraspis*, sp. n.

♀. Head and thorax greyish ochreous suffused with red-brown; palpi irrorated with dark brown; abdomen greyish ochreous with some rufous at base of dorsum; pectus, legs, and ventral surface of abdomen white tinged with ochreous brown. Fore wing greyish ochreous suffused with red-brown, the cell and areas just below and beyond it irrorated with darker red-brown; a curved postmedial series of small red-brown spots in the interspaces from below costa to above vein 1; a terminal series of minute black spots defined on inner side by slight white spots; cilia with slight red-brown lines near base and at middle. Hind wing pure white. Underside of fore wing white tinged with rufous.

Hab. MADRAS, Coimbatore (*B. Fletcher*), 1 ♀ type. *Exp.* 30 mm.

Larva in sugar-cane.

(26 *k*) *Argyria coniorta*, sp. n.

♂. Head and thorax pale red-brown; antennæ ringed black and white; palpi whitish mixed with dark brown; abdomen whitish dorsally suffused with brown towards base; pectus, legs, and ventral surface of abdomen white tinged with red-brown. Fore wing pale red-brown mixed with some whitish and irrorated with blackish especially along median nervure and in and beyond end of cell; a medial series of slight black spots, incurved below the cell and sometimes reduced to the spot below the cell; two blackish discoidal points, the lower one sometimes defined on outer side by white; a more or less distinct curved blackish subterminal line; a terminal series of minute black spots defined on inner side by minute white spots; cilia with two dark lines through them. Hind wing almost pure white. Underside of fore wing and the costal area of hind wing suffused with reddish brown.

Hab. SIKKIM, 1800′ (*Dudgeon*), 1 ♂; BENGAL, Behar, Pusa, 4 ♂ type. *Exp.* 20–26 mm.

Larva in sugar-cane.

(26 *l*) *Argyria argyropasta*, sp. n.

♂. Head and thorax red-brown mixed with some dark brown; abdomen white tinged with brown; pectus, legs, and ventral

another above inner margin; the inner area and that surrounding the lower part of the reniform, which is white and ochreous, suffused with chestnut and olive browns; some black streaks on the oblique apical fascia, and some on outer margin below the middle; cilia chestnut, intersected by pale teeth. Hind wing white, the veins and outer area suffused with fuscous.

Hab. Amur; Japan; throughout India and Ceylon. *Exp.* 38 millim.

*1942. **Auchmis ? cashmirensis**, *Moore, P. Z. S.* 1881, p. 358; *C. & S.* no. 1821.

Pale ochreous; palpi and legs marked with black. Fore wing with the base speckled with black, and with a black streak above inner margin; a medial ochreous band, bordered by indistinct sinuous double black lines; the orbicular, reniform, and claviform blackish, with ochreous edges; some dentate discal black marks; a patch above outer angle and marginal series of black and white specks. Hind wing pale fuscous; the outer area darker, with crenulate marginal line.

Hab. Kashmir. *Exp.* 32 millim.

Genus **NONAGRIA**.

Nonagria, *Ochs. Eur. Schmett.* iv, p. 82 (1816).
Microsemyra, *Butl. P. Z. S.* 1883, p. 155.

Type, *N. dissoluta*, Tr., from Europe.

Range. Palæarctic region and throughout India, Ceylon, and Burma.

Eyes naked and without lashes; proboscis thin. Palpi obliquely porrect, the 2nd joint evenly scaled, the 3rd prominent; thorax and abdomen without tufts; tibiæ without spines. Wings with the cilia not crenulate; fore wing with the apex somewhat acute; hind wing with veins 6 and 7 stalked or from cell.

SECT. I. Hind wing with veins 3 and 4 from the cell.

A. *Antennæ of male ciliated, of female simple.*

1943. **Nonagria inferens**, *Wlk. Cat.* ix, p. 105; *Moore, Lep. Ceyl.* iii, pl. 145, fig. 3; *C. & S.* no. 1654.
Leucania proscripta, *Wlk. Cat.* ix, p. 106; *C. & S.* no. 1655.
Sesamia fraterna, *Moore, Lep. Atk.* p. 103; *C. & S.* no. 1653.

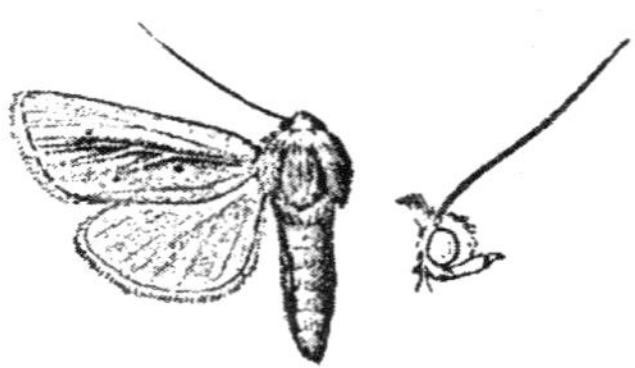

Fig. 153.
Nonagria inferens, ♂. 1/1.

Ochreous. Fore wing with a red-brown suffusion along median nervure and veins 2 to 5; a marginal dark line; cilia paler. Hind wing white.

Hab. Sind; Bombay; Mhow; Ceylon Burma; Flores. *Exp.* 28 millim.

(61) *HYPSOTROPA BIPARTELLA.

Peoria bipartella Rag. N. Am. Phyc. p. 19 (1887); id. Rom. Mém. viii. p. 386, pl. 40. f. 33; Dyar, Cat. Lep. N. Am. p. 439.

U.S.A., N. Carolina.

Auctorum.

Peoria albidella Hulst, Can. Ent. xxxiii. p. 175 (1900); Dyar, Cat. Lep. N. Am. p. 439 .... U.S.A., California.
Hypsotropha laropis Turner, Pr. R. Soc. Queensl. xxiv. p. 173 (1913) .... Queensland.
,, *neurica* Turner, Pr. R. Soc. Queensl. xxiv. p. 113 (1913) .... N. Australia.
,, *castella* Pag. Zoologica, xxix. p. 163 (1900). Bismarck Arch.

Genus RAPHIMETOPUS, nov.

Type, *R. spinifrontella*.

Proboscis aborted and minute; palpi downcurved, extending about three times length of head and thickly scaled; maxillary palpi minute and filiform; frons with long flattened corneous plate, somewhat trifid in front and produced to short lateral processes, a corneous plate below the frons; antennæ of male laminate. Fore wing narrow, the apex rounded, the termen obliquely curved; vein 2 from close to angle of cell; 4 absent; 5 from above angle; 6 from below upper angle; 8, 9 stalked; 10, 11 from cell. Hind wing with vein 2 from just before angle of cell; 3 and 5 stalked, 4 absent; 6, 7 shortly stalked; 8 closely approximated to or anastomosing with 7.

(1) RAPHIMETOPUS SPINIFRONTELLA.

Anerastia spinifrontella Rag. Nouv. Gen. p. 48 (1888); id. Rom. Mém. viii. p. 399, pl. 40. f. 2; Hmpsn. Moths Ind. ii. p. 56.

BOMBAY, Sind.

(2) RAPHIMETOPUS ABLUTELLA.

Anerastia ablutella Zell. Isis, 1839, p. 17; Herr.-Schäff. Eur. Schmett. iv. p. 109, Tin. f. 89; Rag. Rom. Mém. viii. p. 403; Hmpsn. Moths Ind. iv. p. 56; Staud. Cat. Lep. pal. ii. p. 12.

†*Anerastia stigmatella* Rag. Nouv. Gen. p. 49 (1888); id. Rom. Mém. viii. p. 403, pl. 40. f. 4.

Anerastia bimaculata Rag. Nouv. Gen. p. 49 (1888); id. Rom. Mém. viii. p. 404, pl. 40. f. 5; Hmpsn. Moths Ind. iv. p. 56.

Anerastia majorella Roths. Nov. Zool. xx. p. 138 (1913).

SPAIN; SICILY; ALGERIA; N. NIGERIA; CONGO, EGYPT; DAMARALAND; TRANSVAAL; ORANGE R. COLONY; ADEN; ASIA MINOR; PERSIA; W. TURKESTAN; JAPAN; KASHMIR; PUNJAB; MADRAS.

brown beyond it; a marginal series of specks. Hind wing rather paler.

Hab. Simla (*Harford*). *Exp.* 32 millim. Type in B. M.

4312. **Polyocha depressella,** *Swinh. P. Z. S.* 1885, p. 876, pl. 57. fig. 5; *C. & S.* no. 4733.

Polyocha variegatella, *Rag. Nouv. Gen.* p. 39; *id. Mon. Phyc.* pl. xxxvi, fig. 16; *C. & S.* no. 4618.

♀. Head pale pink; thorax brown; abdomen ochreous. Fore wing pink, with a white costal fascia slightly suffused with pink and narrowing to base and apex; the area below the costal fascia broadly suffused with brown; the veins of outer area streaked with brown. Hind wing white with a slight yellow tinge.

Hab. Dharmsála; Poona; Nilgiris; Pulo Laut. *Exp.* 22–36 millim.

*4313. **Polyocha detritella,** *Rag. Nouv. Gen.* p. 39; *id. Mon. Phyc.* pl. xxxvi, fig. 14; *C. & S.* no. 4617.

♂. Yellowish white. Fore wing with a subcostal brownish streak. Hind wing white.

Hab. Dharmsála. *Exp.* 23 millim.

*4314. **Polyocha carnatella,** *Rag. Nouv. Gen.* p. 39; *id. Mon. Phyc.* pl. xxxv, fig. 25; *C. & S.* no. 4616.

♀. Yellowish pink. Fore wing with the costa narrowly edged with white. Hind wing yellowish white, with pinkish marginal line.

Hab. Lahore. *Exp.* 24 millim.

Sect. II. Antennæ of male with uniseriate branches; fore wing with vein 4 stalked with 5, and 10 with 8, 9; hind wing with vein 3 stalked with 4, 5.

4315. **Polyocha leucania,** *Feld. Reis. Nov.* pl. 137, fig. 13; *C. & S.* no. 4684.

Polyocha vesculella, *Rag. Nouv. Gen.* p. 39; *C. & S.* no. 4619.

Head and thorax pink; abdomen ochreous. Fore wing grey suffused with pink; the interspaces streaked with pink; the veins white defined by pink and usually also by fuscous. Hind wing pale fuscous.

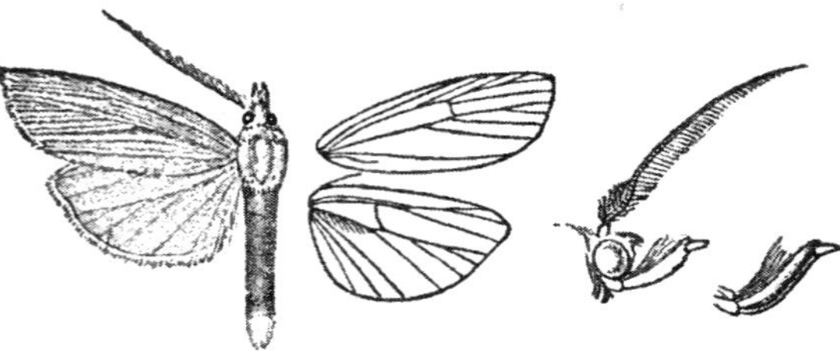

Fig. 42.—*Polyocha leucania*, ♂. ½.

The Sikhim male has the branches of antennæ shorter than in the Southern form.

Hab. Dharmsála; Sikhim; Nilgiris; Ceylon; Burma. *Exp.* 22–40 millim.

Hampson (1918)

(7) †Emmalocera depressella.

Melissoblaptes depressella Swinh. P. Z. S. 1885, p. 876, pl. 57. f. 5; Hmpsn. Moths Ind. iv. p. 63; id. Rom. Mém. viii. p. 324, pl. 36. f. 12.

Polyocha saccharella Dudgeon, J. Bomb. Nat. Hist. Soc. xvi. p. 405 (1905).

Aden; Punjab; Bengal; Bombay. The larva feed on the roots of sugar-cane.

(8) †Emmalocera strigicostella.

Polyocha strigicostella Hmpsn. P. Z. S. 1896, p. 270; id. Rom. Mém. viii. p. 315, pl. 31. f. 18.

Aden.

B. (*Emmalocera*). Antennæ of male with short uniseriate branches.

(9) †Emmalocera leucocincta.

Crambus leucocinctus Wlk. xxvii. 169 (1863); Hmpsn. Rom. Mém. viii. p. 316, pl. 36. f. 9.

Emmalocera crenatella Rag. Nouv. Gen. p. 38 (1888).

Singapore; Borneo; Philippines.

C. (*Papua*). Antennæ of male laminate, serrate towards base.

(10) †Emmalocera sanguifusalis.

Polyocha sanguifusalis Hmpsn. P. Z. S. 1910, p. 493, pl. xl. f. 9.

N. Rhodesia.

(11) †Emmalocera aurifusella.

Crambus aurifusellus Wlk. xxxv. 1756 (1866); Hmpsn. Moths Ind. iv. p. 62; id. Rom. Mém. viii. p. 317, pl. 36. f. 18.

Kashmir; Punjab; Bombay; Madras.

(12) Emmalocera bifidella.

Polyocha bifidella Wileman, Trans. Ent. Soc. 1911, p. 357, pl. 31. f. 22.

Japan.

(13) †Emmalocera polychroella, sp. n.

Head and thorax ochreous mixed with fiery red; abdomen ochreous; antennæ of male with the tuft blackish on inner side; palpi, pectus, legs, and ventral surface of abdomen ochreous tinged with red-brown. Fore wing with narrow creamy white costal fascia leaving the costal edge red-brown towards base, defined below by a fiery rufous streak to beyond middle, the area below it red-brown to median nervure and vein 4; the inner

PYRALIDÆ CRAMBINÆ.

Chilo auricilia, n. sp.

Antennæ of the male flattened by coalescing lamellæ separated distinctly.

♂. Brownish ochreous, irrorated with fuscous; a raised dark-metallic spot in the cell, another at the origin of veins 4 and 5, one above and one below vein 2; a few golden scales in and beyond the cell; a postmedial row of black dots incurved towards the costa with golden scales upon them; a marginal row of black dots; cilia golden. Hindwing brownish white; cilia slightly silvery.

♀. Differs from the male in being larger and in having the ground colour pale ochreous with the metallic spots smaller. Antennæ simple.

Habitat.—Burogah, Bengal; Sikhim and Bhutan. *Exp.* Male 18 mill., female 25-27 mill.

Type sent to British Museum.

Described from one male received from Mr. Maxwell-Lefroy sent him by Mr. Mackenzie and from five females in my own collection, one of which had been marked "*Platytes argentisparsalis*, Hmpsn." by Sir Geo. Hampson. Mr. Mackenzie has found this insect doing damage to sugarcane in company with *Chilo simplex*, and both species have been reared by him.

ANERASTIINÆ.

Polyocha saccharella, n. sp.

Section II, B. Differs from Sect. II, Hmpsn. in Moths of India, Vol. IV, in vein 10 of the forewing not being stalked with 8 and 9. Antennæ of the male with a large tuft of bicoloured scales in the sinus at the base of the shaft and uniseriate rather short branches. Palpi obliquely upturned in the male, rostriform in the female.

♂. Head, thorax, abdomen and forewing uniform pale ochreous brown more or less suffused with fuscous in and beyond the cell. Hindwing white.

♀. Differs from male in wanting the fuscous suffusion except slightly at the apex of the forewing.

Habitat :—Burogah, Bengal. *Exp.* 32-34 mill.

Types sent to British Museum.

Several specimens of this species were sent me by Mr. Maxwell-Lefroy reared by Mr. Mackenzie. It is said to be destructive to sugarcane in Behar.

Lefroy (1906)

MOTHBORER IN SUGARCANE, MAIZE AND SORGHUM IN WESTERN INDIA.

By H. MAXWELL-LEFROY, M.A., F.E.S., F.Z.S

Entomologist to the Government of India.

WHEREVER sugarcane is grown as a staple crop, it is infested with the caterpillars of moths, known under the general name of "Mothborer." These insects belong to a large family of moths, *Pyralidæ*, and fall chiefly in one subfamily—the Grassmoths or *Crambidæ*.

In different parts of the world, different species occur, and though these are to some extent spread over large areas, those of the East are in the main distinct from those of the West. Java has several species not as yet known in India, nor is the common mothborer of the West Indies, *Diatræa saccharalis*, definitely known in India, though it is recorded in Ceylon.

The mothborers of India are of several species ; the one discussed in these pages, *Chilo simplex*, is certainly the only species widely spread as a pest in Western India. Quite recently it has been found that one of the mothborers of Bengal is *Chilo auricilia*, a species extremely easily confused with *C. simplex*. Though the latter is stated to be the common mothborer of all India, this statement may now be taken with caution, and until more is known of the mothborers of different places, the identity of the particular species is better left in doubt. In addition to the *Chilo*, there are such borers as *Scirpophaga auriflua*, *S. excerptalis*, *Nonagria uniformis* etc. While these do not occur commonly in Western India, they are known in Bengal and their range is a matter of doubt. It has, therefore, been thought best to entitle *Chilo simplex* the mothborer of Western India, with the reservation that it may prove to be the most common mothborer of Bengal, Madras, the United Provinces, Punjab, etc.

Years of work will be required to ascertain the range of the various species, and the mothborers of cane in Bengal should be investigated and worked out as the next step in this enquiry. It may be pointed out that though the difference between *Chilo simplex* and *Chilo auricilia* may be one that appeals only to Entomologists, the differences in their habits may be so great as to make the treatment of each quite distinct. The differences between *Chilo simplex* and *Diatræa saccharalis* are small, but the best remedy for the latter is useless for the former, so that the conclusions arrived at from a study of *Chilo simplex* in Western India cannot be applied offhand to similar insects in other parts of India. It will be necessary to obtain larvæ and rear them for determination. When it is known what species occurs in each province, the remedies for the proper insect can be applied.

LIFE HISTORY IN SUGARCANE.

(*A.*) *Eggs.* The eggs (Plate X, fig. 1) are laid by the moth in clusters on the leaves. Each egg is oval, flattened, about one millimetre long ; the upper surface is very finely marked and is not flat but slightly convex. The eggs overlap, in two rows as a rule, (Plate X, fig. 2) and give the appearance of nearly flat oval cakes of dough laid down one after another in two rows. The whole cluster is fastened to the leaf, sticking to it. The number of eggs forming each cluster is very variable ; as a rule it lies between 10 and 30, the average being about 14. In young canes the egg clusters are found nearer the top or on the centre leaves that stand straight up. The upper side is the most usual and near the mid-rib or on it, but there is no great regularity in this. The eggs when first laid are a dull creamy white ; this colour lasts for some period and towards the time for hatching the eggs turn orange and a black spot appears in the middle. The caterpillar can be seen curled up inside, the black head in the middle. It has only to bite through the thin shell and emerge. The empty egg-shell is white. In canes that are forming "joints" and in ripe canes, the eggs are laid on the leaves, usually on the apical half on the upper surface. It is not easy to find eggs on old canes owing to the great leaf expanse to be examined, nor is it always easy to see eggs on young canes. The colour is so inconspicuous, the size so small, and the clusters so like the spots, splashes of excreta and the like on cane leaves, that the eggs very readily escape notice.

Black eggs. It is not unusual to find that a proportion of the eggs are of a deep brown or black colour. These eggs contain parasites, which feed upon the substance of the egg and then make a tiny round hole in the centre of the egg for emergence. Such eggs may be found before and after the parasite has emerged ; the proportion of black eggs to normal eggs may be very large, as much as 75 per cent ; also as the black eggs that have hatched stay on the leaf, whilst normal eggs that hatch (and which are white) soon fall off and are not found, one may find large numbers of black eggs from which the parasites have emerged. There are, therefore, five kinds of egg clusters: (1) Light yellow eggs, freshly laid. (2) Orange coloured eggs, which will soon hatch caterpillars. (3) White eggs, eggs from which caterpillars have hatched. (4) Black eggs, whole, which contain parasites. (5) Black eggs with a small puncture, from which the parasite has escaped.

(B.) *Caterpillars.*—The caterpillar as it emerges from the egg is a tiny worm-like creature, some two millimetres in length, of an orange colour with black head. The body is slender, the head large: there are bands of deeper colour round each segment and small dark hairs that give it a spiny appearance. It walks actively about the plant, sometimes letting itself down by a thread from the edge of the leaf. The epidermis of the leaf is eaten in little spots, and the caterpillar often burrows into the mid-rib and tunnels down towards the centre. In a short time it reaches the heart of the shoot and lives in the outer leaves for some days. Cane shoots are often found with small patches of epidermis eaten off, fine dust lying on the centre leaves and a general appearance of having been eaten by caterpillars; this is due to the newly hatched caterpillars and a sure sign of the recent hatching of an egg-cluster.

After the caterpillar has become larger (¼-inch long) it tunnels into the heart of the cane and feeds on the rolled up centre leaves or the growing point of the stem. As a rule only one large caterpillar is found in each shoot; the others leave the shoot and attack separate shoots. The centre leaves are cut through and die; as they wither the dead shoots are readily seen and in such shoots (called "dead hearts") the caterpillar can be found. It the shoots are small, one caterpillar will eat several, eating its way in and out of each at the base, at the level of the ground. If the shoots are large, one caterpillar will find sufficient food for its whole life, and there may be two in one shoot. Whenever the growing point is eaten the shoot dies. In older canes, the caterpillar tunnels in the joints and does not necessarily destroy the growing point and kill the shoot: as soon as joints begin to form, the "dead heart" is not seen and fewer borers are found. The borer is seldom found in the ripening cane to any extent: it leaves the canes after the joints form and then prefers other crops.

The full grown caterpillar (Plate X., figs. 3-4) is one inch long, the head black or dark brown, the body dirty white. There is a conspicuous dark patch behind the head, and as a rule two lines of dark colour running down the body from head to tail on each side. On each segment there are small spines or hairs set on dark spots; the usual three pairs of jointed legs and five pairs of sucker feet are present.

The actual colour of the caterpillar is extremely variable. A caterpillar living in a cane shoot and feeding actively has the black spots on each segment well marked ; as it becomes full-fed, purple brown pigment is found in the tissues, showing as a longitudinal line of colour on each side. When resting prior to pupation, the colour may entirely disappear. Hibernating larvæ frequently show no spots, the colour of the hairs and spots being the same dirty white as that of the skin ; the purple brown pigment may be largely developed, not in longitudinal lines but in curved lines in each segment. (Plate X., figs. 5-6.)

It is difficult to be certain of the identification of the larvæ of the mothborer ; it is closely similar to the larva of the *Nonagria* that lives in wheat and juari ; it also resembles closely the borer in brinjal stems, (*Solanum melongena*) several borers in paddy, &c. Microscopic examination of the mouth parts, spiracles, hairs, and the like, is necessary to differentiate these larvæ, and the best plan is to rear the larva in captivity until the moth emerges.

The caterpillar can produce silk and uses it chiefly for preparing the burrow for pupation. Sometimes the whole burrow is lined with closely woven silk ; often the caterpillar lies in a long tube of silk and dust. The duration of the caterpillar stage is exceedingly variable ; in hot weather in young cane the larva is full fed in about four weeks ; during the winter in dry sorghum stalks the caterpillar hibernates and this period may last from six to eight months. It is not hibernation proper, that is, it is not dependent on temperature, since a part of this period (March to May) is in the hot weather. Probably the duration of the larval life is simply a matter of food, moisture and season. The caterpillar can pass to the pupa stage four weeks after hatching : it can rest as a full-grown insect for six months or longer if it wishes to or if the conditions render it necessary.

(*C.*) *The Chrysalis.*—At the end of two days' quiescence, the caterpillar sheds its skin and appears as a small brown chrysalis. (Plate XI., fig. 1.) The male chrysalis is smaller (½-inch) than that of the female (¾-inch), though both are otherwise identical in appearance. The chrysalis remains motionless in the burrow, with its head to the opening ; it can make convulsive movements with its abdomen but has no means of locomotion, and during this period of quiescence the tissues of the moth are built up from the tissues of the caterpillar. At the end of six days the male moth emerges, the chrysalis skin breaking open in front to allow the moth to escape. The female chrysalis requires one day longer, in some cases two. This is the usual period when the weather is warm or during the rains ; chrysalides are sometimes found during the cold weather with hibernating larvæ, and the insect evidently can hibernate as a chrysalis. In this case the chrysalis period may be very long, but apparently when the insect hibernates as a chrysalis, it emerges as a moth when the weather becomes hot. In this case it is true hibernation dependent upon temperature; the larvæ that hibernate can remain as dormant larvæ during the hot weather succeeding the winter, the pupæ apparently cannot.

(*D.*) *The Moth.*—The moth is a small yellowish grey insect with prominent palpi projecting in front of the head like a beak; its appearance is best realised from an examination of the figures on Plate XI. The male is distinctly smaller than the female and is frequently darker in appearance; the hind wing is also slightly darkened, and not, as in the female, white. The wings when spread measure nearly one inch across. In repose the wings are laid on the back, one overlapping the other, as if wrapped round the abdomen.

The moths fly in the dusk and at night. During the daytime they hide among the leaves or on the ground and come out as dusk falls. In captivity they are not long lived, a week being usually the longest period; it is probable that being unable to feed (the proboscis being absent) they live only long enough to mate and lay eggs, the male dying after mating, the female after all the eggs are deposited. It would be interesting to know if the moths can remain dormant when there are no plants on which to lay eggs; moths emerging during the hot weather before the rains may not find any food-plants and it seems likely that they can remain dormant probably in the ground until the rains bring fresh crops.

FOOD PLANTS.

In addition to cane, this insect has been found in maize (*Zea mais*) and juari (*Andropogon sorghum*), and rarely in bajri (*Pennisetum typhoideum*).

In cane it principally attacks the young shoots, but can attack the stem at any stage of growth. In maize, besides the young plants, it attacks the stem, the cob and the tassel; the succulent stems are often riddled through by the tunnels of these caterpillars; the young caterpillars feed on the leaves and then pass down to attack the stem. The tassel is sometimes eaten up by them; great numbers especially young ones may be found, and they like to feed on the tassel before burrowing into the stem. The cob is also eaten, the worm eating through the sheathing leaves and burrowing in the cob or simply eating the grain. In sorghum, the young plants are attacked and also the stems of older ones. In all stages of growth the stem is attacked and as a rule the grain is untouched. The following is quoted from the report of the Cawnpore Farm for 1901-1902 :—

"This borer is not the only one, but that which about here apparently does most damage to sugarcane, maize and juari. Last year the sugarcane

Appendix III (a)

Cotes (1889)

2.—THE SUGARCANE BORER MOTH.

Diatraea saccharalis,[1] Fabr.

Plate II, fig. 2 a *moth,* b *larva* (*dorsal view*), c *larva* (*side view*), e *pupa* (*all natural size*) ; *fig.* 2 f *pupa* (*enlarged*), g *piece of sugarcane* (*natl. size*) *to show tunnels.*

The larvæ of this moth bore into the stalks of sugarcane, often thereby setting up putrefaction, so that the whole stalk becomes worthless.

The zoological position of the insect.

It has not yet been satisfactorily determined whether the sugarcane borer, found in different parts of the world, belongs exclusively to a single species, or whether there are several closely allied species, all of which damage sugarcane by boring into it. Until, therefore, this question has been definitely settled the Fabrician name of *saccharalis* may continue to be applied[2] to the pest wherever it occurs, and the insect may be determined zoologically as a Pyralid moth, belonging to the genus *Diatraea* and to the species *saccharalis* of Fabricius.

Occurrence of the pest in India.

Sugarcane, in different parts of the world has, for at least the last hundred years, been known to be subject to the attack, either of this pest or of others so closely allied to it as to be scarcely distinguishable from it, and during the last year information has been sent to the Museum of damage done to sugarcane in several parts of India, where the pest would seem to have long been known, though but little has been recorded concerning it.

In 1857 Babu Joykissen Mukerji described[3] the total destruction, by the pest which he calls '*dhosah,*' of an imported variety of sugarcane (known as the Bombay or red sugarcane) in the districts of Rungpur, Hooghly, and a portion of Burdwan. The cultivation of this variety had been carried on for some years, and had proved very profitable, but when the pest appeared its cultivation had to be entirely given up, as it was found to be very much more subject to attack than the country varieties of cane.

In the Indian Museum are specimens of the pest, which were received in 1885, with the information that the insect had done great injury to sugarcane in Dhulia.

In 1888, the Personal Assistant to the Director of Land Records and Agriculture, North-Western Provinces, wrote[4] that the pest, which ap-

[1] The thanks of the writer are due to Mr. R. Blechynden, of the Agri-Horticultural Society of India, for help in hunting up the previous history of this pest.

[2] This is the view taken by Dr. Riley in his paper on the pest in the Entomology Report of the U. S. Department of Agriculture, 1880, page 240.

[3] In a paper published by the Agri-Horticultural Society, India, Volume IX, page 355 1857).

[4] In a letter dated 1st September.

pears in dry seasons, had destroyed as much as a fourth of the sugarcane crop in the neighbourhood of the Cawnpore Experimental Farm.

The Special Manager of the Dhaukora Wards Estate writes [1] that the pest, which is known locally as *Mandaruah*, has this year (1888) done injury to the sugarcane crops in several parts of the estate.

The Collector of Ganjam also notices [2] injury done by this insect, which is known as *Monjikila purugu*.

The Agricultural Officer of Burdwan and Seebpur, writes [3] that the loss occasioned by the pest, sometimes amounts to fully one-fourth of the cane crops of a neighbourhood.

The pest in other parts of the world.

With regard to injury by the pest in other parts of the world :—

In the year 1750 Hughes mentions [4] injury done to the cane in Barbadoes by the larvae of small moths which are likely to have been the same as the borer.

Porter notices [5] that the pest was found fatally destructive in Guadaloupe in 1785 and 1786.

Beckford notices [6] the presence of the borer in Jamaica in 1790.

In 1828 the Revd. L. Guilding wrote [7] an account of the borer in St. Vincent, describing it as *Diatraea sacchari.*

Westwood mentions [8] the pest as destructive to sugarcane in Jamaica in 1841.

About the year 1856 the insect did great damage to cane in Mauritius, [9] into which island it was supposed to have been introduced from Ceylon.

In 1857 the borers were very abundant along the Lower Mississippi in the United States.[10]

About the year 1879 the borer did great damage to the sugarcane crop in British Guiana.[11]

[1] In a letter dated 23rd September to the Collector of Mymensingh that was forwarded by the Director of Land Records and Agriculture Bengal.

[2] In a report forwarded by the Revenue and Agricultural Department.

[3] In a letter dated 4th July which was forwarded by the Director of Land Records and Agriculture, Bengal.

[4] See Roth's Animal Parasites of the Sugarcane, page 12, (1885).

[5] "Nature and properties of sugarcane," by G. R. Porter, London, (1830).

[6] See Roth, Animal Parasites of the Sugarcane, p. 8, (1885).

[7] Trans. Soc. Arts., Vol. XLVI, p. 143, (1828).

[8] *Gardener's Chronicle*, 5th July 1856, page 453.

[9] Bojer's Report of the Select Committee appointed to examine the extent of the damage done by the cane-borer in Mauritius, 1856 ; reprinted in the *Sugarcane* for 1873, (see Roth's Parasites of the Sugarcane).

[10] See Dr. Riley's paper in the Entomology Report of the U. S. Department of Agriculture, 1880, page 240.

[11] See papers by Miss Ormerod in Proc. Ent. Soc. Lond. 1879, pp. 33 and 36 ; and 1880, page 16.

In 1880 Dr. Riley reported [1] on the insect as injurious in the United States.

In 1885 Mr. H. Ling Roth described [2] the insect as occasionally very destructive to sugarcane in Queensland, Australia.

The Agricultural Officer of Burdwan and Seebpore writes [3]—

Nature of the injury in India.

"The sugarcane planting season extends from the beginning of February to the end of May. If there be no rains [4] in April or May, and if the cane fields are not frequently irrigated, which, from the scarcity of water at this time, is hardly possible, the pest makes its appearance. The pest first shows itself by the drying of the middle stalk of the plant, and is hence called by the ryots the *Majera* (a Bengalee term meaning relating to the middle); on pulling, the stalk now easily comes out, and its lower end is found to have become a rotten mass. Very soon the whole plant dies away, and from the root stock a number of smaller plants make their appearance to be in their turn attacked by the worm. If the rains hold off a long time, or if the fields are not thoroughly irrigated, three or four generations of plants are in this way attacked and destroyed. At last, when the rains set in, the fields become free from the insect, and a number of sickly-looking cane plants shoot out, but these make very little progress and never attain the proper size of the cane plant. If only one generation of plants is lost, and if this happens at an early stage of the growth of the plant, the damage done is not much."

Life history.

The life history of the insect has not yet been fully studied in India, but what has been observed agrees so closely with the observations made on the corresponding sugarcane pests of other parts of the world, that we may safely infer the rest, and the following account therefore is taken from Dr. Riley's paper [5] on the pest in America, where, however, the insect is likely to take rather longer to pass through the various stages of its existence than in the warm climate of India.

The parent moth lays her eggs upon the leaves of the young cane near the axils, and the young borer, hatching in the course of a few days, penetrates the stalk at or near the joint, and commences to tunnel through the soft pith. The eggs are flat and circular, one twenty-fifth of an inch in diameter, and are white when first deposited, turning yellow as they

[1] Riley, l. c.

[2] See his 'Animal Parasites of the Sugarcane.'

[3] In a letter dated Calcutta 4th July, which has been forwarded by the Director of Land Records and Agriculture, Bengal.

[4] It is noticeable that while considerable injury by the pest is almost universally supposed to take place only when moisture is deficient, Ling Roth, in his account of the pest in Australia notices particularly that the pest occurs in "wet springs." (See his 'Animal Parasites of the Sugarcane')

[5] Report, U. S. Department of Agriculture, Entomology, 1880, p. 240.

approach the hatching point. The growth of the borer worm must be very rapid, less than thirty days being probably occupied in the larval state. The borers are quite active, and occasionally leave their burrows and crawl about upon the outside of the stalk, seeking another place to enter. The full-grown borer is about an inch long, rather slender, nearly cylindrical, and cream white in general colour, but speckled [1] with black spots with a yellow head and black mouth-parts. Upon attaining its full size, it bores to the outside of the cane and makes a large round hole for its future exit—a hole which is usually at least one-fifth of an inch in diameter. It then retires into its burrow and transforms, a short distance from the opening, into a slender brown pupa,[2] three quarters of an inch long. The pupa state lasts but a few days and then the moth makes its exit. The moth has a spread of wings of about an inch and a quarter, and is of a light, greyish-brown color. With the female moth the hind wings are of nearly the same color with the fore wings, but with the male the former are silvery white. There are several broods in the course of the season, and the insects hybernate almost exclusively in the larval or "worm" state. During the winter they are to be found most abundantly in the seed cane, but also in the discarded tops, and to a slighter extent in the stubble.

Food plants of the worm other than sugarcane.

The Agricultural Officer of Burdwan and Seebpore notices [3] that he has seen the Kash plant (*Saccharum spontaneum*), attacked by the insect in the same way as sugarcane is attacked. Specimens of what appear to be the larvæ of the sugarcane borer have been received from the Collector of Ganjam, who writes [4] that they destroy paddy and brinjal plants, besides sugarcane. Similar specimens have also been received from the Agricultural Officer of Ranchi, who reports them as injurious to brinjal; and from Mr. Woodrow, of Poona, where they bore into jowari stalks and are said to make the plant poisonous to cattle. (See p. 28.) Dr. Riley notices [5] a very closely allied, if not identical, insect that bores into corn (maize) stalks in America.

Parasites and natural enemies of the pest.

In British Guiana [6] the pest is subject to the attack of ants which live in the cane fields, and are supposed to wage continual warfare against the borers; and in Mau-

[1] Dr. Riley found that the black speckles were not always present, but the specimens sent to the Museum have them plainly visible.

[2] The insect which damaged sugarcane in Mauritius about the year 1856 is said (see Bojer's Report) to have spun itself up in the leaves of the plant instead of transforming into a pupa in its burrow. In the case, however, of the insects reared in the Museum, the pupæ were formed in the burrows in the cane.

[3] In a letter dated Calcutta, 4th July, forwarded by the Director of Land Records and Agricultnre, Bengal.

[4] In a report forwarded by the Revenue and Agricultural Department.

[5] Report of U. S. Department of Agriculture, Division Entomology, 1880.

[6] See Miss Ormerod's paper in Proc. Ent. Soc., Lond., 1879, p. 33.

ritius the chrysalids of the borer were found to be subject to the attack of mites; no evidence has yet been obtained of any effect produced in either of these ways upon the numbers of the sugarcane pest in India, though the writer has observed large numbers of a chalcid fly which is parasitic upon the allied, or identical, jowaree borer. (See p. 29.)

Remedies.

A large number of remedies have been proposed for the pest, and it seems to be pretty well established that it can be to a great extent controlled by the burning or burying all the discarded tops, and clearing the fields of all waste sugarcane stalks after the crop has been taken; for, as the insect passes the winter as a larva inside the sugarcane, if these are destroyed, there are no moths in the spring to lay the eggs which produce the next year's "borers." The waste tops, however, should be carefully gathered together and removed from the field before being burnt, for if they are burnt carelessly, on the field itself, many predaceous insects will be liable to be destroyed, which take shelter in the ground and assist in reducing the numbers of the pest.

The following may be noticed among the remedies that have been suggested :—

Guilding recommends[1] that all the dry and useless leaves, under which he says the moth lays its eggs, should be stripped off: he claims that this treatment has been found effective in removing the pest.

Porter quotes[2] the practice of "introducing a pinch of quicklime into the heart of the young cane" for the destruction of the pest.

Westwood notices[3] that in Jamaica in 1841, the ravages of the borer were to a great extent checked by allowing the refuse to accumulate upon the grounds, and burning them there, the old roots subsequently throwing up more vigorous shoots.

Miss Ormerod,[4] in writing of the pest in British Guiana about the year 1879, quotes the practice of cutting back the cane below the surface of the ground, covering the plant with mould, and adding a handful of lime. The cutting out of the affected canes was tried on one estate over 246 acres, the result being considered satisfactory. In this case the canes cut out were put through the mill, and sufficient rum and megus obtained from them to pay expenses. Miss Ormerod also quotes the practice of steeping the cane for 48 hours in water before planting it, a treatment which was thought on one plantation, where it was tried, to destroy the hybernating larvæ without injury to the cane. It appears from the inquiries instituted in British Guiana, that it is a mistake to burn the refuse sugarcane on the fields themselves, as this destroys the ants, which, when unmolested, rendered valuable assistance in keeping down the pest. The plan therefore approved was to burn the refuse cane after collecting it in heaps outside the fields.

Dr. Riley recommends[5] burning all "tops" during the winter so as to destroy the larvæ which hybernate in them; selecting seed cane from the least infested portion of the plantation and laying it down in furrows during the winter, covered with earth as deeply as should be found possible without inducing decay, and only uncovering it as

[1] Trans. Soc. Arts, Vol. XLVI, p. 43 (1828).

[2] Nature and Properties of Sugarcane, by G. R. Porter, London, 1830.

[3] *Gardener's Chronicle*, 5th July 1856, p. 453.

[4] Proc. Ent. Soc., Lond. 1880, p. 16; and 1879, pp. 33 and 36.

[5] In his report on the insect.

history of this pest. It probably takes some weeks to reach the perfect state, or possibly some months. The perfect insect is about half-an-inch in total length, the head produced into a long snout upturned at the tip, the eyes black and prominent, coloration pale wainscoat-brown, the fore wing with some small scattered black dots hardly to be seen except under a lens, the hind wing transparent, white, immaculate. The female has a large mass of white flocculent material at the end of the body, with which she covers the eggs when laying them.

This insect, if it attacks the sugarcane in large numbers, must be a serious pest, as it must greatly reduce the vitality of the plant by sucking up the sap through its proboscis or beak, piercing the cuticle of the leaf for that purpose. The remedy for it is simple if taken in time. The egg-masses being so very conspicuous, it would be extremely easy to collect and destroy them. Children could do it. They should each be provided with an empty kerosine oil tin, and made to go over the lines of young sugarcanes, examining the underside of each leaf carefully. Any leaf with an egg-mass attached to it should be cut off and put into the tin to be subsequently burnt. As the eggs drop off at the slightest touch, the collecting receptacle should not be a basket or other porous article through which the loose eggs can fall. If the children went over the fields once a week until no more eggs were found, the pest would be eradicated for that season, and with a little care and attention in subsequent years it could be practically exterminated. Should it be found advisable to attack the insect after it has become a hopper or larva, kerosine emulsion (made from two-thirds kerosine oil and one-third soap, added to ten times its bulk of water) should be sprayed over the plants on which the insects are at work.

At Muzaffarpur I found eggs commonly at Mr. Roland Hudson's estate at Ottur. The insect is probably widely spread, and doubtless is found on other plants besides the sugarcane.

Mr. J. M. Hayman, Deputy Director of Agriculture, Experimental Station, Cawnpore, has found freshly-laid eggs of this insect up to the end of May, so that it appears probable that there are no well-defined broods, but that the egg-laying goes on for a considerable period.

2. *Dinoderus minutus*, Fabricius. Family *Bostrychidæ*; *Læmotmetus ferrugineus*, Gerst., Family *Cucujidæ*, and *Sitophilus oryzæ* Linnæus, Family *Curculionidæ*; all Order *Coleoptera*.

In June, 1900, Captain A. T. Gage, of the Royal Botanic Gardens, Sibpur, near Calcutta, sent to the Museum for identification three species of beetles which were found to destroy the tissues of sugar-cane. They bore small tunnels through the tissue of the cane, and reduce it to powder, leaving the vascular system practically untouched.

One beetle was pronounced by M. P. Lesne, of the Natural History Museum, Paris, to be the cosmopolitan warm region species *Dinoærus minutus,* Fabricius, Family *Bostrychidæ;* the second beetle was pronounced by Mr. M. A. Grenville to be *Læmotmetus ferrugineus*, Gerst., Family *Cucujidæ;* and the third beetle was identified by M. J. Desbrochers des Loges to be *Sitophilus oryzæ,* Linnæus, Family *Curculionidæ,* which he says is found all over the world.

The first species, *Dinoderus minutus,* Fabricius, has been referred to in "Indian Museum Notes," vol. i, n. 1, p. 43 (1889); in vol. ii, n. 6, p. 150, n. 21 (1893); in vol. iii, n. 3, p. 123, process block of imago dorsal and lateral views, antenna, fore, mid, and hind legs (1896); in vol. iv, n. 1, p. 37 (1896); and n. 3, p. 135 (1899) as in all cases being destructive to bamboo in various parts of India. It is figured in Bulletin No. 7, 1900, Agricultural Series, No. 6, Department of Land Records and Agriculture, Bengal, on "Saccharum, sugar, diseases, pests, etc., to which the sugarcane is subject," by Lieutenant A. T. Gage, M.A., M.B., B.Sc. This paper deals with sugar-cane diseases in Bengal. Dr. Gage records the following varieties of sugarcane as affected by the shot-borer, *Dinoderus minutus,* Fabricius:—"Chinia" or "Tikh" or "Ketari" variety at the Soupoul sub-division of Bhagalpur; "Mango, Bhoorli, and Bansahi" varieties of cane at Beheea, East Indian Railway; "Belati, Bheula, Mukhi and Nori" varieties at Bogra; "Kunri, Shamsara, Puri, Kajooli and Basta" varieties at Burdwan; "Auck, Ganna or Gandari, Kooshur," etc., varieties at Dacca; "Reora" and "Bhuli" varieties from Saran and Darbhanga districts respectively, where the disease is known as "Pilna;" "Mugi, Saheban, Kheri, Kajla and Banisha" varieties at Dinajpur; sugarcane from Faridpur; a variety known as "Nola" from the Jhenida sub-division, and "Dhulsundur and Wajli" varieties from the Narail sub-division, both in Jessore; "Shamsara, Bombai, Bhoori and Kajli" varieties from Midnapore; "Shamsara" variety in the Tamluk sub-division of Midnapore; "Bhooli" variety from the Begusarai sub-division, and "Mango, Maneria and Rounda" varieties in the Jamni sub-division, both in Monghyr; "Chinia, Ketar and Mango" varieties from the Barbigha police station in the Jamni sub-division; "Sakor Chinia, Maneria, Mango and Rounda" varieties from the Sikandra police station, also in the Jamni sub-division;

two on each side, each spot smooth, shining, raised, oval and blackish; below this are two lateral series of similar spots, one below the other, placed obliquely, the upper spot nearer the head the larger; the eleventh, twelfth and thirteenth segments with one spot only in each series; below these again are the similarly formed and coloured spiracles; true legs, abdominal legs and oval claspers concolorous with the body, the abdominal legs and claspers with black hooks; the body bears a few scattered hairs.

PUPA reddish-brown, smooth, shining, length ·75 of an inch, head rather pointed, spiracles prominent, black.

At Seeraha I did not succeed in finding any eggs of this moth, nor any young larvæ. Mr. Cotes in Ind. Mus. Notes, vol. i, n. 1, p. 24, says that the eggs of the Amercian sugarcane borer are laid upon the young cane near the axils. In the West Indian cane borer, *Diatræa saccharalis*, Fabricius, the eggs are at first light yellow in colour, within thirty-six hours a tinge of orange appears, which deepens gradually becoming orange-brown; the eggs are laid in clusters of from four to fifty-seven, usually between ten and thirty, on the green leaves (*H. Maxwell-Lefroy*). In Java the allied *Chilo infuscatellus*, Snellen, lays its eggs in clusters on the under surface and near the base of the leaves, although sometimes they are laid on the upper side and then almost always just on the midrib (*Dr. L. Zehntner*). It is probable that our species also lays its eggs on the leaves, where they would be very conspicuous.

On emergence the young larva probably at once tunnels into the heart of a shoot of cane, and gradually eats its way downwards, killing the shoot in doing so. At intervals it drives holes to the outside of the cane, presumably for air. The central growing shoot of the cane dies and grows no more; it turns yellow, and it is by this dead, yellow central leaf (called a "dead heart") that the presence of a borer in a shoot or stem of cane is at once demonstrated. If this "dead-heart" is gently pulled it comes out at once, its base being rotten and turned black. The entire tunnel or burrow made by the larva also turns black; it is full of moisture, smells badly, is in a generally very insanitary state, and is the home of many small insects that live on the putrescent matter, especially a small grub of some fly. When full-grown the larva turns to a pupa as stated above and emerges in about ten days as a dingy grey or drab-coloured moth.

As regards remedies for this pest, the following notes may be of use. When the cane is very young children in gangs in charge of some intelligent man should go over the canes row by row, and pick off the eggs which are probably conspicuous and laid in rows or clusters on the leaves. The eggs of *Chilo simplex* not having as yet been discovered, it is not certain what their appearance is, but they are probably white, and laid as described above. The eggs with the leaves on which they are laid should be placed in kerosine oil tins or some other unperforated vessel (a basket is not a good thing to use, as the eggs may get rubbed off the leaves, fall through the holes in the basket and ultimately hatch in the cane field, quickly finding their way on to the canes), the leaves being burnt or fed to cattle. The rows of canes should be carefully searched for eggs once a week, until no more eggs are found. Later on all diseased canes should be cut out close to the bottom of each near the ground, such canes being destroyed and not allowed to remain on the field. If left where cut, the caterpillars will crawl out and attack a neighbouring growing cane. If the canes are not cut low down, the larvæ may be in the piece of the cane remaining uncut and so escape. A few affected canes can be at once detected if the borer is of sufficient age by the presence of the "dead heart." If the larva is very young the cane will not show a "dead heart" at this early stage of its existence. If the "dead hearts" are carefully removed week by week so long as any are to be found, together with the eggs, the borer will be entirely eradicated. When the canes grow large it is almost hopeless owing to the size of the canes, their numerous and large leaves, and the density of the fields, to search for either eggs or caterpillars, but if care be taken when the cane is young to eradicate the insect, there should be no borers left when the cane grows large. In one field of "Burali" cane, which is a small variety, at Seeraha, I found many bored canes, but on cutting these out I found no borers whatever at any stage. The reason appears to be this. The cane is planted (sown) in trenches, the earth from the trenches being heaped up on each side. When the trenches are filled up after the cane has grown to a foot or so in height, the earth covers the stems of the cane above the hole in the cane made by the borer for breathing purposes, and the borer is killed in the cane. It might be advantageous to level off the fields by filling up the trenches at as early a date as possible, so killing any borers there may be in the canes; but this would be unnecessary if all the eggs have been previously collected and destroyed and all "dead hearts" cut out. In old and large cane if "dead hearts" are searched for and found, it is not necessary to cut out the stems close to the ground, thereby wasting several feet of sound cane, but the cane should be cut within eight inches or a foot of its growing point. The cane being at this stage very large and thick, the borer will be found to have penetrated down the stalk but a short distance. There is no loss to the

N

cultivator in cutting out diseased canes, as in the case of young cane the bored shoot invariably dies, and only throws out thin, straggling, lateral shoots which, when full grown, will give but little or no produce in the mill.

It has often been recommended to burn the trash (outside leaves of the cane in the field after the cane is cut for milling. This appears to me to be a wasteful process, as the trash if allowed to rot is of great value for manure. I would rather advocate the trash being at once ploughed into the ground and allowed to rot there, by this means killing any eggs there may be in the leaves, or collecting and burying it till thoroughly rotted, subsequently spreading it over the fields.

It might be found advantageous to catch the perfect moths. The following apparatus should be used. Stout stakes should be driven into the fields at intervals to be settled by experience, a few traps being experimented with at first. On the top of the stake a shallow wooden tray, tarred inside to make it water-tight, should be fixed, and the tray filled to a depth of about half an inch with thin treacle or molasses. The moths will probably be found to come in numbers to feed on the molasses, to which they will stick. A kerosine-oil lamp might be experimented with to attract the moths. It should be suspended close to the tray of sweets, or placed on a stand in the middle of the tray. It remains to be seen whether the molasses alone would attract the moths, or whether a lamp in addition to the sweets proves to be more efficacious. A common tin lamp such as natives commonly use, with a piece of cotton twisted up for a wick, would probably give sufficient light. These lamps, however, would be useless on windy nights, and a hurricane or other wind-proof lantern would have to be substituted.

Plate xiv, Fig. 1, *male imago life-size, shows both surfaces of the wings, the left-hand side the upper side, the right-hand side the under side;* Fig. 1 *a, the upper side of a female imago life-size in natural position when at rest;* Fig. 1 *d, larva full size seen from the side;* Figs. 1 *b* and 1 *c, young larva, dorsal and lateral views found breeding in maize stalks in Calcutta,** Fig. 1 *e, pupa seen from the side.*

5. *Scirpophaga auriflua,* Zeller. Family *Crambidæ.* Sub-order *Phalænæ.* Order *Lepidoptera.*

Scirpophaga auriflua, Zeller, Monograph *Chilo et Crambus,* p. 2 (1863); Walker, List Lep. Ins. British Museum, vol. xxx, p. 968 (1864); Moore, Proc. Zool. Soc. Lond, 1867, p. 666; Lep. Cey., vol. iii, p. 387 (1884-7); Cotes and

* These larvæ died young, so were not bred up to the imago stage, but as there is only one maize borer known, they are probably *Chilo simplex.*

Swinhoe, Cat. Moths, India, p. 684, n. 4655 (1889); Alcock, Ind. Mus. Notes, vol. v, p. 43, pl. vi, figs. 1, *female twice enlarged;* 2, *section of sugarcane showing cocoon* (1900).

This insect has only once been reported to damage sugarcane, the caterpillar having been sent down to the Indian Museum, Calcutta, from Kushtea, the moths bred, and the species figured in "Indian Museum Notes," vol. v. In May, 1901, Mr. H. Thorp of Sikta, Betteah, on the Nepal frontier, sent some caterpillars to the Museum, which proved to be this species.

In June, 1901, I visited Sikta, and found that quite fifty per cent. of the shoots of the "khari" variety of cane grown by Mr. Thorp was bored by this insect; the "reonda" variety of cane was also badly affected. The "khari" cane was obtained from the Burdwan Experimental Farm. The symptoms of an affected cane are very similar to those attacked by *Chilo simplex*, Butler, a "dead heart" is always present if the larva has been in the cane some while, but the presence of *S. auriflua* can always be told by the base of the leaves from the growing point of the cane having a series of holes across each leaf at right-angles to the length of the leaf bored with a parallel row, which were made by the young larva when it first penetrated into the cane after hatching; these holes seem never to be present in a shoot attacked by *C. simplex*. As a rule the leaves are bent backward and fall over at the point of perforation, so an affected cane can at once be detected by this feature as well as by the "dead heart." The line of holes arise from the fact that the larva penetrates into the heart of the cane when the leaves are tightly rolled up before they unfold, so that each leaf is penetrated in several places. The falling over of the leaves is due to the wind acting on a weakened spot. In all other respects the injury done by this insect is similar to that of *C. simplex*; the larva bores right down the centre of the shoot, the inside turns black and rotten, and harbours many insects which feed on the decayed mass Each attacked shoot stops growing, the growing point being killed. When full-grown the larva has nearly reached the root of the cane; it bores a hole at right-angles to the tunnel in which it has hitherto lived to the outside of the cane, lines this hole with silk, and fits a neat lid or operculum across the hole of exit, through which the moth on emergence from the chrysalis leaves the cane. The larva then retreats about an inch up its tunnel, spins a flimsy silk cocoon of white silk, and turns to a pupa, its head downwards and towards the hole of exit. A light placed out in the fields over a tray of molasses only produced one moth, so this trap does not appear to be very effective, probably the moths fly but little, especially in windy

N2

weather. In the "Sewari" variety of cane which is a very hard and tough variety, no borers were found at Sikta.

LARVA when full-grown about an inch in length (the males a third smaller), pale yellow in colour, without any markings, the dorsal pulsating "heart" often dark and prominent. The caterpillar is extremely soft and flaccid, with a very thin integument; the constrictions between the segments deep; the head is very small (smaller than in any of the other sugarcane borers), very pale ochreous in colour, with a similarly coloured plate on the dorsal area of the first segment; the spiracles inconspicuous, concolorous with the body; the true legs small; the abdominal legs and anal claspers not furnished with hooks; the body hairless. The larva when removed from its tunnel is very inactive and lethargic.

PUPA of the male ·50 of an inch, of the female ·75 of an inch in length. In both sexes it is extremely soft and flaccid, and is very easily injured. Colour cream or very pale yellow, without markings, but when reaching maturity and just before the emergence of the moth, the wing-covers and the legs are very prominent, the hinder pair reaching beyond the outer edge of the wing-covers; the head is narrow, square in front, the black eyes and the cases of the antennæ prominent; the abdomen of the male tapers to a sharp point, in the female the anal segment is very large, thicker than the segment in front of it, ferruginous in colour.

The larva turns to a chrysalis within the tunnel in which it has lived as a caterpillar. Before doing so it bores to the extreme outside of the shoot, forming a short tunnel at right angles to its original bore. This tunnel it closes on the outside by a neat circular operculum or cover made from a piece of the thin bark of the sheath of the cane fastened at the edges with silk. It then retires back into its tunnel, as it goes spinning "bulk-heads" or partitions of silk across the tunnel as it retreats, one behind the other at short distances (about 1 or 2 mm.) apart. I have counted as many as twelve of these bulk-heads, but five or six is a common number. Lastly it spins a flimsy cocoon of white silk within the tunnel, and turns to a pupa, head downwards towards the opening. The operculum and bulk-heads are doubtless constructed to keep out ants and other predaceous insects which would otherwise enter and eat up the defenceless pupa.

By shutting up both sexes of the moth I obtained the eggs of *S. auriflua* at Sikta. They are laid in groups of three or four, each group being covered with the long red hairs with which the anal segment of the female moth is furnished. The eggs are large for the size of the moth, oval in shape, pale green in colour, smooth and without structure. The eggs are laid close to the base of the penultimate whorl of leaves at the growing point of the shoot. The female probably climbs down between the two outermost leaves, posterior end foremost, and lays the eggs as low down as possible where they are entirely hidden from view, the eggs being coated with the "fur" from the mother's body to protect them from the ants which pervade the canes. As soon as the young larva hatches, it penetrates into the central shoot, which dies.

Later on at Seeraha I found larvæ, pupæ and moths of *S. auriflua* in all the five varieties of cane grown there.

Dr. F. Moore in Lep. Cey. describes the moths thus :—

"Silky-white. Wings without markings. Body white; abdomen with broad cinereous-grey bands above and beneath, anal tuft luteous [in the female only]; palpi and fore-legs with cinereous-grey bands. Expanse: 1·4 inch."

Zeller's description I have not seen. Dr. Moore records it from Calcutta as well as from Ceylon; there are specimens from Kushtea, in Bengal, in the Museum, and I have found it in two places in the Champaran District of Behar. In my specimens the male measures 1·0 inch in expanse across the open wings, the female 1·3 to 1·6 inches. Both sexes have the wings long and very narrow, and are exactly like white satin, the cilia or fringes to the wings long, especially those on the hind wing. The male is much smaller than the female, especially the abdomen, which in the male has no large, luteous or pale ferruginous tuft of hairs at the end; the broad ashy-grey bands on the abdomen are often wanting. As regards remedies for this pest, it seems hopeless to search for the eggs, as they are hidden away out of sight and are not at all conspicuous when found. Cutting out affected canes and possibly catching the perfect moths in trays of molasses, as recommended for *C. simplex*, seem to be the only possible remedies to exterminate the pest.

In July, 1901, the Collector of Rangpur, Bengal, forwarded some insects reported to be destructive to Sugarcane, together with a copy of letter No. 356, dated 16th idem, from the District Engineer, Rangpur. On examining them they proved to be larvæ of *S. auriflua*. The District Engineer writes :—

"I have the honour to state that another kind of insect called "Banga" is destroying the sugarcane in the district. These insects attack the canes from the top and go downwards, eating up the pith and then the flesh. It is said that when drizzling rains continue for a few days these insects cease to attack the canes in large numbers."

A considerable number of large hymenopterous parasites were bred in Calcutta from the larvæ of *S. auriflua* sent to me in July by Mr. Thorp of Sikta. One *Bracon nicevillii*, Bingh. has black antennæ, longer than the body ; the head black in the middle yellow in front and behind ; the thorax yellow ; abdomen yellow tipped with black above ; the legs yellow ; both the wings hyaline, pale yellow, margined irregularly with fuscous ; the fore wing with a large medial black spot stretching across the wing, a smaller spot beyond it from the costa to the middle of the wing, with a smaller one behind it on the inner margin. It is figured on plate vii, fig. 2, *female*. The other *Pimpla predator* Fabr. var. has the wings entirely hyaline, the body yellow, with lateral black spots on the abdomen. It is figured on plate vii, fig. 3, *male*. This borer is also attacked by 2 species of Hymenopterous parasites belonging to the family BETHYLIDÆ.

They have been pronounced by Mr. William H. Ashmead to be new species and have been named by him *Goniozus indicus* and *Apanteles scirpophagæ*.

Plate xiv, fig. 2, *shows a male imago ; 2 a, a female imago, upper and under sides ; 2 b, a female imago in its natural position when at rest ; 2 d, the larva, side view, life size ; 2 e, male pupa ; 2 f, female pupa ; 2 g, a bored cane stem with tunnel made by larva, with a pupa* in situ, *and thirteen partitions separating the pupa from the hole of exit ; 2 c, a batch of eggs laid on a sugarcane leaf.*

6. *Nonagria inferens,* Walker. Family *Noctuidæ*. Sub-order *Phalænæ*. Order *Lepidoptera*. Plate xii, Figs. 9, ♀ *imago ; 9 a, larva, lateral view ; 9 b, larva, dorsal view ; 9 c, pupa, ventral view.*

Leucania inferens, Walker, List Lep. Ins. British Museum, vol, ix, p. 105, n. 65 (1856) ; *Seramia inferens,* Moore, Lep. Cey., vol. iii, p. 8, pl. cxlv, fig. 3 (1884). Swinhoe, Proc. Zool. Soc. Lond., 1885, p. 447, n. 6 ; 1886, p. 441, n. 68 ; Butler, Proc. Zool. Soc. Lond. 1883, p. 158, n. 55 ; Cotes and Swinhoe, Cat. Moths, India, p. 261, n. 1654 (1887) ; *Nonagria inferens,* Hampson Fauna Brit. India, Moths. vol. i ; p. 284, n. 1943, woodcut, n. 153, *male,* also *head of female,* (1894).

Leucania proscripta, Walker, List Lep. Ins. British Museum, vol. ix, p. 106 n. 67 (1856) ; Butler, Ann. and Mag. of Nat. Hist., fifth series. vol. xviii, p. 190, n. 55 (1886) ; Cotes and Swinhoe, Cat. Moths India, p. 261, n. 1655 (1887).

Seramia fraterna, Moore, Desc. Lep. Coll. Atkinson, p. 103 (1879) ; Cotes and Swinhoe, Cat. Moths, India, p. 261, n. 1653 (1887).

I reprint the original descriptions of this species by Mr. Francis Walker, Dr. F. Moore and Sir George Hampson, and those of the two synonymic names it has received from Mr. Walker and Dr. Frederic Moore.

Leucania inferens, Walker. *Habitat* : Ceylon. *Description* : "Pale bright fawn-colour ; abdomen whitish-testaceous, much paler than the thorax, fore-wings with three black discal dots and with an intermediate brownish stripe, which accompanies the median veins and its branches; a hardly interrupted blackish line along the exterior border. Hind wings white. Length of the body 4½ lines; of the wings 10 lines." (*Walker,* l. c.).

Seramia inferens, Moore. *Habitat*: Ceylon. *Description*: "Fore wing dull yellowish-ochreous, with a longitudinal ochreous-brown fascia extending from the base along the median vein to the exterior margin; a brown exterior marginal line: hind wing white. Thorax dull yellowish-ochreous; abdomen paler; legs dull brownish-ochreous. Expanse: $1\frac{2}{10}$ inch." (*Moore, l. c.*)

Nonagria inferens, Hampson. *Habitat*: Sind, Bombay, Mhow Ceylon, Burma, Flores. *Description*: "Hind wing with veins 3 and 4 from the cell. Antennæ of male ciliated; of female simple. Ochreous. Fore wing with a red-brown suffusion along median nervure and veins 2 to 5; a marginal dark-line; cilia paler. Hind wing white. Expanse: 28 millim." (*Hampson, l. c.*)

Leucania proscripta, Walker. *Habitat*: East Indies. *Description*: "Pale testaceous. Abdomen whitish. Fore wings fawn-coloured towards the exterior border, with the exterior dots indistinct; a pale brownish discal stripe, and a brown line along the exterior border, which has whitish-testaceous cilia. Hind wings, white. Length of the body, 5 lines; of the wings, 10 lines." (*Walker, l. c.*)

Seramia fraterna, Moore. *Habitat*: Dharmsala, Western Himalayas. *Description*: "Very similar to *S. proscripta* and *S. inferens* Fore wing with a slightly darker shade longitudinally through the median veins, and a small distinct black spot below the cell: hind wing white. *Expanse* ♂ $1\frac{1}{10}$. ♀ $1\frac{3}{10}$ inch" (*Moore, l. c.*).

LARVA when full-grown 1·3 inches in length when walking; cylindrical, tapering rather suddenly towards the head, less rapidly towards the anal segment, which is rounded; rather stout, smooth, shining, with a few colourless bristles about ·15 of an inch in length; the segments broadly and shallowly pitted; flesh-coloured, the pulsating dorsal vessel (heart) dark and rather prominent; spiracles small, black; head cordate, small, ferruginous, the mouth-parts black; legs concolorous with the body, short; the abdominal legs (claspers) on the seventh to tenth segments inclusive counting the head as the first; the fifth and sixth, eleventh and twelfth segments without true legs or claspers. The larva is very active when removed from its tunnel in a cane, walking briskly.

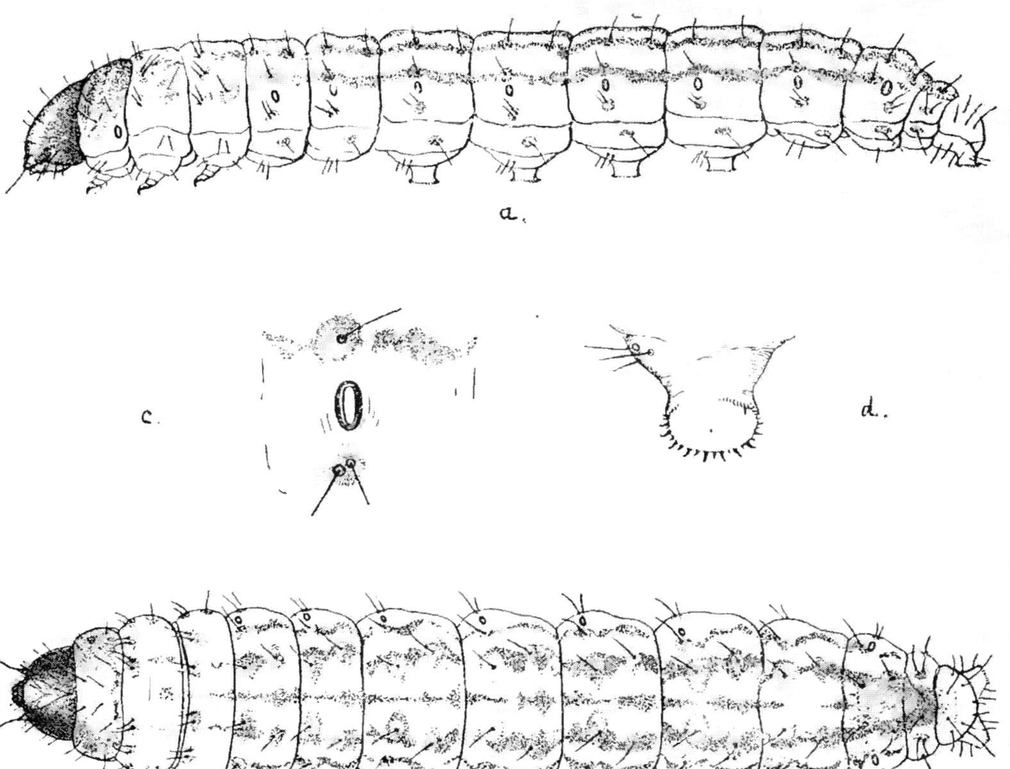

Fig. 1.—*Diatræa auricilia* (C. S. 1574).

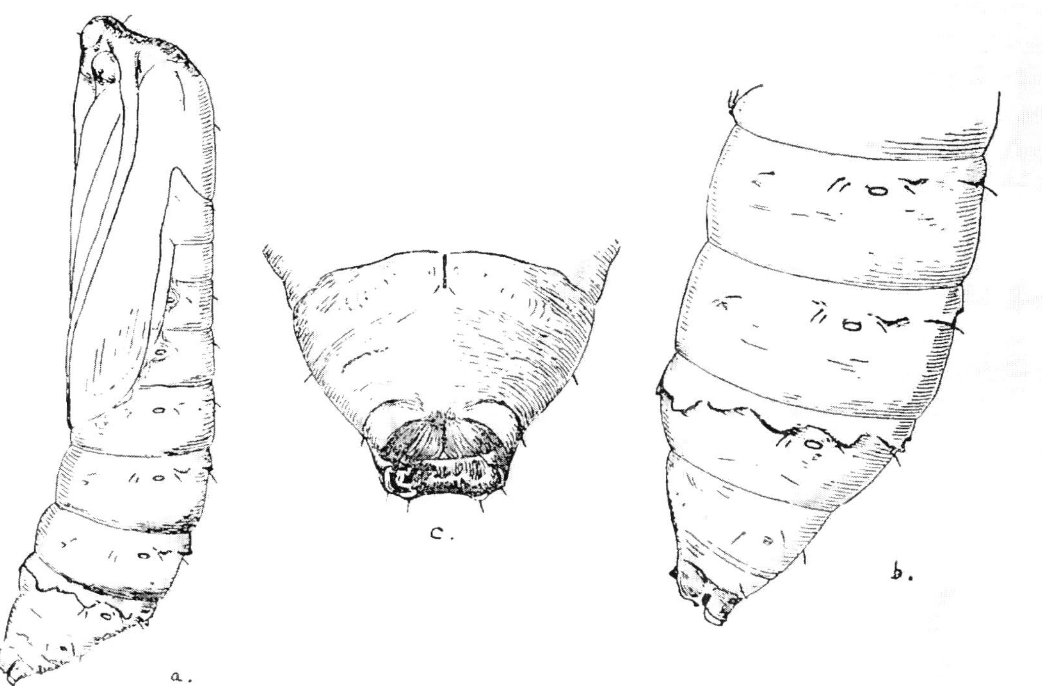

Fig. 2.—*Diatræa auricilia* (C. S. 1560).

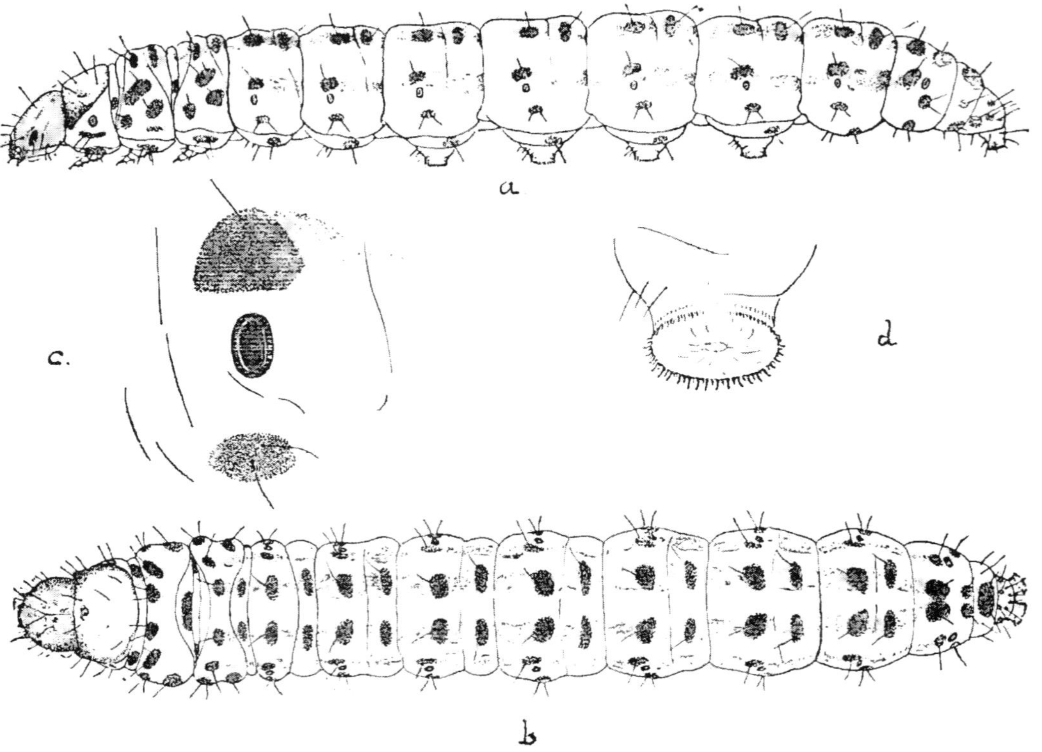

Fig. 1.—*Diatraea venosata* (C. S. 1635). *a*. lateral, and *b*. dorsal view of larva ×4 ; *c*. detail of spiracle of fifth segment, more highly magnified ; *d*, details of first proleg, seen from below, more highly magnified.

Fig. 2.—*Diatraea venosata* (C. S. 1607) ; *a*. pupa ×5 ; *b*. posterior segments of pupa, lateral view, more highly magnified ; *c*. anal segment of pupa, ventral surface, more highly magnified.

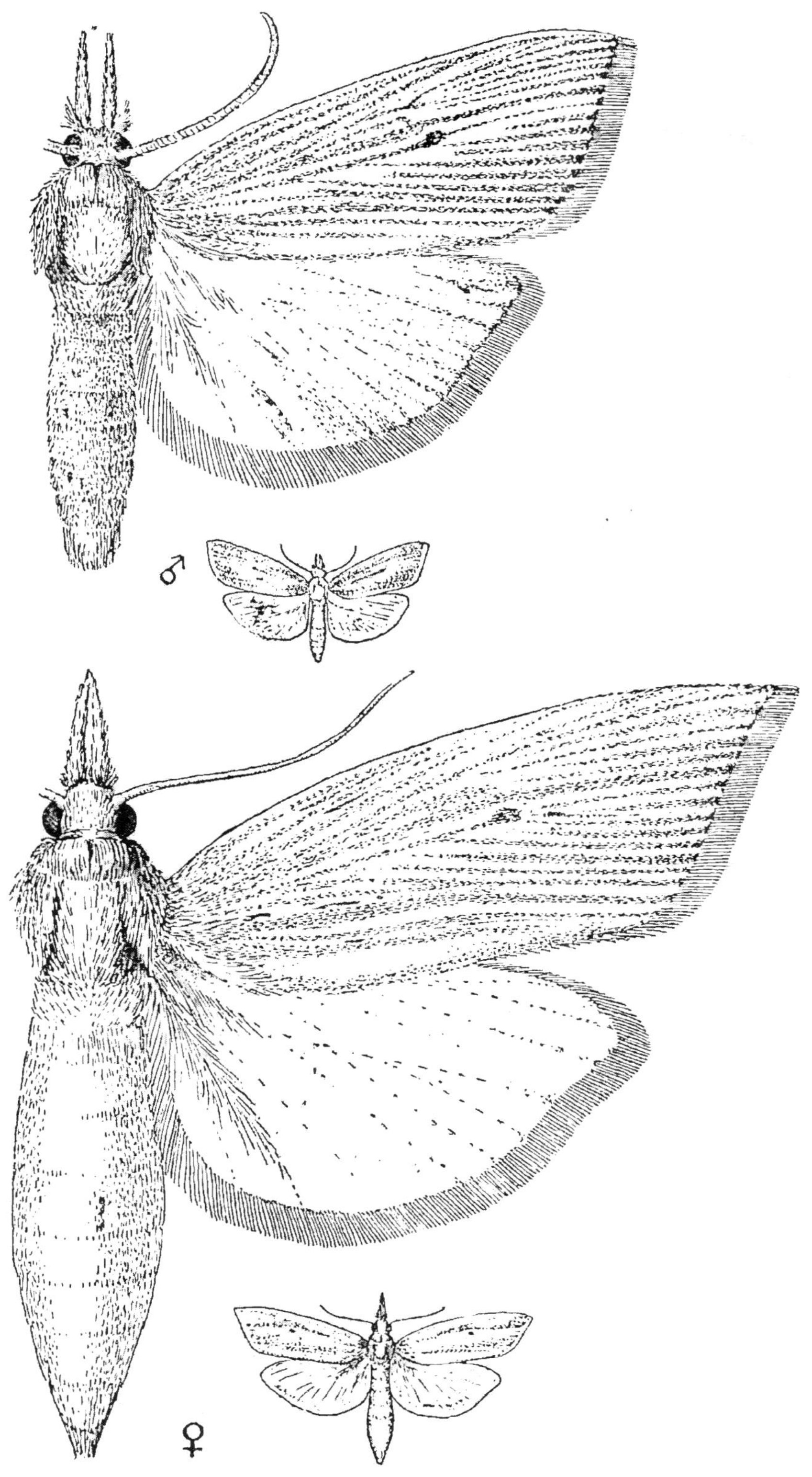

Diatraea venosata, moths × 5. The smaller figures show the natural sizes.

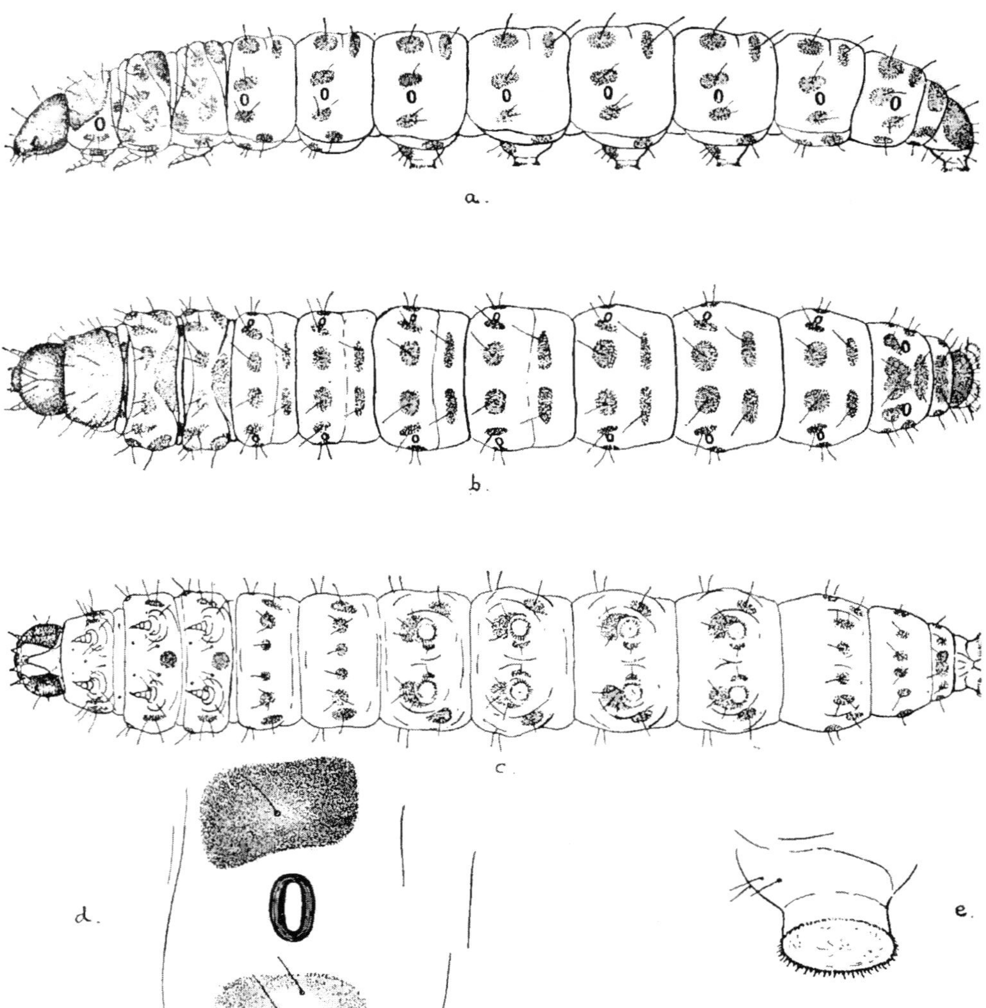

Fig. 1.—Argyria tumidicostalis (C. S. 1610).

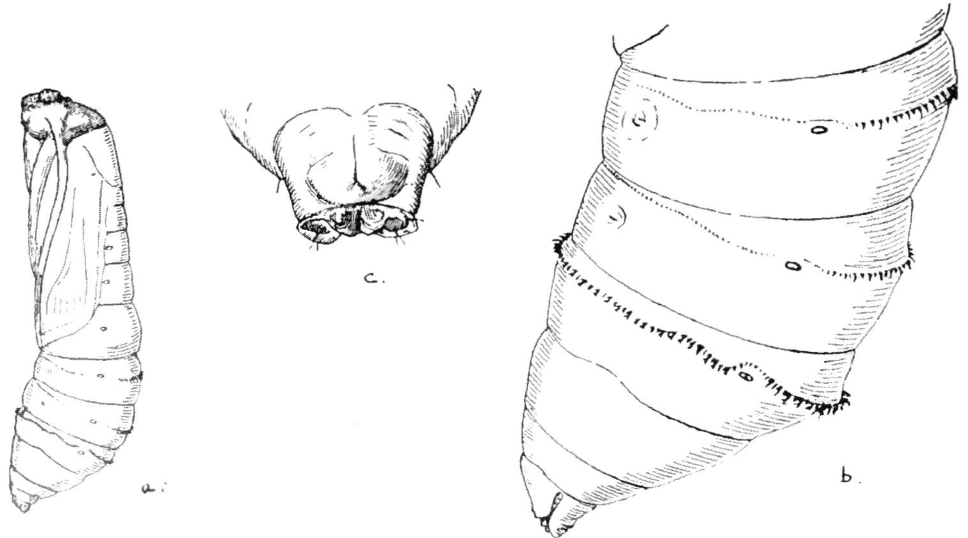

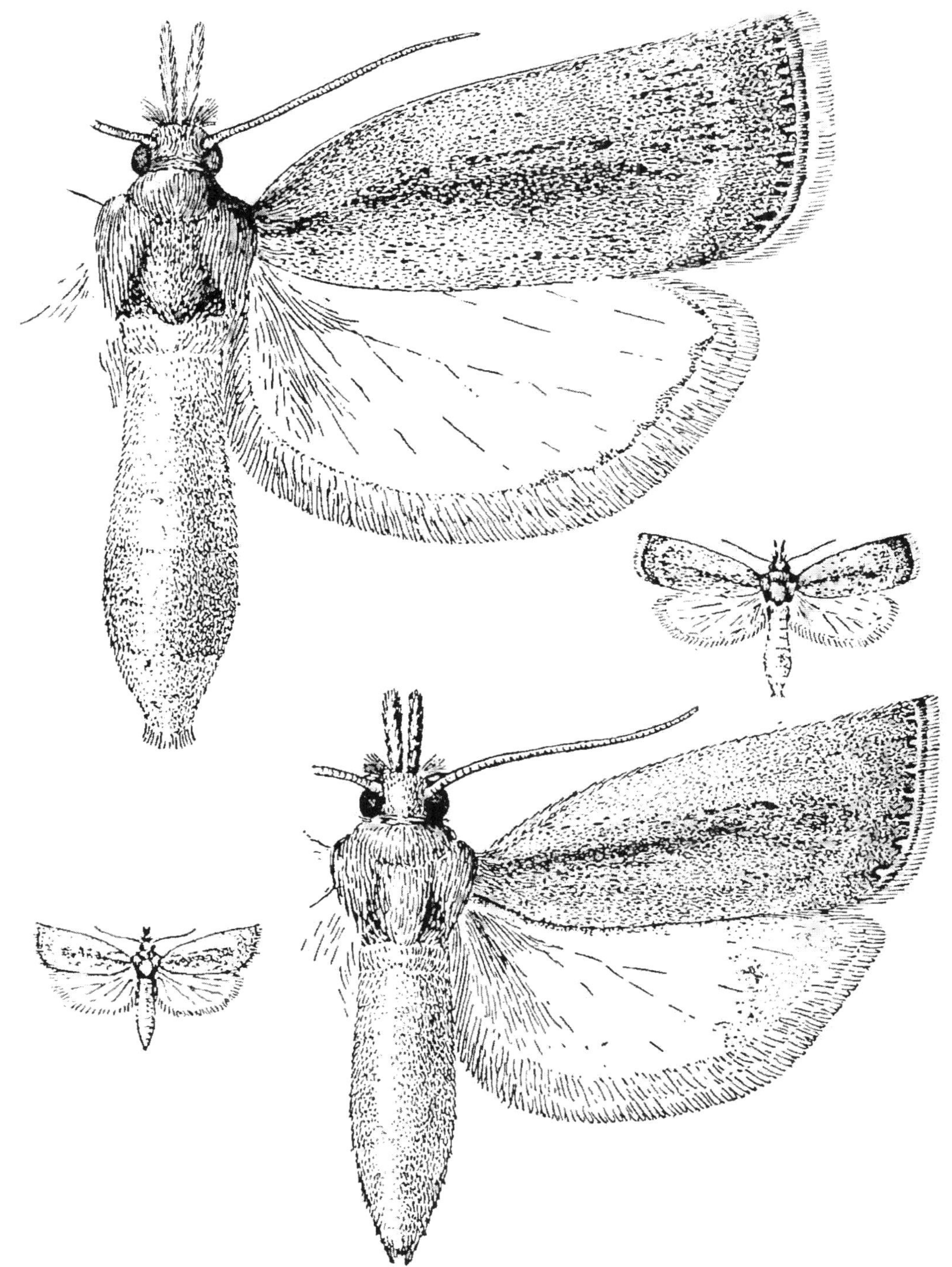

Argyria tumidicostalis (C. S. 1610). Moths ×5. The smaller figures indicate the natural sizes.

Raphimetopus (*Anerastia*) *ablutella* (C. S. 1801)

Plate 44, fig. 1.

Foodplant—Sugarcane.

In the course of the two seasons 1917 and 1918, the caterpillars of this species were observed to occur in very small numbers in young sugarcane in April 1918 and to cause "dead heart" like the other borers.

The full-grown larva is about 16 mm. long and about 2 mm. across the middle of the body which is cylindrical and slightly tapering at the ends. The head is yellow-brown, glossy. The prothorax green; the meso-and metathorax and the abdomen copper-coloured dorsally and greenish ventrally. The spiracles are round, yellow-brown. The hairs are brown, thin and longish. The hooklets on prolegs are in a complete circle.

One pupated on 3rd May and emerged as moth on the 11th May. The pupa is about 8 mm. long and tapering hindwards, the hind end having a few long circinate hairs. The colour is greenish brown.

This insect is stated by Messrs. Mackenzie and Lefroy to have been found in great numbers at Siripur, Hathwa, Saran, in young sugarcane when about three weeks to a month above the ground. (*Agri. Journ. Ind.* III, p. 104). The larva observed by them was completely green.

Chilo simplex, Butl.

Plates 45—47.

Foodplants—

- Maize—Pusa; Mirpurkhas; Sukkur; Poona; Jhalrapatan; Lyallpur.
- *Juar* (*Andropogon Sorghum*)—Pusa; Mirpurkhas; Landhi; Larkana; Sukkur; Jalaun; Jhalrapatan; Babugarh; Lyallpur.
- Sugarcane—Pusa.
- Rice—Pusa; Konkan.
- *Bajra* (*Pennisetum typhoideum*)—Pusa; Sukkur; Mirpurkhas.
- *Sama* (*Panicum frumentaceum*)—Pusa.
- *Marua* (*Eleusine coracana*)—Pusa.
- Sudan grass—Pusa.
- Job's Tears (*Coix lachryma-jobi*)—Konkan.

In this list of foodplants the localities cited are those from which the larvæ have been collected in the course of the last two years and definitely identified as those of *Chilo simplex*.

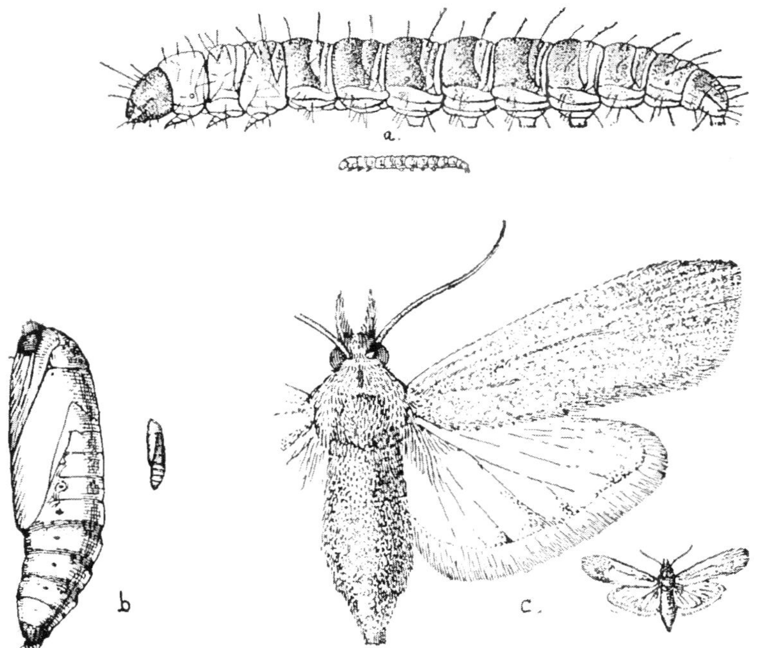

Fig. 1.—*Raphimetopus ablutella* in sugarcane (C. S. 1801).
a. Larva, ×5.
b. Pupa, ×5.
c. Imago, ×5.
The smaller figures show the natural sizes.

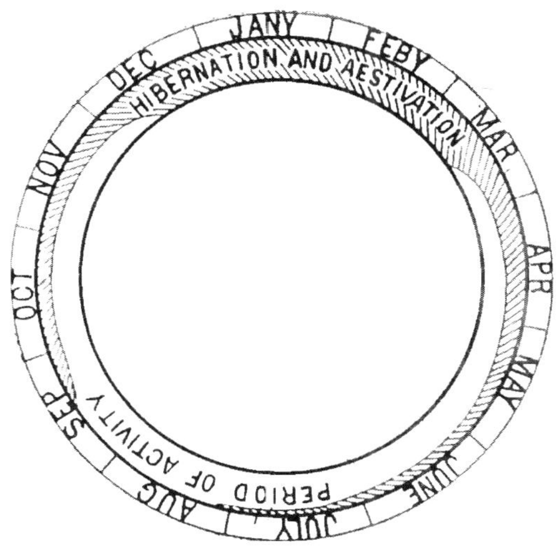

Fig. 2.—Annual life-cycle of *Chilo simplex*.

? Pyralid borer.

Plate 64, fig. 1.

Foodplant—*Rarhi* and *Batri* (*Saccharum spontaneum*).

These caterpillars, which we have not yet been successful in rearing, occur in very large numbers. The hard stems of grown-up plants are converted into hollow tubes throughout their length, a single larva usually occurring in individual plants. The heart shoot dies and dries only when the larva bores near the top.

The insect has only one generation in the year. Young larvæ are found about July and they feed up to about the early part of November, by which time they become fullgrown. Then they rest through the winter and the hot weather, pupating and emerging as moths about June-July. The larvæ have been observed to rest in the Insectary up to the middle of August and then die.

The fullgrown larva is about 35 mm. long. The shape is cylindrical or rather semicylindrical being slightly compressed dorsoventrally. The head is glossy-yellow. The segments of the body have an elongated and slightly chitinized and glossy appearance. The hairs are small and black. The spiracles are elongated oval, brown with a dark narrow rim. The five pairs of prolegs are short, equally developed and have their hooklets in complete circles.

? Pyralid borer. (Plate 64, fig. 2).

Foodplant—*Ikri* (*Saccharum fuscum*).

A caterpillar with a glossy yellow-brown head and the body having a green colour on the ventral surface and the sides and dorsally a brownish tinge on the prothorax, pink on the mesothorax and deep pink or rather red on the rest of the segments, was observed to occur in large numbers in March. They fed on sugarcane stems in the Insectary but could not be reared.

? Pyralid borer.

Foodplant—*Batri* (*Saccharum spontaneum batri*).

A rather slender-bodied, pale-yellow caterpillar, with a yellow head. reddish-brown patches in the subdorsal region of the segments of the body arranged in longitudinal rows on the back and a very prominent visible broad brown tracheal tube running from the thoracic to the last abdominal spiracle on each side, was collected in small numbers in June. It could not be reared.

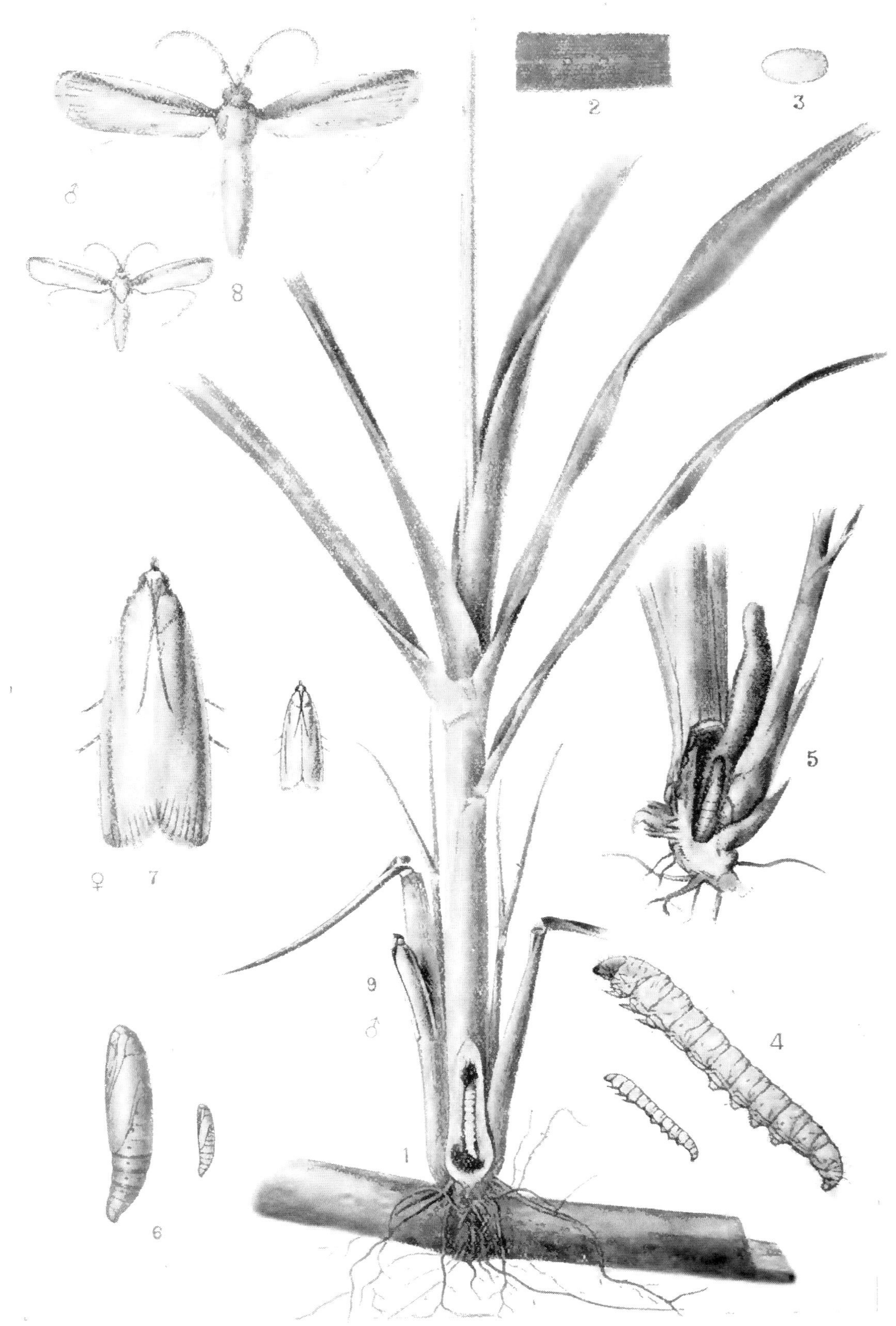
2
3
♂
8
5
♀ 7
9
♂
4
1
6